Handbook of Water Quality Management Planning

Handbook of Water Quality Management Planning

Edited by

Joseph L. Pavoni

Vice-President
TenEch Environmental Consultants, Inc.
Louisville, Kentucky

VAN NOSTRAND REINHOLD ENVIRONMENTAL ENGINEERING SERIES

VNR VAN NOSTRAND REINHOLD COMPANY
NEW YORK CINCINNATI ATLANTA DALLAS SAN FRANCISCO
LONDON TORONTO MELBOURNE

Van Nostrand Reinhold Company Regional Offices:
New York Cincinnati Atlanta Dallas San Francisco

Van Nostrand Reinhold Company International Offices:
London Toronto Melbourne

Library of Congress Catalog Card Number: 77-21601
ISBN: 0-442-23282-9

Manufactured in the United States of America

Published by Van Nostrand Reinhold Company
450 West 33rd Street, New York, N.Y. 10001

Published simultaneously in Canada by Van Nostrand Reinhold Ltd.

15 14 13 12 11 10 9 8 7 6 5 4 3 2 1

Library of Congress Cataloging in Publication Data
Main entry under title:

Water quality management planning.

(Van Nostrand Reinhold environmental engineering series)
Includes bibliographical references and index.
1. Water quality management–United States.
I. Pavoni, Joseph L.
TD223.W295 333.9'1'00973 77-21601
ISBN 0-442-23282-9

To
Gina Marie and Joseph Vincent

Contributors

Ronald G. Alderfer, Harland Bartholomew and Associates, St. Louis, Missouri.

T. Scott Cantine, Harland Bartholomew and Associates, St. Louis, Missouri.

Edward L. Findley, Roy F. Weston, Inc., Wilmette, Illinois.

Edward G. Foree, Civil Engineering Department, University of Kentucky, Lexington, Kentucky.

Marva Gay, Schimpeler-Corradino Associates, Miami, Florida.

D. Joseph Hagerty, Civil Engineering Department, University of Louisville, Louisville, Kentucky.

John E. Heer, Jr., Civil Engineering Department, University of Louisville, Louisville, Kentucky.

Eldridge Lovelace, Harland Bartholomew and Associates, St. Louis, Missouri.

George Noble, Roy F. Weston, Inc., Wilmette, Illinois.

Joseph L. Pavoni, TenEch Environmental Consultants, Inc., Louisville, Kentucky.

Jerry R. Perrich, Olin Corporation, Brandenburg, Kentucky.

Adelbert L. Roark, A. L. Roark and Associates, Lexington, Kentucky.

Charles C. Schimpeler, Schimpeler-Corradino Associates, Louisville, Kentucky.

John S. Tapp, Commonwealth Technology, Inc., Lexington, Kentucky.

Van Nostrand Reinhold Environmental Engineering Series

THE VAN NOSTRAND REINHOLD ENVIRONMENTAL ENGINEERING SERIES is dedicated to the presentation of current and vital information relative to the engineering aspects of controlling man's physical environment. Systems and subsystems available to exercise control of both the indoor and outdoor environment continue to become more sophisticated and to involve a number of engineering disciplines. The aim of the series is to provide books which, though often concerned with the life cycle—design, installation, and operation and maintenance—of a specific system or subsystem, are complementary when viewed in their relationship to the total environment.

The Van Nostrand Reinhold Environmental Engineering Series includes books concerned with the engineering of mechanical systems designed (1) to control the environment within structures, including those in which manufacturing processes are carried out, and (2) to control the exterior environment through control of waste products expelled by inhabitants of structures and from manufacturing processes. The series includes books on heating, air conditioning and ventilation, control of air and water pollution, control of the acoustic environment, sanitary engineering and waste disposal, illumination, and piping systems for transporting media of all kinds.

Van Nostrand Reinhold Environmental Engineering Series

HANDBOOK OF ADVANCED WASTEWATER TREATMENT (Second Edition) by Russell L. Culp, George Mack Wesner and Gordon L. Culp

ADVANCED WASTEWATER TREATMENT, by Russell L. Culp and Gordon L. Culp

ARCHITECTURAL INTERIOR SYSTEMS–Lighting, Air Conditioning, Acoustics, John E. Flynn and Arthur W. Segil

SOLID WASTE MANAGEMENT, by D. Joseph Hagerty, Joseph L. Pavoni and John E. Heer, Jr.

THERMAL INSULATION, by F. Malloy

AIR POLLUTION AND INDUSTRY, edited by Richard D. Ross

INDUSTRIAL WASTE DISPOSAL, edited by Richard D. Ross

MICROBIAL CONTAMINATION CONTROL FACILITIES, by Robert S. Runkle and G. Briggs Phillips

SOUND, NOISE, AND VIBRATION CONTROL, by Lyle F. Yerges

NEW CONCEPTS IN WATER PURIFICATION, by Gordon L. Culp and Russell L. Culp

HANDBOOK OF SOLID WASTE DISPOSAL: MATERIALS AND ENERGY RECOVERY, by Joseph L. Pavoni, John E. Heer, Jr., and D. Joseph Hagerty

ENVIRONMENTAL ASSESSMENTS AND STATEMENTS, by John E. Heer, Jr. and D. Joseph Hagerty

ENVIRONMENTAL IMPACT ANALYSIS: A New Dimension in Decision Making, by R. K. Jain, L. V. Urban and G. S. Stacey

CONTROL SYSTEMS FOR HEATING, VENTILATING AND AIR CONDITIONING (Second Edition), by Roger W. Haines

HANDBOOK OF WATER QUALITY MANAGEMENT PLANNING, edited by Joseph L. Pavoni

Preface

The Federal Water Pollution Control Act Amendments of 1972 (Public Law 92-500) is perhaps the most comprehensive program that the Congress has ever established for solving the nation's water quality problems. The Act charters a ten year program whose goal is the achievement throughout the country, by 1983, of waters in which we can swim and fish. Obviously, this task is an enormous one, which can best be achieved by properly structured and coordinated phases of water quality management planning. A very important aspect of the Act is that it recognizes the complex technical and institutional problems of water quality protection vary so widely across the country that long-term solutions to these problems will necessarily depend on decentralized management subject to Federal overview. Consequently, the fundamental responsibility for planning and implementing the requirements of the Act should reside with State and local governments.

In the five years since the Act became a law, State and local governments have embarked on a variety of water quality management planning efforts including:

- River Basin Planning (Section 303e)
- Areawide Waste Treatment Management Planning (Section 208)
- Facilities Planning (Section 201)
- River Basin Planning Strategy (Section 209)

These planning processes are interdependent and multidisciplinary in nature, and are developed, reviewed, and implemented by a wide spectrum of individuals including regulatory personnel, governmental officials, planners, engineers, citizens, industrial representatives, farmers, and legislators. Therefore, it would seem mandatory that the water quality management processes currently underway be completely understandable to the various segments of our society. The accomplishment of the immediate and long-range objectives of water quality management planning relies to a great extent on the involvement of the above segments of our society in the continuation of the planning processes in which adopted plans are periodically updated.

This manuscript is intended to acquaint individuals interested in water quality management planning with the various aspects of the planning process. The chapters of this manuscript have been written by professionals engaged in specific phases of water quality management planning activities since the passage of Public Law 92-500. Consequently, each chapter has been authored by individuals experienced in the specific aspect of water quality management planning which that chapter addresses. Chapter 1 outlines the development of water quality management planning in the United States and the purpose, scope, and approach utilized in various planning studies. Chapter 2 discusses the intricate relationships which exist between water quality management planning and land use planning. Chapter 3 describes wastewater treatment alternatives that can be used in water quality management studies and a method of evaluating these treatment alternatives on a cost effective basis. Chapter 4 discusses the use of computer modeling in water quality management planning studies, models available for various types of planning studies, and the relationship of modeling results to the decision making process. Chapter 5 outlines the management and institutional requirements and constraints that must be considered in water quality management planning. Chapter 6 discusses the requirements for public participation in the planning process, methods available for public participation and alternatives evaluation, and the applicability of these methods to various planning studies. Chapter 7 describes the role of environmental assessment in various water quality management planning studies and various assessment methodologies available to the planner.

No seemingly interminable project such as this manuscript comes to fruition through the efforts of the editor alone. The various authors have genuinely interacted and lent support to the editor from the conceptual development of this text through its various stages of revision and production. The authors must be commended for their interest in developing a water quality management planning manuscript focused at the wide spectrum of individuals involved in the planning process. The editor also wishes to acknowledge the able assistance of Jan A. Tatum, and a long time collaborator, Judith E. Pavoni, for secretarial assistance, proofreading, and preparation of the index.

JOSEPH L. PAVONI

Contents

Handbook of Water Quality Management Planning

1

Development of water quality management planning in the United States

George Noble and Edward L. Findley

Roy F. Weston, Inc.
Wilmette, Illinois

1-1 THE DAWNING OF ENVIRONMENTAL AWARENESS

Contrary to popular belief, the dawning of environmental awareness is comparatively recent. Only in the area of water pollution was there any indication of general concern before the turn of the century, and this concern usually related to the riparian user and the protection of his particular interests.

In the case of Tennessee Coal, Iron & R. R. Co. v. Hamilton, 100 Ala. 252, 260, 14 So. 167, 170 (1893), the judge placed on record some of these early concerns:

> "The exigencies of the great industrial interests must be kept standing in view; the property of large and useful interests should not be hampered or hindered for frivolous or trifling causes. For slight inconveniences or occasional annoyances, they ought not to be held responsible. . . .
>
> It is certainly true that owing to the want, if not necessities, of the present age . . . some changes must be tolerated

in the channels in which water naturally flows and in its adaption to beneficial uses. Reasonable diminution of its quality, in gratifying and meeting customary wants, has always been permitted. So, its temporary detention for manufacturing purposes, followed by its release in increased volume, is a necessary consequence of its utilization as a propelling force. Nor must we shut our eyes to the tendency—the inevitable tendency—of these and other uses in which water is an indispensable element, to detract somewhat from its normal purity. These modifications of individual right must be submitted to, in order that the greater good of the public be conserved and promoted. But there is a duty to yield, to this claim and right to expect and demand. The water course must not be diverted from its channel, or so diminished in volume, or so corrupted and polluted as practically to destroy or greatly impair its value to the lower riparian proprietor."

As might be expected, the general development of town and industry at that time tended largely to ignore this admirable judicial constraint, and before the turn of the century many of our major rivers and streams were seriously polluted. It is reported by Fuller and McClintock, in their treatise on "Solving Sewage Problems" published in 1926, that at one time one of the branches of the Chicago River (appropriately called "Bubbly Creek") became covered with such a thick scum "that people were able to walk upon it quite safely."

Similarly in England, where the industrial revolution was taking its toll of the environment far ahead of American experience, the River Irwell was reported at one time to be so thick and solid that birds walked on it without sinking (Ref. 1-1).

The following lines were reported found on a board room table after a meeting of the Mersey and Irwell Joint Committee about the year 1901:

"If with a stick you stir well
The poor old River Irwell,
Very sick of the amusement
You will very soon become;
For foetid bubbles rise and burst
But that is really not the worst
For little birds can hop about
Dry-footed on the scum" (Ref. 1-2).

1-2 EARLY LEGISLATIVE ATTEMPTS

1-2.1 The Rivers and Harbors Act of 1899

No legislative attempts to handle the problems of increased pollution of the natural water courses emerged in this country until well into the twentieth century, except for the Rivers and Harbors Appropriations Act of 1899. (See Figure 1-1 for a summary of major legislative history.)

1972 Federal Water Pollution Control Act Amendments

1966 Clean Water Restoration Act

1965 Water Quality Act

1961 Water Pollution Control Act Amendments

1956 Water Pollution Control Act Amendments

1948 The Water Pollution Control Act

1924 The Oil Pollution Act

1899 The Rivers and Harbors Act

Fig. 1-1. Summary of major legislative events in summary pollution control legislation.

The emergence of this Act as a tool for controlling pollutant discharge to rivers and streams is widely held to be anomalous. The Act itself was chiefly designed to prevent interference with navigation. Section 9, for example (now 33 U.S.C. §401), expressly forbids the construction of bridges or dams across navigable waters without Congressional consent and Corps of Engineers' approval. Section 10 forbids the creation of any obstruction not affirmatively authorized by Congress to the navigable capacity of any of the waters of the United States, and requires Corps' approval of wharfs, piers, excavations, etc.

However, under Section 13, 33 U.S.C. §401, appears the following provision:

> "It shall not be lawful to throw, discharge or deposit, or cause, suffer, or procure to be thrown, discharged, or deposited either from or out of any ship, barge, or other floating craft of any kind, or from the shore, wharf, manufacturing establishment, or mill of any kind, any refuse matter of any kind or description whatever other than that flowing from streets and sewers and passing therefrom in a liquid state, into any navigable water of the United States, or into any tributary of any navigable water from which the same shall float or be washed into such navigable water; and it shall not be lawful to deposit, or cause, suffer, or procure to be deposited material of any kind in any place on the bank of any navigable water, or on the bank of any tributary of any navigable water, where the same shall be liable to be washed into such navigable water, either by ordinary or high tides, or by storms or floods, or otherwise, whereby navigation shall or may be impeded or obstructed: *Provided*, . . . That the Secretary of the Army, whenever in the judgment of the Chief of Engineers anchorage and navigation will not be injured thereby, may permit the deposit of any material above mentioned in navigable waters, within limits to be defined and under conditions to be prescribed by him. . ."

Given the overwhelming concerns of the Act to limit obstruction to navigation, it is not surprising that the above section was rarely exercised. However, in 1960, the Supreme Court rendered a decision which turned the modest Section 13 provision into an apparent giant-killer. In United States v. Republic Steel Corp., 362 U.S. 482, the complaint was that industrial wastes containing suspended solids were discharged to the river where they settled to form shoals which interfered with navigation.

Though this case was clearly still concerned with physical interference with navigation, it turned out to be an appropriate platform from which to launch an appeal that all discharges of "refuse" were covered by this Act. The next case was United States v. Standard Oil Co., 384 U.S. 224 (1966), in which the defendant was charged under Section 13 for accidentally discharging gasoline into the water. It was variously argued that since the gasoline was valuable, it could not be considered "refuse" within the meaning of the statute. However, the Supreme Court argued for a broader reading, stating that:

> "Oil is oil and whether usable or not by industrial standards it has the same deleterious effect on waterways. In either case, its presence in our rivers and harbors is both a menace to navigation and a pollutant."

As in the Republic case, it was argued that navigation was the central concern, since gasoline in the harbor could certainly create a fire hazard that menaces navigation. However, the court went further to suggest a much broader field of operation for Section 13:

> "It is plain from . . . the legislative history that the "serious injury" to our watercourses . . . sought to be remedied was caused in part by obstacles that impeded navigation and in part by pollution—"the discharge of sawmill waste into streams" . . . and the injury of channels by deposits of ballast, steam-boat ashes, oysters and rubbish from passing vessels."

Possibly all this was intended to imply that pollution as well as "obstacles" was prohibited if it interfered with navigation, but it certainly encouraged the argument that pollution control was an independent goal of the statute.

By 1973, the Supreme Court reported that, in spite of a long history of contrary interpretation by the Corps of Engineers, there was almost universal agreement:

> "that Section 13 is to be read in accordance with its plain language as imposing a flat ban on the unauthorized deposit of foreign substances into navigable waters, regardless of the effect on navigation."

The court went on to say that the Section 13 limitation requiring interference with navigation applied only to the separate offense of placing material on the bank of the stream and that the prohibitation against discharge of refuse:

> "is set off from the second clause by a semicolon, (which) contains no language of its own limiting its prohibition to navigation–impeding deposits." (Ref. 1-3)

This interpretation lead to a flood of actions attempting to use the statute as a basis for controlling pollution, and the situation was for a time further complicated by the revival of the virtually forgotten permit clause at the end of Section 13. This lead to the view that the 1899 statute need not be viewed as an absolute ban on pollution, but merely a restriction on discharge without a permit to do so. Unfortunately, the Corps had omitted to develop an appropriate permit program and in 1970, pursuant to Executive Order (35 Fed. Reg. 19627), the Corps duly set up such a permit system, based on the assumption that Section 13 could be used as a means of enforcing the water quality standards.

A number of problems followed this action, not the least of which was the contention in Kalur v. Resor, 335 F. Supp. 1 D.D.C. 1971, that the permit authority only applied to navigable waterways. Thus it was argued that Congress must have intended a complete ban on all discharges to nonnavigable waterways.

These problems were, fortunately, more or less resolved in the most recent law. However, this brief history of the Refuse Act stands as an indicator of the confusion surrounding water pollution legislation in this country until fairly recently. More importantly, the substantial efforts to bend existing statutes to achieve more stringent pollution controls are clear evidence that the old body of law which existed until fairly recently was generally considered to be altogether too slight to meet the challenge of the twentieth century.

1-2.2 The Oil Pollution Act of 1924

The first water pollution control statute knowingly enacted by Congress was the Oil Pollution Act of 1924, 43 Stat., 604. The Act was designed to control oil discharges into coastal waters. The Act specifically made it unlawful, except in grave emergency or unavoidable accident, to discharge oil into waters within the ebb and flow of the tide.

1-2.3 The Water Pollution Control Act of 1948 (P.L. 845)

Between 1924 and 1948 (with the exception of the war years), a number of attempts were made to introduce water pollution control legislation. A number of different approaches were put forward. It was suggested that the Federal role should be strongly regulatory with wide enforcement powers. At one stage, a federal permit system was proposed. In another bill, it was proposed that the Federal government should be prohibited from purchasing paper from any manufacturers who discharged wastes into a stream.

On three occasions–specifically, in 1936, 1938, and 1940–comprehensive water pollution control legislation narrowly missed approval. After World War II, renewed efforts in Congress finally resulted in passage of the 1948 Act by the 80th Congress. In many respects, the 1948 Act was a very modest step toward a comprehensive water pollution control program. Initially, it was intended to be a temporary 5-year program though it was later extended for 3 years by P.L. 82-579.

The 1948 Act provided for federal research into water pollution control strategies. Federal grants were also provided to support state and local water pollution control programs. Limited federal loans were also provided to help in the construction of municipal treatment works.

Recognizing the problems of state control of interstate waters, Congress authorized the setting up of interstate compacts to act as joint control agencies. The pollution of "interstate waters" was declared a "public nuisance" if it endangered health or welfare.

Unfortunately, the enforcement provisions were very weak. Formal notice of a violation had to be made both to the polluter and to the state agency in which the infringement occurred. If two formal notices failed to produce results, the federal agency was authorized to call a "public hearing." The public hearing was intended to produce recommendations for abating the nuisance. In the event that recommendations were not heeded, the federal authorities, according to the statute, were permitted to bring suit (but only with the permission of the State in which the discharge occurred). Even pollution, as such, was not specifically defined by the act (everything being left up to the court to decide) it being required to:

> "enter such judgment . . . as the public interest and the equities of the case may require (after) giving due consideration to the practicability and to the physical and economic feasibility of securing abatement."

1-2.4 The Water Pollution Control Act Amendments of 1956 (P.L. 660)

Comprehensive water pollution control legislation came closer to a reality as a result of the amendments of 1956.

The amended Act was administered by the Surgeon General of the Public Health Service under the supervision and direction of the Secretary of Health, Education, and Welfare.

The most important provisions of the 1956 Act as amended were:

- Affirmation of the congressional policy to recognize, preserve, and protect the primary responsibilities and rights of the states in preventing and controlling water pollution.
- Authorization of increased technical assistance to states in addition to an increase in research by nongovernmental research organizations.
- Authorization of collection and dissemination of basic data on water quality relating to water pollution prevention and control.
- Direction of the Surgeon General to continue encouragement of interstate compacts and uniform state laws.
- Modification and simplification of procedures governing federal abatement actions against interstate pollution.
- Appointment of a Water Pollution Control Advisory Board.
- Establishment of a program specifically directed toward the control of pollution from federal installations.

The 1956 amendments represented significant progress toward development of comprehensive legislation and established a firm foundation for later developments.

1-2.5 The Water Pollution Control Act Amendments of 1961 (P.L. 87-88)

In July 1961, President Kennedy signed into law a new round of amendments to the Water Pollution Control Act. These amendments improved and strengthened the Act, particularly in the following areas:

- Extension of federal authority to enforce abatement of intrastate as well as interstate pollution by making "navigable" waters subject to enforcement.
- Increase in the dollars authorized for financial assistance to municipalities for the construction of waste treatment works.
- Intensification of research toward more effective methods of pollution control and authorization of the necessary funds.
- Authorization of storage being included as a valid concept to regulate stream flow, for the purpose of quality control in the planning of federal reservoirs and impoundments.
- Designation of the Secretary of Health, Education, and Welfare to administer the Act.

Once again, the Act affirmed the primary responsibilities and rights of the states in preventing and controlling water pollution. The federal functions were specifically designed to be carried out in cooperation with state and interstate agencies and with local public and private interests.

1-2.6 The Water Quality Act of 1965 (P.L. 89-234)

The Federal Water Pollution Control Act was strengthened considerably in 1965 when Congress passed P.L. 89-234—the Water Quality Act. This Act removed the administration of the Water Pollution Control Act from the Public Health Service and established a new and independent agency—the Federal Water Pollution Control Administration (FWPCA). A completely new four-year program of grants and contracts was also approved for the development of new and improved methods for controlling waste discharges from storm sewers and combined storm and sanitary sewers.

The concept of water quality standards was introduced into water pollution control legislation for the first time with the passage of the Water Quality Act. The Act required that States adopt water quality criteria together with a plan for implementation and enforcement. The criteria and plans were to be submitted by June 30, 1967 to the Secretary of the FWPCA, whose job it was to establish whether the proposals met the terms of the Act. Upon approval, the criteria and the plan, taken together, were designated the state water quality standards.

Prior to this Act, enforcement was entirely by way of the conference method. That is to say, the Secretary would be asked to call for a conference, by a state that felt an adjacent state was causing pollution (sufficient to endanger the health and welfare of the people). Cumbersome though this system may appear, it seemed remarkably effective once a case was introduced.

With the passage of the 1966 Act, a new enforcement procedure came into being (though the conference procedures continued in many states due to delay in establishing standards). Under Section 10(c) (5), a discharge of matter that reduced water quality below standards was subject to abatement by an action brought by the United States Government. The Secretary was required to give 180 days notice of such proposed action. The written consent of the Governor of the state in which the violation occurred was also required.

1-2.7 The Clean Waters Restoration Act of 1966 (P.L. 89-753)

Based primarily on the evidence of the 1961 Act, this new statute recognized the need for comprehensive basin planning. Congress authorized grants (not to exceed 50 percent of cost) to enable the states to set up a planning agency to develop effective water quality control and abatement plans on a river basin basis.

The comprehensive plan was intended to be consistent with the applicable water quality standards. It was also required that the plan should recommend treatment works and sewer systems capable of providing the most effective and economic collection storage and treatment of wastes for both municipal and industrial installations.

The 1966 Act increased the Federal contribution to 40 percent of the project cost (if the State agreed to match at least 30 percent of the cost of all federally assisted projects in a single fiscal year). If the State agreed to match only 25 percent, but was also willing to establish adequate and enforceable water quality standards, the Federal contribution was increased to 50 percent.

Grants for research and development were also included in the new Act. For programs designed to demonstrate new or improved methods of controlling storm water discharges, the Federal share was increased from 50 to 75 percent.

Grants to States or local government compacts to demonstrate advanced treatment methods were increased to 75 percent. Grants to individuals for waste treatment research were also increased substantially.

The Clean Waters Restoration Act of 1966 also directed the Secretary to undertake a comprehensive study of the effect of pollution in the estuaries and estuarine zones of the United States.

1-3 THE FEDERAL WATER POLLUTION CONTROL ACT AMENDMENTS OF 1972 (P. L. 92-500)

In October 1972, the above statute was enacted over the President's veto, and House Bill 247-23/Senate Bill 52-12 became P.L. 92-500.

The new bill, the result of extensive conference proceedings, was nearly three years in the making. As early as 1969, work began on a senate bill to amend the Water Pollution Control Act. Those amendments (S.2770) passed only in the Senate. They were intended to authorize a concerted attack on industrial and municipal polluters. They specifically set a goal of ending all discharges of pollutants into all water bodies by 1985.

However, in March 1972, the House passed a different bill (H.R. 11896) which provided a lower level of action, and called upon the states to play a greater role in cleaning up the nation's waterways. In September of 1972, after much disagreement over the importance of thermal pollution in the projected bills, a joint House-Senate Conference report was finally issued.

The most profound philosophical distinction between P.L. 92-500 and all previous law is the proposal to abandon the case by case basis upon which treatment needs were traditionally established. Hitherto, states were permitted to determine water quality standards on a case by case basis depending upon the projected uses of the water body under discussion. In turn, the projected uses were themselves based upon ambient water quality. Thus, the worst water quality was rarely improved significantly above standards since no higher use was assigned greater than current levels of usage.

To reduce the administrative difficulties of relating effluent quality to ambient water quality, P.L. 92-500 establishes minimum effluent limitations for each category of discharger, regardless of receiving water requirements. Where water quality standards cannot be reached by imposition of a standard level of treatment alone, then and only then will the conditions of the receiving waters dictate more stringent controls.

Prior to the 1972 statute, water pollution control legislation rested heavily on a de facto goal of making the waters of the state clean enough to support specified uses (fishing, swimming, boating, water supply, farming needs, and industrial needs). The new law rejects these distinctions and establishes a clean water goal. To achieve this objective, the law establishes the goal of elimination of pollutant discharges by 1985. In the interim, whenever attainable, the law establishes the goal of achieving, by the year 1983, a water quality standard sufficient to protect and propagate fish, shellfish, and wildlife and for recreational uses.

The new law does not specifically mandate achievement of these goals. Even the no-discharge goal by 1983 is only mandated for certain classes of nonmunicipal discharger for which it is "technologically and economically achievable." The law requires that effluent standards be established for all discharges "which can reasonably be expected to contribute to the attainment or maintenance" of the water quality needed to achieve the 1983 interim goal.

The basic regulatory requirement of the new law is that the discharge of effluent will be controlled by a new permit system. (The specific need for such a permit system grew out of the difficulties encountered over the interpretation of the Refuse Act.) Basically, all point sources (industry, municipal treatment facilities, feedlots, and other discrete sources) must, prior to discharge, obtain a permit which specifies the quantities of each pollutant which may be discharged.

The new law encourages states to administer the permit program, to be known as the National Pollutant Discharge Elimination System (NPDES). A state can qualify to administer the program by adopting enforcement authorities, public notice procedures, and certain other authorities specified by EPA. Even after a state program is approved, however, each permit will be subject to EPA review to ensure compliance with law.

Permits are required to be consistent with the specific effluent guidelines which are issued by the EPA for major classes and categories of industry or publicly-owned treatment facilities. The application of the technology–based effluent guidelines is to take place in two phases, as illustrated in Figure 1-2. By 1977, municipal plants are required to provide "secondary treatment" (a level of treatment to be defined in more detail when EPA develops detailed regulations). By 1977, industrial facilities are required to establish what is defined by EPA as "best practicable control technology currently available" (BPT).

According to the law the second phase ends in the year 1983 by which time municipalities must provide "best practicable waste treatment technology" (BPT) and industries must provide the "best available technology economically achievable (BAT)." This will result in "reasonable further progress" toward the goal of eliminating the discharge of pollutants.

There is a separate provision in the new law that allows for even more stringent effluent limitations for individual industries or municipalities when necessary "to assure protection of public water supplies, agricultural and industrial uses, and the protection and propagation of a balanced population of shellfish, fish

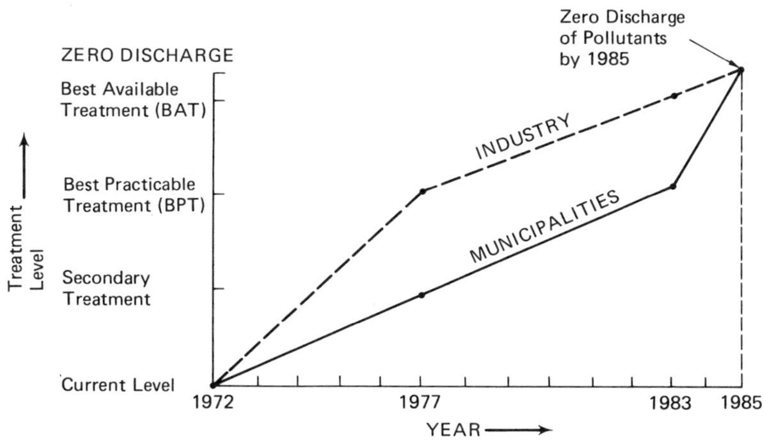

Fig. 1-2. Summary of treatment goals established by P.L.92-500.

and wildlife, and to allow recreational activities in and on the water." The more stringent limitation will not be applied where the discharger can show that there is "no reasonable relationship" between the economic and social costs and benefits to be obtained from the application of the more stringent level of treatment. The EPA is also required, under the Act, to set special effluent standards for new industrial point sources, based on best available demonstrated control technology.

The Act requires that the Administrator of the EPA publish a list of toxic pollutants and effluent limitations or prohibitions for their discharge. Toxic pollutants are defined as those which, when assimilated either directly from the environment or indirectly by ingestion through food chains, will cause death, disease, behavioral abnormalities, cancer, genetic mutations, physiological malfunctions, or physical deformities in any organism or its offspring.

The Administrator is also required to issue pretreatment standards for industrial facilities discharging into municipal sewage treatment systems. These pretreatment standards will require industry to pretreat effluent before discharge to the municipal system so that the resultant effluent does not interfere with the operation of the municipal system nor pass completely through the system without adequate treatment. It is estimated that about 50 percent of all industrial facilities discharge their wastes into municipal systems, so it is easy to see why pretreatment standards are essential to achieve control over industrial effluent.

Under the new law, all states are required to develop a comprehensive and continuing planning process for water quality management. Plans are required to include controls for point sources, diffuse land runoff, and other nonpoint sources. Starting in 1975, all states are required to submit annual reports to the EPA which inventory all point sources, assess existing and anticipated water quality, and propose programs for nonpoint source control.

Enforcement under the Act is greatly simplified over earlier statutes. Put simply, it is unlawful to discharge pollutants from point sources except in compliance with a permit. The EPA has authority to enforce the law through either administrative or judicial channels. When a violation occurs, (either a permit violation or any other provision of the law), the Administrator may notify the polluter and issue an administrative order prohibiting further violation, or he may request the Attorney General to seek relief in the Federal Court.

Under the Act, private citizens may seek judicial relief against a polluter for violating an effluent limitation or an administrative order. Private citizens may also institute proceedings against the Administrator should he fail to perform as required by the law.

A considerably expanded Federal grant program is also an important element of the new Act. For the purposes of municipal treatment improvement, up to $18 billion was authorized to be made available for fiscal years 1973 through 1975. The Federal share was to be 75 percent. The remainder was to be matched by either municipalities or states.

A major concern of the new Act of particular importance, considering the level of appropriation envisioned, was the industrial use of municipal facilities in receipt of Federal construction grants. Under prior law, those industries using municipal plants automatically received a Federal subsidy for their waste treatment costs. Since industrial use of municipal treatment facilities is often more cost-effective for the industry, the new law did not foreclose the option, though it did lay down arrangements by which industry would pay their way. Industry using the municipal treatment option is required to pay a share of construction costs and user charges.

Finally, the financial portions of the Act reflect the congressional concern that funding for treatment plants should not be considered a continuing Federal handout. Some attempt is made to move municipalities towards self-sufficiency in maintaining and replacing facilities.

Viewed in comparison with earlier legislation, there is no question that P.L. 92-500 is far superior. There is considerable room for improvement, which is perhaps to be expected in such a complex piece of legislation. It is to be hoped that in the development of regulations, and in the process of court interpretation and congressional amendment, some of the more complex and convoluted provisions may be straightened out and rendered more serviceable.

1-4 THE HIERARCHY OF WATER QUALITY MANAGEMENT PLANNING

Planning and management programs are heavily stressed by the new Act in order to achieve and maintain clean water. The Act has established several new, distinct, interrelated levels of planning. Each with its own scope and purpose, and each dependent upon one another for ultimate effectiveness.

1-4.1 The Basin Plan (Section 303e)

The Basin Plan specifically deals with a river basin (the area drained by a river and its tributaries). The plan establishes, for the entire basin, the nature and extent of the pollution of its waters, in order to arrive at a strategy for bringing substandard waters up to water quality standards and to prevent the degradation of high quality waters. From this assessment, the allowable volume of pollutants which can be discharged into each segment can be determined, and loads can be allocated to each point source as a basis for permits. The Basin Plan is also concerned with establishing priorities for the construction and modification of treatment plants needed throughout the basin. Since areawide planning and facilities planning are both carried out within the boundaries of the river basin, both of these planning processes must conform to the management strategy and constraints of the basin plan.

1-4.2 Areawide Planning (Section 208)

According to the Act, areawide planning is generally confined to those areas of the basin where there are complex urban/industrial and nonpoint source water quality problems. The Governor of each State designates areas for the areawide planning program. The program specifically stresses planning and management by local governments (with State guidance and assistance) in finding and implementing solutions to water quality management problems. Section 208 also calls for the development of a special management structure to carry out the planning, and provides grants to help local governments cover their planning costs. Areawide planning also deals with urban nonpoint pollution sources, and attempts to integrate the water quality management program with other environmental and developmental activities.

1-4.3 Facilities Planning (Section 201)

Narrowing the scope still further, the 201 Facilities Plan deals with the actual facilities needed to achieve water quality improvement and maintenance. A 201 plan is needed to construct or modify a sewage treatment plant and its related facilities. The resulting highly detailed plan examines the problems experienced by the individual community and evaluates a number of alternative methods or techniques of wastewater treatment and disposal. The alternative chosen on the basis of economic, social, and environmental factors, achieves the most benefit for the least cost.

1-4.4 River Basin Planning Strategy (Section 209)

On a basinwide scale, the scope of a water quality management basin plan is broadened by a water resources basin plan (level B plan) prepared under Section

209 of the Act. This plan is intended to establish a broad strategy for long-term river basin management. Besides water quality, it includes considerations of water supply, transportation, recreation, and others. The 209 plan needs to build upon the 303(e) Basin Plan, as well as all 208 areas and specific facilities within the basin.

1-5 THE RIVER BASIN PLANNING PROCESS

The complexity of the water quality control problems in the United States requires that each State have an effective water quality management program. The Program should be capable of providing the basis for making centralized, coordinated water quality management decisions for the State, consistent with the national strategy of the U.S. Environmental Protection Agency (EPA).

The State planning process is directed toward establishing water quality standards designed to achieve the goals set forth in the Act, and is intended to be an effective and dynamic management tool at the State level to assist in implementing Federal and State programs designed to achieve clean water.

The broad goals of the 303(e) Continuing Planning Process are threefold:

- To provide the States with water quality assessment and program management information necessary to make centralized, coordinated management decisions.
- To encourage water quality objectives which take into account overall State policies and programs, including those for land use and other related natural resources.
- To furnish the strategic guidance for developing the annual State program required under Section 106 of the Act.

The State strategy contains a Statewide assessment of water quality problems and their causes (including nonpoint pollution sources), a listing of the geographical priorities of these problems, and a description of the State's approach to solving the problems identified. This data is included in the State's Annual (305b) Water Quality Report. Priorities and scheduling of permits, construction grants, basin and areawide plans, and other program actions needed to solve water quality problems are also included. Each State is also required to establish both a State municipal and a State industrial discharge inventory, and to rank significant dischargers.

A Basin Plan is a water quality management document which identifies the water quality problems of a particular basin and sets forth a remedial program to alleviate those problems. The value of a basin plan is that water quality management decisions can be made on a basinwide scale. Basinwide planning decisions can take advantage of the economies of large scale planning and also coordinate pollution control efforts on all waters of the common drainage basin.

It is an essential element of the State water quality management program, and is prepared under the State continuing planning process.

A Basin Plan provides for orderly water quality management. In each plan, water quality is assessed in terms of applicable standards, considering both point and nonpoint sources of pollution. Water quality and abatement needs are assessed in order to identify segment priority rankings and discharge inventories. Where permits have been issued, compliance schedules are set forth. Target abatement dates are established where permits have not been issued.

Needs and priorities are identified for Section 201 Facilities Plans and Section 208 Areawide Plans within the planning basin.

1-5.1 Specific Objectives of the Basin Plan

The elements of a Basin Plan vary with the water quality problems and decisions to be made in a particular basin. All waters within the planning basin must be classified as either water quality or effluent limited, the specific contents of the plan being dependent upon this designation.

A Water Quality Limited Segment (WQLS) is a segment where it is known that water quality does not meet applicable water quality standards, and which is not expected to meet water quality standards even after application of the effluent limitations required by the Federal Water Pollution Control Act, as amended in 1972. These effluent limitations are base level limitations consisting generally of Best Practicable Control Technology Currently Available (BPCTCA) for industrial point sources, and secondary treatment for municipal sources.

An Effluent Limited Segment (ELS) is a segment where water quality is meeting and will continue to meet applicable water quality standards, or where there is adequate demonstration that water quality will meet applicable water quality standards after the application of the effluent limitations required by the Federal Water Quality Act, as amended in 1972.

Basin plans have six major objectives:

1. Setting and recommending revisions of water quality standards, including the antidegradation statement, applicable to each classified segment of water in the basin.
2. Point source management provisions, including significant discharge inventories and data assembly.
3. Schedules of compliance or target abatement dates.
4. Wasteload analysis in water quality limited segments. For each parameter in violation of standards, point source load allocations are established to assure attainment of applicable, instream water quality standards.
5. A recognition of nonpoint sources in water quality limited segments.
6. The assessment of needs for municipal wastewater treatment facilities.

This assessment is used to develop a detailed cost estimate of future needs submitted annually to Congress through the EPA. The annual report forms the basis for allocations to the various States of Federal construction grant assistance.

1-5.2 Major Activities Conducted to Prepare the Basin Plan

Segment Delineation. River mile points, showing the beginning and the end of the main stem of the segment, should be identified. All the land area which is drained to a stream segment should be identified. Location of the segment in the basin should be described with respect to:

- Drainage areas directly upstream and downstream. This will establish the influence of this segment on other areas and the influences from upstream areas that are most important.
- Major urban centers or landmarks telling the reader that the segment surrounds city(ies) or notable landmarks will give him a better reference on the importance of this area to him as a worker or resident in the basin.

Basin Planning Data. General statements should identify the population category containing the majority of the population, and the general trend in the growth of this category during the planning period. Reference can be made to the size of this group with respect to the total for the basin, when this category represents 10 percent or more of the total for that category in the basin. Specific statements can identify the cities in the urban category. Trends in this area must be identified. For example, if a nonfarm city grows to urban status by 1995, this should be specified.

Housing units need not be discussed unless a major urban center is in the basin. Housing and industrial trends constitute a major nonpoint source problem. Therefore, any information on housing, i.e., type of residential housing, typical densities, and soil suitability, should be stated.

Employment data should be identified for the counties wholly or partly included in this segment. The total for manufacturing and nonmanufacturing categories should be given, and the specific labor markets in or adjacent to the segment can be identified. This will establish a basis for employment-related projections of industrial activity. The proportion of land use in each category—urban, industrial, residential, open space, etc.—should also be specified. The trend in each group is also important. Generally, urban land use will grow and the other categories will decrease (with stable trends in agriculture and silviculture). Areas where resource extraction occurs or is anticipated should also be identified.

A *major* land use activity in or adjacent to the segment may also indicate future water quality issues. Obvious activities in this category include major highway construction or airports. Probably the most important future project in a basin is the area affected by a proposed water resource project.

Hydrology. The character of flows in a segment should be identified. This can be done by identifying the difference in flows varying with: location in the segment, the seasons, and the influence of man-made forces. The discussion of variation in flow by location in the segment should be limited to the critical low flow event. The streams having some natural or augmented flow should be named.

The variation in flow by seasons should be limited to those streams where gaging stations exist. Therefore, the mean flow and the critical low flow can be identified for specific streams in the segments. Likewise, a statement that no gaging records are available for flow on nonaugmented streams in the segment may also be in order.

The variation of flows with respect to man-made forces is more difficult to determine. As the result of point source inventory, a tabulation of total waste flows for every stream in the basin is available. The values should be identified along with augment flows for each stream.

Obviously the first segment in the basin will receive the entire flow of the sub-drainage basin. In other segments tributaries may drain to a main stem, and the hydrology of such segments may change significantly at the confluence of each tributary. These changes should be stated in the hydrology section. Obviously, natural low flows can be very small. Therefore, these statements should address the average flows coming into the segment from the segments immediately adjacent to them, and the contribution of waste flow from *all* upstream segments.

Water Quality Data. Whereas, a dual approach in and out of segment is appropriate for discussing hydrology data, specific EPA guidelines exist for addressing water quality in the segment analysis.

Specifically, this data is divided into three categories:

- Water quality monitoring data.
- Water quality data used in modeling.
- Water quality standards and degradation.

Water quality monitoring data of specific interest in the segment analysis is contained in standards violations. This information can be augmented, in a basin plan, by statements concerning the value of the data presented in the table and the relationship of stations to one another. Specifically, the quantity of data available on a station should be identified, and the years of record for each of the criteria violated. For example, the concentration of dissolved oxygen (D.O.) may have been measured over a 30-year period on a monthly basis. Only one sample out of that period may show a violation.

The principal reason for locating stream monitoring stations is to identify their linear relationship to one another. This can best be done by describing the station where trends may appear, i.e., the station downstream from one or more stations in the segment may indicate persistent water quality character.

Information from special monitoring surveys should also be reported. Specifically, the data from the EPA Lake Eutrophication Studies. Also, in the event of fish kill incidents, the data and the stream in which they occur are specified.

Point Source Discharges. EPA guidelines require the following issues to be addressed:

- Each point source contributing to a Standard Violation must be identified.
- An inventory of point sources in the segment, giving existing facility and permit information, must be prepared.

A tabulation of existing discharge characteristics can provide all the specific information required for point source inventory. Since municipal and private discharges which are "significant" are to be highlighted, the discussion (text) can provide three levels of information:

- The number of point sources in the basin.
- The waste character from different types of point sources.
- The contribution of each point source type to the total wasteload in the segment.

The number of point sources in the basin will describe the magnitude of point source management required. Three important categories of point sources should be identified: municipal, private domestic, and industrial process. There are significant and nonsignificant levels of discharge within this group.

The waste character from different point sources basically requires the identification of industrial process water or cooling water contributed to a segment. Where information is available, an attempt should be made to identify nonsewage wastes received by municipal plants in the basin. For example, the basin plan should specify that discharge character estimates are based only on raw sewage character in the influent, and that the flow values are generally design values, because data on actual flow is not available.

In light of these constraints on the development of current wasteload character, the identification of waste generated in each segment may seem less valid. However, the data should be available to identify the contribution of waste from significant versus nonsignificant sources.

Finally, some indication must be made of the discharges contributing to standards violation at water quality monitoring stations. This will be limited to discharges in each segment unless it is immediately obvious that the source of a violation is a discharge in an upstream segment. The basis must be laid for establishing effluent limits by relating standards violations and secondary treatment requirements to specific point source(s).

Nonpoint Source Discharges. EPA regulations require the Basin Plan to address the following nonpoint source issues:

- Description of the nonpoint source.
- Identification of the streams affected by nonpoint pollution.
- Estimate of the significance of the nonpoint load, relative to point loads to the affected waters.
- Discussion of the relationship between any identified nonpoint pollution and existing water quality.

Specific discussion of nonpoint sources should identify the affected streams by name, and identify the extent of the problem. Nonpoint sources such as erosion may remain uncontrolled and, therefore, present a problem to more than one segment. The most important point to address in the basin plan is the affect this nonpoint pollution will have on meeting water quality standards in the segment.

All statements about nonpoint problems should be stated in terms of their current status. For example, if erosion from agriculture is a problem, the streams where it has been observed must be identified. If erosion from agriculture is only thought to be a problem, then the reason must be given. A similar type of statement can be made regarding unreclaimed strip mined land.

Wasteload Allocations and Priority. The Basin Plan must establish effluent goals for each segment and for every discharge in it. This requires identification of the values used in calculating the segment priority number and the significant municipal facility priority number.

In order to meet water quality standards in Water Quality Limited segments, a total maximum daily load (TMDL) must be spedified. This will prevent further degradation of these waters after the best available treatment technology is applied. The TMDL represents the level of waste disposal allowable throughout a segment for each water quality parameter currently violated.

Municipal Needs. EPA regulations require an assessment of municipal needs based on the following factors:

- Load reduction to be achieved by each significant municipal facility.
- 20-year service area population.
- Treatment and disposal improvements required to meet standards.
- Cost estimates for these treatment needs.

Likewise, schedules of compliance or target abatement dates must be identified for *all significant point sources*. Schedules of compliance are the treatment processes and construction timetable identified in approved NPDES permits. Target abatement dates are chronological points specified for the point source to:

- Get NPDES permit.
- Install or improve facilities.
- Apply for construction grants.

A table of compliance schedules should be identified in a basin plan as the proposal for specific water quality management actions in each segment for significant point sources. Discussion in the accompanying text can comment on the absence of permits and the specific knowledge used in specifying target treatment. Specifically, if metals removal or nitrification is needed, reference should be made to the appropriate information in the standards violation table, to water quality standards or the effluent limitations. For example, streams standards require ammonia-nitrogen (NH_3-N) levels to be less than 2 mg/l. Therefore, for discharges to low flow streams, some form of nitrification is required.

1-5.3 Content of the Basin Plan

To summarize, the EPA requires that a basin plan contain the following information.

A. Basinwide Information
   1. Maps
      a. Basin location in the State.
      b. Stream quality segment delineation.
   2. Water Quality Standards
      State criteria and water use classifications.
   3. Segment Classification
      A listing of segments of each body of water where water pollution abatement will be limited by effluent standards or water quality standards.
   4. Priorities
      a. A ranking of stream segments in the basin.
      b. Explanation of ranking system.
      c. Explanation of deviations from national priorities.
   5. Industrial Discharge Inventory
      a. Inventory of industrial dischargers.
      b. Ranking in order of abatement priority.
   6. Municipal Discharge Inventory
      a. Inventory of municipal dischargers.
      b. Ranking in order of abatement priority.
      c. Municipal facilities investment needs.
   7. Permit Issuance Target Dates for NPDES

B. Individual Stream Segment Information
   1. Existing Water Quality Data
      a. Water quality data.

   b. Hydrologic data.
   c. Hydraulic data.
   d. Note applicable water quality.
2. Inventory Existing Dischargers
   a. Location.
   b. Quantity, characteristics, and current treatment of each discharge.
3. Estimates of Existing Population, Employment and Land Use
   a. Population number and distribution.
   b. Employment number and distribution.
   c. Land use data.
4. Baseline Projections (at 5-year intervals of 20 years)
   a. Population.
   b. Employment.
   c. Land use.
5. Specify Water Quality Criteria Applicable in this Segment
6. Description of Existing Water Quality and Related Data
   Disaggregate this from general basin section.
7. Inventory Categorize and Rank Existing Dischargers
8. Nonpoint Sources
   General evaluation.
9. Determine Total Maximum Daily Loads (Water Quality Limited segments only)
   a. List violated WQC.
   b. TMDL for violated or expected violation parameters in the WQ segment
10. Wasteload Allocations
   a. Show coordination with:
      – "201" facilities plans.
      – "208" areawide waste treatment management plans.
      – NPDES permits.
   b. Base allocation on:
      – Accommodation of future growth.
      – Nonpoint sources.
      – Upstream contribution.
      – Levels of uncertainty/margin of safety.
11. Effluent Limitations
   Specify effluent target limitations for each significant point source in lb/day/parameter.
12. Treatment Plant Residuals
   Show quantity estimates and specific disposal sites.
13. Assess Municipal Facilities Requirements
   Municipal investment requirements.
14. Schedules for Significant Dischargers
   Compliance schedule for each discharger.

15. Monitoring
    Show monitoring program for each water quality limited segment.

1-5.4 Approaches to Implementing the Basin Planning Process

Water Quality Planning in Pennsylvania. Prior to the passage of the 1965 Amendments to the Federal Water Pollution Control Act, Pennsylvania used a system of stream classifications and effluent limitations as the basis for its water quality planning program. A water quality standards program was implemented after the passage of the Federal Act and water quality standards were established for all interstate waters, prior to the June, 1967 deadline specified in the Act.

Following the adoption of standards for the interstate waters, (required by the 1965 Water Quality Act), the water quality standards program was expanded to cover all intrastate waters.

All waters must support a broad spectrum of uses including water supply and fish, and other aquatic life. In carrying out the water quality standards phase of the planning program, Pennsylvania attempts to obtain the best input possible from conservation agencies and organizations as to the water uses that should be protected for a given stream segment.

In establishing water quality standards, the Commonwealth solicits input from various agencies and organizations. Public hearings are held to obtain as much information as possible to aid the Environmental Quality Board in amending its regulations to incorporate water quality criteria.

In order to carry out the overall planning process, the Federal Water Pollution Control Act Amendments of 1972 assign each level of government–local, state, and federal–a responsible and needed planning task. The States are expected to play a role in all elements. Their major task; however, is one of developing and implementing regulatory standards and criteria, along with certain ancillary tasks as set forth in Section 303(e). Due, however, to previous federal requirements and to state pollution control and water resources statutes, Pennsylvania was already extensively involved not only in the activities expressly required under Section 303(e), but also activities that serve as the basis for statewide compliance with the areawide waste management planning requirement.

Five broad planning areas are included in Pennsylvania's Comprehensive Water Quality Management Planning Program (COWAMP):

- Areawide municipal and industrial wastewater management.
- Control of sewer overflow and agricultural runoff pollution.
- Maintenance of groundwater quality.
- Control of acid mine drainage.
- Disposal of sludges and other treatment-process residuals.

Pennsylvania's Clean Streams Law and the 1972 Federal Water Pollution Control Act (P.L. 92-500), allow COWAMP to serve four major state-level functions:

1. Define future water pollution control regulations, policies, legislation, and enforcement actions.
2. Evaluate sewerage project priorities and eligibility for Federal and State funding.
3. Review water resource and land development projects affecting surface and groundwater quality.
4. Meet U.S. EPA water quality management planning requirements.

As these statements indicate, Pennsylvania is complying with the intent of the Federal law in its current planning efforts. However, a direct and rigid adherence to Basin Planning Guidelines is absent.

Water Quality Planning in Kentucky. Kentucky has followed 303(e) Guidelines more closely. In order to achieve river basin planning objectives, the State of Kentucky elected to produce basin plans by contracting with outside consultants. During FY 1974, "Water Quality Planning in the State of Kentucky" was prepared under contract with Roy F. Weston, Inc. This publication included a detailed analysis of Area Development Districts water and sewer facility plans as they relate to the requirements in 303(e) Basin Planning and a scope of work which outlines Federal requirements under 40 CFR 130 and 131 for river basin planning.

Recognizing the importance of this effort, a state coordinating and review team was assembled which represented the following interests: facility planning, land use, commercial and industrial development, legislative authority, and implementation of the river basin plan. During 1974, a major output was obtained, in the form of a wasteload allocation procedure as developed under contract with the University of Kentucky and the University of Louisville. Through the procedure developed in this contract, the State of Kentucky had an accurate method of assigning wasteloads to any water quality limited segment which involved dissolved oxygen. The method employed is on the safe side; in other words, it provides for a lower strength factor than would be needed to satisfy water quality standards.

The river basin plan also used the "NEEDS" Survey to interface with the plans. The river basin plans can be updated as construction costs indexes for sewer and sewage treatment plant change.

By July 1975, River Basin Plans had been prepared for all areas in Kentucky. The format for each plan, as well as the planning principles used in developing all plan outputs, were similar. (An example of the format is provided in Figure 1-3.) This presents the strictest compliance with 303(e) Basin Planning Guidelines.

Introduction and Summary

Project Scope: A description of past and present water quality management planning in Kentucky including the basis of the plans.

Report Scope: A description of the principal outputs of the River Basin Planning process and the location of specific items in the plan.

Description of the River Basin

Introduction: Describe the contents of this section.

Basin Physiography: A description of the major geohydrologic sub-divisions of the river basin, including data on the character of surface and subsurface soils, topography and groundwater table.

Basin Land Uses: A description of the major urban, industrial and undeveloped resources of the basin, including the location and density of population centers, the location and productivity of developed resources, and the location and potential of developable resources.

Basin Population: A description of all incorporated population units in the river basin including projections, distribution, migration and settlement trends.

Basin Economic Resources: A description of the major economic resources of the basin including extracted resources, manufacturing resources, recreation resources and employment resources.

Basin Wastewater Control: A description of the most common methods of wastewater collection, treatment and disposal for various population densities, and industrial and agricultural operations in the basin.

Basin Wastewater Control Needs: A description of the wastewater collection, treatment and disposal alternatives available to various types of wastewater generators to meet the requirements of Federal Law and Kentucky's Water Quality Standards.

Water Quality Management in the River Basin

Introduction: Describe the contents of this section.

Segment Classification: A tabulation of stream segments and supporting material including:

- Method for Water Quality/Effluent Limited Classification,
- Method for formulation of Segment Priority,
- Location of each segment,
- 1975 Segment Classification, and
- 1975 Segment Priority.

Significant Municipal Discharges: A tabulation of significant municipal discharges and supporting material including:

- Method for Significant Discharge Designation,
- Method for formulation of Abatement Priority,
- Owner of discharging facility,
- Geographic location of discharge,
- Segment in which discharge occurs,
- 1975 abatement priority, and
- NPDES Permit or Permit Application No.

Fig. 1-3. Format of river basin plans developed in Kentucky.

Significant Industrial Discharges: A tabulation of significant industrial discharges and supporting materials including:
- Method for Significant Discharge Designation,
- Method for formulating Abatement Priority,
- Owner of discharging facility,
- Geographic location of each discharge outfall,
- Segment in which discharge occurs,
- 1975 abatement priority, and
- NPDES Permit or Permit Application No.

Significant Non-point Sources: A tabulation of hydrologic subdivisions in which dispersed sources of pollution represent a significant water quality management problem. The information presented should include:
- Geographic area contributing to the non-point pollution load,
- Segment in which the pollution is observed,
- Principal pollutant, and
- Federal, State or local programs which can respond to the non-point source management needs.

Land use Policies and Controls: A summary of official policy and legislative controls available under Federal, State and local statutes for the control of projected growth in the river basin, including:
- Regulatory implications of the River Basin Plan, and
- Sufficiency of State and Local statutes to comply with requirements of P.L. 92-500.

Wastewater Residuals Management: A tabulation of types and sources of wastewater residuals and management procedures, including:
- Character of residuals from various sources in basin,
- Methods for ultimate disposal of residuals in basin,
- Projected residuals generated by all sources,
- Sites licensed to receive wastewater residuals, and
- Projected capacity of licensed disposal sites.

Water Quality Monitoring and Surveillance: A tabulation of instream monitoring requirements and point source monitoring, including:
- Location of monitoring stations,
- Segment monitoring parameters,
- Sampling and analyses frequency,
- Length of previous records,
- Self-monitoring requirements, and
- Proposed changes to monitoring and surveillance program.

Planning Coordination: A description of pertinent Federal, State and local activities in the river basin, including:
- Ongoing water resource projects of the Corps of Engineers and Soil Conservation Service,
- Ongoing water projects of the Kentucky Department for Natural Resources and Environmental Protection, and
- Significant Plans and Studies of Water and related Land management in the river basin.

Bibliography: An inventory of all plans, reports, studies, graphics, maps and data consulted in the preparation of the plan for this river basin.

Fig. 1-3 (Continued).

Appendices

Appendix: Segment Analyses

Introduction: Describe the contents format of the segment analyses which follow.

Segment Analysis: _ _ _ _ _ (Code number developed to uniquely describe the segment) This analysis should provide information supporting the segment classification and abatement priority, including:

- Geographic location of the segment.
- Major political subdivisions within the segment.
- Segment Classification.
- Segment Priority.
- Standards violations in WQ segments, including:
  Sampling location,
  Parameters sampled, and
  Dates of violations.
- Discharge information, including:
  Owner of discharging facility,
  Facility code number (may be used to identify facility on segment map),
  Geographic location of discharges,
  1975 abatement priority,
  Treatment and disposal process,
  Effluent characteristics,
  Effluent characteristics proposed,
  Load reduction proposed,
  1980 treatment & disposal investment needs,
  NPDES permit or application no., and
  Statutory compliance date.
- Maximum allowable daily load in WQ segments.
- Influence of non-point sources on the segment classification.
- Alternative Water Quality management plans for Water Quality segments.
- 201 planning areas.
- Segment map showing:
  Political sub-divisions,
  Streams,
  Segment boundary,
  Significant discharges, and
  Monitoring stations.

Appendix: Summary of Public Participation

Introduction: Describe the contents of this appendix.

Public Information/Education: A description of the measures taken to provide public awareness of the planning process and the means for public responses.

Public Response: A compilation of formal responses of individuals and organizations to the discretionary elements of the proposed Basin Plan, including the text of all public hearings.

Modification of the Basin Plan: A presentation of the rationale used to modify the Basin Plan or support the status quo in response to formal public response.

Fig. 1-3 (Continued).

Graphics

– Basin Map Showing:
Basin boundary,
Political subdivisions,
Segment boundaries,
1975 Segment classification, and
1975 Segment Priority.

Fig. 1-3 (Continued).

1.6 THE 208 PLANNING PROCESS

The 208 program stresses planning by local governments. Local governments in a particular planning area are invited to work together to achieve solutions to their common, water quality management problems. The 208 program gives local planning agencies a means of jointly solving these problems, instead of the simple application of a standard approach to waste treatment, which would be less likely to succeed. The 208 program helps local governments to find integrated solutions that deal with both environmental and management issues. Finally, and most importantly, the program provides grants to help local governments cover planning costs.

The program sets up a management structure to insure that the 208 plan can be carried out effectively. The result is a two-part approach to solving water quality problems.

- Systematic planning to find workable solutions.
- Follow-up management to insure that the solutions are carried out properly.

This reliance on management as an essential component of areawide planning applies nationwide.

Areawide planning brings together a number of varied aspects of water pollution control under one planning program. The treatment of municipal and industrial wastes is an important part of a program that includes the development of a discharge permit system which requires industries to reduce their discharge of pollutants over a specified period. It is intended that planning should ultimately develop mechanisms for dealing with "nonpoint" sources of pollutants, such as stormwater runoff from cities and farms, erosion at construction sites, and the seepage of chemicals from mines and areas that rely on septic tanks for treating wastes. In addition to tying these disparate aspects of pollution control together under one planning system, 208 planning also provides a structure to coordinate the water program with other environmental activities, such as air quality improvement and solid waste management.

Areawide planning also looks at the issue of development in terms of its environmental implications. In the long run, the program should enable planners to consider the environmental implication of such related issues as land use and development by both the public and private sectors.

In many respects, areawide planning provides a unique opportunity to solve pollution problems and make progress in the overall improvement in water quality. In an urban area, an all-inclusive plan could be put together that might cover a wide variety of problems affecting water quality. The plan (once approved) could be used by state and local governments to manage the area's waters in the future, as well as to solve the current problems.

Areawide planning expands the scope of a 201 facilities plan. A 208 planning area can include one or more 201 facilities planning areas and, in such cases, the areawide approach can build upon the contents of completed facilities plans. In short, the 208 plan permits a more comprehensive view of the problem and its possible solutions.

1-6.1 Specific Objectives of 208 Planning

According to the EPA, information included in an areawide waste treatment management plan should identify:

- All treatment works needed to maintain acceptable water quality over a 20-year period. The list will be revised annually to reflect changing conditions. In addition, the plan establishes priorities for building or modifying treatment plants, and offers time schedules to put those priorities into effect.
- Mechanisms to coordinate all waste treatment planning activities within the area in question. This insures that all such planning contributes to an organized and integrated solution to the problem. To do this, these mechanisms regulate the location, modification, or construction of all facilities within the area. They also insure that industrial wastes are properly pretreated to remove incompatible or toxic substances before they are discharged into a municipal treatment system.
- The management structure to be responsible for implementing the plan. This may range from a single agency to a number of agencies and government organizations. It must have the legal authority to take whatever actions are necessary to carry out the plan, either directly or through contractual agreements. It will also manage all included waste treatment works and other related facilities. Inclusion of this management provision is one of the features that gives 208 planning its unique potential. Therefore, the management system identified in the plan must be specifically approved by EPA before the plan can be put into effect.
- Means of defining and dealing with "non-point" sources of pollutants. This is especially important in relation to agricultural, forestry, mining, and construction activities, as well as saltwater intrusion into groundwater supplies due to reductions in flow of fresh water from any cause.

- Other information needed to carry out the plan. This includes data on financing arrangements, time requirements, the anticipated cost of implementing the plan, and its probable economic, social, and environmental effects on the planning area.
- Methods to dispose of the residues and treated wastewater that remain after treatment. This involves defining the problems associated with land and subsurface disposal of these wastes. Doing this insures that the many safe uses that have been found for sewage sludge and treated wastewater will be given full consideration.

Collecting such diverse and detailed information allows planners to solve a particular area's water quality control problems in a way that is socially acceptable and economically sound. Of course, the plan does more than just offer solutions; it also provides the management structure that will be needed once the plan has been approved (Ref. 1-4).

1-6.2 Major Activities Conducted to Prepare the 208 Plan

The purpose of the 208 planning process is to evaluate various alternative methods of achieving water quality goals and formulate a plan that can be implemented by a management agency. The technical portion of the planning process involves:

- The identification of priority water quality problems in the area.
- Recognition of any constraints in dealing with the problems.
- Development of alternatives to achieve water quality goals.

Management planning (the selection of a management agency or agencies and development of appropriate institutional arrangements for plan implementation) should be conducted in coordination with technical planning. Management planning should identify water quality management problems, and evaluate the ability of existing agencies to carry out the management requirements of the 208 program. Institutional problems (lack of authority, or lack of financial capacity) should be identified in the plan. Alternative means to acquire proper authority, financial capacity, and effective institutional arrangements should also be evaluated. Finally, alternative management agency(s) and institutional arrangements should be evaluated and a single alternative selected.

The development of alternative technical and management portions of the plan, and the final selection of an areawide waste treatment management plan requires public participation throughout the planning process.

Sources of Information for 208 Planning

303(e) Basin Plans and Facilities Plans. Information from 303(e) basin plans and 201 facilities plans provide the basic inputs for 208 planning. 201 Facilities plans under Title II of the Act, or preceding facility plans under 18 CFR and

Sec. 3(c) of the Water Quality Act of 1965, should be coordinated with 208 planning.

Update information on discharges into navigable water should be available through the National Pollutant Discharge Elimination System (NPDES). The specific terms and conditions of any permits already issued to dischargers should be accounted for in formulating pollution control strategies in the 208 plan.

Related Water Management Information. Much of the information necessary for developing an effective 208 plan should be available from any one of the following related water management programs and studies:

- Basin Studies conducted under the Water Resources Planning Act.
- Urban Studies (U.S. Army Corps of Engineers).
- Floodplain Information Studies (Bureau of Housing and Urban Development, U.S. Geological Survey, and U.S. Army Corps of Engineers).
- State and Local Water Supply Studies.

208 Goals Other Than Water Quality. While the 208 plan is concerned with water quality, it should be recognized that selection of a final plan could affect other community goals. It is important in the planning process to establish an understanding of community goals and plans, especially with respect to housing, economic development, transportation, education, recreation, and other environmental goals. The relationship between these broad planning goals on the one hand, and water quality and environmental goals on the other, should be understood from the outset of the planning process. It is in this area of emphasis upon integrated planning that the 208 program may have the greatest impact upon the comprehensive planning process.

Water Quality Problems and Standards. The objective of 208 planning is to achieve the 1983 water quality goals. An assessment should therefore be made of those existing and projected water quality problems which should be solved to meet those goals. Water quality standards which correspond to the goals should then be specified in the plan. It is the responsibility of the states to set standards and the responsibility of the EPA to review and approve standards. The 208 planning agencies can also develop proposals for water quality standards revision.

The 208 planning guidelines suggest that an initial assessment of water quality conditions should be made on the basis of existing data and a general familiarity with the planning area. Knowledge of industries, waste treatment, nonpoint source pollution generation, land use, and receiving water quality all aid in this evaluation.

The guidelines go on to state that water quality standards must be reviewed by the states at least once during each three-year period, beginning with the passage of the Act. Upon review of standards, recommendation for their revision may be considered and adopted. The 303(e) basin plans should then be revised accordingly during the next basin plan review.

Since 208 planning is concerned with pollution control over a 20-year planning period, it is rightly required that control levels should be based on the water quality standards applying over the planning period. Thus, in order to develop 208 plans, it is necessary to make assumptions concerning the change in standards, and any such assumptions should be made only after consultation with the state.

Source Inventory. Discharge inventories provided in 303(e) basin plans should be reviewed and compared with information provided by the NPDES Program in order to identify and locate all significant dischargers. Special attention should be given in the plan to the adequacy of information on storm and combined sewer discharges, as well as any point source discharges to groundwater.

Nonpoint Sources. Any existing nonpoint source information should be used to identify monitoring and modeling needs. Where nonpoint sources are a significant pollution problem, they should be identified by use of generalized prediction techniques. Some monitoring of water quality responses to various land uses should be conducted to establish the reliability of these techniques.

Technical Planning. The purpose of technical planning is to develop a coordinated water quality management strategy for abatement of water pollution in areas unable to meet 1983 water quality goals. The strategy may be a combination of the following:

1. Municipal wastewater treatment systems.
2. Industrial wastewater pretreatment or treatment.
3. Residual waste management.
4. Urban stormwater management.
5. Nonpoint source management.

The plan is expected to implement these abatement measures through the use of existing regulatory procedures, such as local ordinances, capital construction of wastewater treatment facilities, and improved management of stormwater.

In a similar manner, technical planning should emphasize water quality problems that are capable of solution with existing technology and current sources of funding. The water quality problems that should receive priority are municipal and industrial point source problems. Next in importance should be those nonpoint source problems that can be dealt with through better management practices. Lower priority should be placed on nonpoint source and stormwater problems that require large capital investments. For each type of water quality problem, there should be an overall preference for regulatory approaches which will prevent problems from occurring in the first place.

Water Quality Modeling

The Basis for Modeling. Modeling allows description of the physical, chemical, and biological processes which cause changes in water quality, resulting from the introduction of point and nonpoint sources of pollution.

Priorities for modeling should be those pollution problems that cannot be solved with existing technology. The detail of analysis required for these different water bodies and sources of pollutants should be identified early in the planning process. This will require identifying:

- The type of analytic techniques.
- Need for additional data collection.
- Need to consider innovative (nonstructural) pollution control methods.
- Management alternatives to implement the controls.

In general, these pollution problems should be analyzed in sufficient detail to enable a cost-effectiveness analysis (Ref. 1-5) of reasonable alternatives.

The EPA has developed a system for classifying waters according to the level of federally mandated effluent limitations required to meet Water Quality Standards in a particular body of water. Each body of water is classified as "effluent limited," unless high levels of abatement are needed to meet the Water Quality Standard, in which case the body of water is classified as "water quality limited."

The use of water quality modeling in 208 planning will help identify how structural controls must change over the planning period under the influence of such factors as the stringency of the standards adopted to meet 1983 water quality goals; or the change from one level of effluent limitations to another as raw waste generation increases.

To initiate the water quality modeling for structural pollution abatement on other bodies of water, design conditions (stream flow conditions) should be established. Design conditions will include range of flow, temperature, and meteorological and seasonal conditions. Design conditions for point sources should consider pollution under low flow, dry weather conditions. For nonpoint sources transported in runoff, design conditions should consider rainfall-related high flow conditions.

Selecting an Analytical Model. A number of analytical techniques exist for a wide spectrum of design conditions. The techniques selected should be able to:

- Consider the relationships of nonpoint sources to water quality.

- Allow an estimate of a single quantity for all nonpoint sources contributing to a given body of water.
- Consider the magnitude of the loads associated with each category of nonpoint sources.

Application of the Model. An inventory of existing water quality data should be prepared in a form suitable for input to the model. Data collected through a monitoring and surveillance program should be used to supplement and strengthen this data base, for areas where specific pollution problems are known to occur.

A calibration procedure should be established to determine the value of the coefficients in the analytical model. These values should then be used to verify a previous condition, event, or time period, which is independent from the conditions used for calibration. The model may then be used with measurable confidence to predict water quality responses to changes in point and non-point source wasteloads. The interpretation of analyses conducted for various load conditions will depend in part on the results of the verification and degree of confidence which exists in the reliability of pollution abatement controls.

Projections. To plan waste treatment and pollution control over a 20-year period requires assumptions about population, employment, and land use. Based upon these assumptions and estimates of the wasteload generation characteristics of different activities, future pollution control needs can be projected.

Using Population, Employment, and Land Use Data. Existing population, employment, and land use data should be used to establish existing patterns of point and nonpoint sources, and as a basis for projecting the amount and spatial distribution of future wasteloads. These projections span 20-years in five-year increments. Considerations which should be used in establishing the projection methods include:

- The effect of the 208 plan itself.
- Local growth policies.
- Plans for attainment and maintenance of air quality.
- Regional plans for transportation.
- Solid waste management.
- Water supply.
- Public investment.

The EPA guidelines for 208 Planning state that "Population data are available from the Bureau of Economic Analysis (U.S. Department of Commerce). This data should be updated with the most recent demographic and economic projections developed by the Bureau of Economic Analysis (U.S. Department of

Commerce) and the Economic Research Service (U.S. Department of Agriculture), and with projections used as a basis for state planning for air quality management. In general, the SERIES E projections developed by BEA or comparable projections developed by the state should be used as overall growth projections for the planning area." EPA requires that the use of any projections that deviate significantly from BEA be justified.

Projecting Wasteloads. EPA 208 planning guidelines require that a materials balance be calculated to relate instream water quality to the contribution and transport of pollutants. For nonpoint sources, a materials balance should be based on average factors for wasteload generation per unit of land area and activity.

By utilizing the baseline projections and factors of wasteload generation per unit of area or unit of activity, future wasteloads can be projected, covering land areas and residential, commercial, industrial, and nonpoint sources. It would be advisable to disaggregate those land uses into more specific categories, where specific point and nonpoint control alternatives are to be considered.

Developing Technical Alternatives. Modeling of wasteloads will identify the wasteload reduction which must be achieved by each control option in the detailed development of alternatives for point and nonpoint sources. For each unique point and nonpoint source a control alternative must be identified. The EPA has suggested that these alternatives be grouped into "subplans," according to the category of sources and design conditions to which they apply. For instance, continuous point sources, such as municipal treatment works and some industrial point sources, may comprise one subplan category while intermittent point sources, such as storm and combined sewer overflows or seasonal industrial wastewater discharge, would be another subplan. All nonpoint sources would generally be included in one subplan. Although the categories chosen for subplans may be as detailed as the 208 Planner sees as necessary, the EPA does require each subplan to specify the following:

- "Wasteload characteristics of each alternative, expressed in appropriate units.
- Total cost of each alternative, expressed at its present value or average equivalent value of capital, and operating costs for the overall alternative and subsystem components.
- Information on the reliability of each alternative and subplan included in each alternative.
- Significant environmental effects of each alternative consistent with NEPA procedures, including a specific statement on future development impact.
- Contribution of each alternative to other water-related goals of the planning area." (Ref. 1-5)

Producing a Plan from Technical Subplans. Since the alternatives in each subplan are numerous, only those that can be implemented by the management agency(s) should be considered. Therefore, technical subplans should be integrated with the results of management planning.

Implementable subplans should be combined into alternative areawide technical plans for final evaluation and selection. This evaluation and selection process should consider public preference, but rely heavily on the comparison of information required in the "cost-effectiveness analysis" of the alternatives.

Management Planning. The purpose of management planning is to develop institutional arrangements through which the 208 plan can be implemented. The standards that should be met to allow the management agency(s) to properly carry out its responsibilities are:

- Adequate legal authority.
- Adequate financial capacity.

To ensure plan implementation, the management agency(s) must be able to handle the following activities:

- Program administration.
- Regulatory program.
- Waste treatment program.
- Program coordination.

One of the most important functions of the 208 management agency(s) is to provide general direction for the implementation of the plan, including:

- Agency supervision.
- Coordination with local agencies.
- Ensuring that the 208 program is being implemented.
- Updating the areawide plan and implementation mechanism, as required by changing conditions; and
- Generating capital resources needed to implement the regulatory and waste management programs.

The EPA guidelines for 208 planning indicate that a regulatory program should be the vehicle for enforcing the abatement measures that have been developed through the technical planning process. A waste treatment program includes all the capital construction responsibilities to carry out a 208 plan, while program coordination represents a formal strategy for establishing needed institutional arrangements. The arrangements include one or more agencies which may be existing or under creation, and an explicit arrangement for ensuring overall supervision and enforcement of the 208 plan.

Initiating Management Planning. The initial step is planning the management process should be an analysis of the area's water quality management resources.

This analysis should evaluate the capability of existing agencies to meet the management requirements of Federal law and identify any authority needed to satisfy these requirements. The EPA 208 Planning guidelines indicate that the following analyses should be provided:

"1. An assessment of the specific legal authority required under Subsection 208(c) (2) and 208(b) (2) (C) to carry out the regulatory and waste management programs, and an approach for acquiring such authority.
2. An evaluation of existing financial arrangements to determine what changes will be necessary to provide affected agencies with the capacity to meet financial needs and obligations for carrying out general management responsibilities, the regulatory program, and waste management program.
3. An assessment of the potential of existing institutions in the area to perform the required functions. An overall assessment should be made of the area's potential for regional water quality management." (Ref. 1-5)

Developing Alternative Management Plans. Upon completion of the management analysis, alternative management plans should be developed. Close coordination with technical planning is necessary to ensure consistency of the management plan alternatives with alternative technical plans. The management alternatives should be evaluated in terms of their feasibility and the viable management alternatives identified.

Producing the 208 Plan

Technical and management planning should produce a series of alternative technical plans which can be implemented by one or more alternative management plans. These alternative technical and management plans should be simply combined to produce several alternative 208 plans. For each alternative areawide plan, the EPA 208 planning guidelines specify that sufficient detail be provided to "enable accurate evaluation of the plan in terms of meeting 1983 water quality goals," (Ref. 1-5) including:

- Capability to achieve water quality.
- Goals of the area.
- Technical reliability.
- Monetary costs.
- Environmental, economic, and social impacts.
- Implementation feasibility.
- Public acceptability.

Selection of the 208 plan must include public involvement in a discussion of the merits of the various alternatives in a formal public hearing process.

To summarize, through Section 208 of the Federal Water Pollution Control Act Amendments of 1972, local areas are provided with an opportunity to plan and manage a program integrating control over such activities as municipal and industrial wastewater, storm and combined sewer runoff, nonpoint source pollutants, and land use as it relates to water quality. The planning approach should focus initially on the most serious water pollution problems and, within time, resolve the remaining water quality issues. Particular emphasis should be placed on nonstructural approaches to pollution control, especially where these can be shown to reduce the large capital costs associated with traditional structural methods. A sound approach to areawide waste management planning should be based on the following planning features:

- The identification of pollution problems in terms of their relative impact on water quality. Similarly, existing institutional problems impeding the solution of water quality problems should be identified.
- The identification of both technical and management constraints.
- The identification of possible solutions to problems relating to all reasonable regulatory and management control methods.
- The development of alternatives integrating technical and regulatory control methods for municipal and industrial wastes, stormwater control, nonpoint source control and growth and development.
- The alternatives should be evaluated in terms of minimizing overall costs, maintaining environmental, social, and economic values, and assuring adequate management authority, financial capacity, and institutional feasibility.
- The selection of an areawide plan should be based on a systematic and publicly approved comparison of the alternatives.

A planning process based on these provisions should achieve the planning objectives of Section 208. In seeking to meet these objectives, planning agencies can use any other formal planning process, as long as that process addresses the major issues of areawide waste management.

1-6.3 Content of the 208 Plan

In providing for the most immediate needs for waste management, the contents of the 208 Plan should be applied to such specific outputs as technical requirements for permits, the location of and treatment levels for facilities, initial construction priorities, and feasible measures leading to the control of nonpoint sources discharge problems. The overall contents of a 208 Plan will include the following outputs:

- Identification of anticipated municipal and industrial treatment works construction over a 20-year period.

- Planning for facilities eligible for grant assistance expected during the five-year period following 208 plan approval.
- Identification of required urban stormwater runoff control systems.
- Establishment of a regulatory program to provide for waste treatment management on an areawide basis and for identification evaluation, and control or treatment of all point and nonpoint pollution sources.
- Identification of agencies necessary to construct, operate, and maintain facilities required by the plan.
- Identification of nonpoint pollution sources.
- Identification of agencies necessary to construct, operate, and maintain facilities required by the plan.
- Identification of nonpoint sources of pollution related to agriculture, mining, construction, and procedures and methods (including land use requirements) to control those sources.
- Processes to control the disposal of residual waste and land disposal of pollutants to protect ground and surface water quality.
- Selection of a management system to implement the plan and identification of the major management alternatives which include enforcement, financing, land use and other regulatory measures and associated management authorities and practices.
- A schedule for implementing all elements of the plan, including identification of the costs of implementation.

1-6.4 Approaches to Conducting the 208 Planning Process

Although there are few completed 208 plans (at the time of this writing mid 1977) the approach taken in the Commonwealth of Pennsylvania is interesting because of its intent to coordinate 208 Planning with ongoing regional water management planning. Although the Commonwealth had been actively involved in all aspects of water quality management for a number of years, there was no formal state program for the planning of waste management systems until 1969. In that year, the legislature enacted the Pennsylvania Sewage Facilities Act (35 P.S. 750.1 et seq.), which requires that all municipalities in Pennsylvania submit to the Department of Environmental Resources (DER)—an Offical Plan for sewage, setting forth the extent of existing public systems, expected additional wastewater disposal needs, and proposed measures to meet expected needs. Among other things, the Act specifies that plans can be prepared jointly by two or more municipalities. As a result, plans have been prepared on a county level in 64 of Pennsylvania's 67 counties.

The federal requirements for this wastewater management planning have now been amplified and modified by Section 208. Consistent with these changes, DER modified its present Comprehensive Wastewater Management

Planning Program so that the planning and management capability required by Section 208 will have been established throughout the state.

The existing HUD sponsored Areawide Planning Organizations (APO) participate directly in studies to be performed in the complex urban-industrial areas of the state, although the state conducts the planning process. Specifically, the APO's assist the Department of Environmental Resources in the preparation and the implementation of the water quality management plan. The APO's role includes technical and coordination responsibilities, and pertain to such matters as:

- Participation in preliminary work plan development.
- Preparation of the land use projections to be used in the study.
- Provision of secretariat services for the Public Participation Program.
- Active participation in the technical and policy level of decision-making process.

At the conclusion, a plan is produced whose maintenance in metropolitan areas will be turned over to qualified APO's. Responsibility for plan maintenance in non-metropolitan areas will remain, as in all other phases of 208 Planning with DER.

1-7 THE 201 FACILITY PLAN

Facility planning is required as the initial step under the construction grant program for municipal and publicly-owned wastewater treatment works. The planning area generally includes a metropolitan area (a municipality and its extraterritorial area) or a special service district.

Planning considerations, although they played a minor role, have been a part of the wastewater treatment design reports compiled during the past. However, current federal requirements and other planning considerations are now an essential element in this process.

The major activities conducted in the 201 facility planning process are outlined in Figure 1-4. It may appear that the activities follow one after the other. However, a large amount of interaction occurs, and the order of these activities can be simultaneous and iterative. Current modifications in federal laws have added an impact assessment and preparation of alternative plans to the wastewater treatment planning process. The impact assessment and refinement of alternatives must also be accomplished in part through public participation.

Each of the activities in Figure 1-4 have been required by federal regulations. The overriding purpose of these regulations is to assure that the rationale used in dispersing federal money for construction of public wastewater treatment facilities conforms with the intent of existing federal environmental and civil rights laws.

- establish water quality management goals
- initiate a public participation program
- develop a detailed inventory of existing waste treatment systems and determine existing flows
- prepare an environmental inventory
- estimate future wasteloads and flows
- develop technical alternatives
- develop implementation arrangements
- prepare a cost-effective analysis
- select a 201 plan from the alternatives

Fig. 1-4. The facility planning process.

1-7.1 Initiating the 201 Planning Process

As a minimum, the Federal government requires that publicly-owned works provide secondary treatment. However, community goals for water quality should also be identified, and more stringent requirements for public wastewater treatment and disposal can be established. These goals should be reviewed through a public involvement process, in which other public goals may be identified or priorities for these goals may be established.

Public participation in the facility planning process is required by federal law. Public participation should be initiated in the early stages, and continue throughout the planning process. It should focus on identifying public interests and providing opportunities for those interests to be expressed to public officials.

Issues that should be defined in the public participation process include:

- Environmental impacts of proposed technical alternatives.
- Economic feasibility of structural and nonstructural wastewater abatement methods.
- The practicality of proposed arrangements to implement the plan.

Finally, a hearing must be held to allow the public the opportunity to make final comments on its acceptability.

1-7.2 Preparing the Background for the 201 Plan

The environmental assessment process required of all public projects by NEPA is an integral part of the planning process. The assessment is based on an inventory of existing environmental conditions and on analyses of expected environmental effects of implementing various technical alternatives. EPA guidelines for 201 planning suggest that the environmental inventory should begin early in the planning process, so that it can have some impact on the initial assumptions, development, and evaluation of alternatives.

The major objective of the environmental assessment is the determination of the extent or frequency of impact caused by facility development. Primarily, this requires anticipating where there will be an interaction between development and sensitive environmental areas.

Anticipating areas of interaction or impact is best achieved by communication with those developing the preliminary facility-design concept (the planners) and engineers preparing the 201 Facility Plan). Before the actual site for treatment and disposal has been selected, many design parameters will be known. Using existing maps showing the facility planning area, a sketch can be made of the configuration and phasing of service area expansion. Alternatives which might alter the site of treatment and disposal should be discussed, and suitable alternative sites within the proposed service area should be delineated. In a similar manner, all developable areas which might change the current land use as the result of service provided by the facility(s) should be identified. The resulting graphics will represent the major zones of influence of the proposed facility development.

As in other planning (303e, 208), forecasts of wasteloads and flows should be based on projection of existing land use, economic and demographic trends, growth constraints, zoning restrictions, and discharge permit conditions, as well as the institution of water reuse and conservation measures. As an initial step in alternative development, the EPA 201 planning guidelines require that the optimal performance of existing facilities establish the basis for each technical alternative. The alternatives themselves may consider methods of improving the collection treatment or disposal of sanitary and combined sewer wastewater and their residuals. These alternatives are identified in terms of their wastewater abatement efficiency, and are subjected to impact assessment and evaluation, in order to select the best alternative.

There are three essential management planning elements of a 201 Plan:

- Institutional arrangements.
- Financial program.
- Abatement schedule.

Institutional arrangements should contain the local government resolutions of assurance that the plan will be carried out. In some cases, they might involve the creation of a new agency to finance, operate, and maintain the structural components proposed by the 201 plan. These arrangements might also consist of contracts among the local units of governments.

The financial program and schedule must identify the nonfederal expenditures for implementing the plan, and identify a method for acquiring those funds. If industrial users are part of the system, a cost allocation between industrial users and other users must be made. A method must be identified which will recover from industrial users that portion of any federal grant amount used to provide treatment for industrial wastes.

There is no prescribed method for choosing the best alternative plan. However, there are criteria for plan selection which have been recommended by the EPA planning guidelines. A minimum criterion is that the plan must meet the applicable regulatory requirements. These include effluent limitations, load allocations, and compliance schedules.

The public participation program and the environmental assessment will establish important issues and evaluative factors that should be known by the time a plan selection is made. Each alternative should be analyzed from the standpoint of each of these factors. The level of detail required in a 201 facility plan must be sufficient to justify the selection of a given wastewater abatement design concept.

1-7.3 Preparation of the 201 Plan

The following list identifies, in roughly chronological order, the detailed planning and engineering activities which must be undertaken during the course of preparing a 201 facility plan.

- Obtain authorization to prepare facility plan, grant application and related documents:
  1. Resolution of council or governing body.
  2. Appointment of official representative for the project.
- Define objectives and show conformance with existing basin plan, areawide plan, and/or other related planning documents.
- Prepare infiltration/inflow analysis of existing collection system related to project.
- Submit infiltration/inflow analysis to EPA, with recommendations concerning excessive infiltration/inflow.
- If infiltration/inflow is excessive, conduct infiltration/inflow field study and evaluation of collection system.
- Submit infiltration/inflow evaluation to EPA.
- Provide program for the handling of infiltration/inflow (EPA Grant).
- Allow for infiltration/inflow in design of the facility, or identify physical correction of infiltration/inflow done simultaneously with construction of the project.
- Establish projections (in 5 year intervals) for the design life of the project of:
  1. Land use patterns.
  2. Population projections.
  3. Projected flows.
- Reevaluate existing basin plan, 201 and 208 plans, and other related planning documents in light of new projections, water quality standards, and minimum treatment requirements. Include all point and nonpoint sources.

- Identify institutional constraints, conflicts of interest, etc., and suggest remedies.
- Develop feasible cost-effective alternatives (including industrial pretreatment) employing: infiltration/inflow analysis and/or evaluation; cost-effective guidelines; pretreatment guidelines; cost recovery guidelines.
- Consider the nature of storm runoff and overflows, and the extent and effect they have on the proposed project. Propose remedies and include preliminary cost-estimates.
- Evaluate alternative flow and waste reduction measures.
- Identify the best practicable waste treatment technology. (Identification to be based on published regulations defining such technology unless water quality dictates a higher degree of treatment.)
- Prepare *draft* environmental assessment statement to include all alternatives.
- Submit draft environmental assessment statement to regional and state clearinghouses, and the regional EPA office for their preliminary review.
- Notify all interested groups and governmental agencies of the meeting.
- Make summaries of alternatives and the environmental assessment publicly available (upon request) 15 days prior to public meeting.
- Hold a public meeting.
- Complete environmental assessment statement and choose optimum plan.
- Forward copies of the transcript of public meeting to EPA and submit the final version of the environmental assessment.
- Prepare the engineering report for the project in accordance with the State and Federal regulations. Identify the proposed plan including cost-estimates and a schedule of all works for the design life of the project. Include facility design data such as:

    1. Results of sewer system evaluation or analysis.
    2. Estimated effluent quality, with respect to all parameters required by applicable effluent limitations.
    3. Proposed method for the ultimate disposal of sludge.

- Prepare a discussion of institutional responsibility, legislation needed, organizational requirements, manpower and training, and relation to other existing plans. Include management plan data such as:

    1. A description of proposed user charge system and capital cost recovery.
    2. A copy of the A-95 review and approval of clearinghouses.
    3. A statement regarding applicant's financial, legal, institutional, and managerial capabilities to insure the proper construction, operation, and maintenance of the treatment works.
    4. A statement regarding the applicants proposed operation and maintenance program and a discussion of the necessary personnel for this.
    5. A statement regarding availability of proposed site (if relevant).

6. A copy of proposed subagreements or an explanation of intended method of awarding subcontracts for performance of any substantial portion of the project work.
7. An informal statement from major industrial dischargers that they intend to comply with the plan.
8. A statement from state that existing discharge permits will be amended compatible with proposed plan.

This process incorporates several key planning elements centered in the environmental assessment; infiltration/inflow (I/I) analysis; and formal documentation of both structural and nonstructural engineering alternatives. All of these items are included under the definition of cost-effectiveness. Heretofore, environmental assessments were not a requirement in the design and construction of wastewater treatment facilities.

There is one aspect of the environmental assessment which deserves additional discussion. Sensitive archaeological and historical areas must be identified. It is advisable to contact the State Historic Preservation Officer at the beginning of the facility planning cycle and provide him with a description of the proposed project activity. Based on this information, the State Historic Preservation Officer may recommend more detailed on-site investigations by qualified professionals.

Work on the I/I analysis should commence at the very early stage of facility planning. Results of the study should be integrated into the Facility Plan, especially where excessive infiltration exists.

When excessive I/I is shown to exist, an estimate of costs needed to perform the second stage of I/I work (sewer system evaluation survey) should be prepared. Under normal conditions, this work must be completed and incorporated into the facility plan prior to final selection of wastewater abatement controls.

- Development and analysis of engineering alternatives.
- Service agreements.

The National Environmental Policy Act of 1969 requires that alternative plans be thoroughly documented, and the rationale used in evaluating alternatives must be shown to be objective and to have considered environmental factors. All feasible alternatives should be described in detail, especially:

- Obvious reasons why any alternative was eliminated.
- Environmental and cost-effective reasons why additional alternatives were found to be unfeasible.
- Detailed engineering concept.
- Environmental and cost considerations of major significance.

If the facility plan requires that a treatment facility serve more than one local jurisdiction, a lead agency must be identified and service agreements must be established. Federal law establishes specific requirements for these service agreements. The ones that are most obvious pertain to industrial cost recovery and user charge system. If two or more municipalities are to provide joint wastewater abatement services, these municipalities should begin working on the service agreements as early as possible, and the local authorities themselves should become acquainted with the specific areas to be covered by the agreement.

1-7.4 Content of the 201 Plan

As can be seen from the topics discussed previously, the facilities plan, by necessity, consists of many elements. To summarize, as a minimum, all facilities plans should have the following components:

A. A statement of the problem and its impact on the environment.
B. A description of the proposed treatment alternatives included in the first stage, for which construction drawings, including cost estimates and implementation schedules, are to be prepared.
C. A description of the complete wastewater management system of which the proposed treatment works is to be a part.
D. Sewer system infiltration/inflow documentation.
E. A cost-effectiveness analysis of the recommended system and the alternatives to that system.
F. An environmental evaluation of the alternatives considered.
G. An identification of effluent discharge limitations, or where permits have been issued, a copy of the permit for the proposed treatment works as required by the national pollutant discharge elimination system.
H. Comments and approvals of local agencies, including compliance with the office management and budget circular A-95.
I. A summary of any public meeting or hearing held to consider the plan.
J. A statement demonstrating that the authorities implementing the plan have the necessary legal, financial, institutional, and managerial resources.

Elements C and E of the plan are to cover a 20-year period. Future system expansions and other modifications, with the appropriate cost estimates, and a program for staged development and implementation, should be included. Steps taken in the planning process, particularly public involvement, must also be documented in the plan.

Planning for facilities in the 201 program is one of the most rigid processes in planning today, being little different in one state than another, because adherence to the EPA prescribed format is linked to approval of future construction grants providing up to 75 percent of costs. Innovation in this process is achieved through presentation of the plan. Graphics, wording, and brevity should be primary planning goals in such a rigid process. Any other planning initiative should be focused on the development of sound background information, specifically the assumptions on which future wasteloads will be based.

REFERENCES

1-1. Turing, H. D., Third Report on Pollution, British Field Sports Society, London, 1949.

1-2. Johnson, J., Reminiscences of 45 years as a Rivers Inspector, *F. Inst. Sew. Purif.*, **1** (1933) 17–19.

1-3. United States versus Pennsylvania Industrial Chemical Corp., (PICCO), 411, U.S. 655.

1-4. Areawide Waste Treatment Management Planning, U.S. Environmental Protection Agency, Washington, D.C., 1974.

1-5. Guidelines for Areawide Waste Treatment Management Planning, U.S. Environmental Protection Agency, Washington, D.C., August 1975.

2

Relationships between water quality management planning and land use planning

Ronald G. Alderfer and Eldridge Lovelace

Harland Bartholomew and Associates
St. Louis, Missouri

2-1. INTRODUCTION

There can be no doubt that land use profoundly affects the quality of water in streams, rivers, lakes, and shallow aquifers. However, the task of finding specific cause-effect relationships between different land uses and water quality problems is seldom easy for several reasons: first, water courses usually assimilate materials from widely dispersed land areas; second, land areas drained by a common waterway nearly always have mixed uses; third, soil supporting given land uses vary significantly in properties affecting infiltration, chemistry, the quality of drainage flows, etc.; and, fourth, movements of materials from land areas to waterways vary greatly with time and local precipitation pattern.

In spite of these difficulties, understanding the interaction between land use and water quality is one of the most important ecological challenges of our time. Only as land owners act on the best knowledge available in this area can our nation hope to stabilize and reverse the trends in water quality which have come with increasingly intensive land use by a burgeoning human population.

This Chapter is organized around three main questions:

1. How does existing Federal legislation on water quality management planning affect land use planning in this country?
2. What is known about the impact of specific land uses on the quality and quantity of water resources?
3. What conclusions can be drawn from this analysis regarding the future course of land use planning properly adjusted to minimize degradation of water resources?

2.2. FEDERAL WATER QUALITY MANAGEMENT PLANNING AND LAND USE PLANNING

The Federal policy on land use planning traditionally has been not to plan (Ref. 2-50). It was assumed that individual states would determine whether or not land use planning of any kind was necessary, and that they would establish appropriate policies and regulations. However, a large fraction of this country's water resources (i.e., large rivers, many large lakes and reservoirs, and groundwater resources) affect and are affected by the lands and waters of two or more states. In the Federal Water Pollution Control Act Amendments of 1972 (PL 92-500), Congress recognized the responsibilities of states in maintaining water quality, but also joined them in recognizing federal responsibility. "It is the policy of the Congress to recognize, preserve, and protect the primary responsibilities and rights of States to prevent, reduce, and eliminate pollution, to *plan the development and use* (including restoration, preservation, and enhancement) *of land and water resources*, and to consult with the Administrator in the exercise of his authority under this Act." (PL 92-500, Section 101 (b)—emphasis added). This section of the Act recognizes (if implicitly) that planning for water quality also requires planning for land use.

The purpose of including land use planning in federal support of water quality planning is twofold. On the one hand, many water pollution sources are land use-specific, particularly non-point sources. Therefore, a thorough analysis of water quality problems in a region will include an identification of these land use-specific, non-point sources, as well as point sources such as municipal and industrial effluents. On the other hand, land use controls (or other nonstructural approaches) may, in certain cases, be the most cost-effective method for controlling one or more pollution sources. This latter possibility makes an understanding of land use-water quality interactions even more imperative. Selective changes in land use may decrease the investment in pollution control which would be required to meet the water quality objectives of PL 92-500.

In its "Report to Congress–1974," the Environmental Protection Agency emphasized that, wherever possible, a research emphasis would be placed on control and prevention of polluting discharges, rather than on collection and treatment (Ref. 2-54). It is also pursuing a research program of land use controls as a means of protecting water quality (Ref. 2-54).

2-2.1 Facilities Planning

(Section 201, PL 92-500). Existing guidelines for "201 planning" make very specific requirements in the area of land use patterns and planning (See Table 2-1). The crucial element of this guidance is the use of projected land use patterns and densities as the basis for determining optimum capacity and location of facilities. Since the availability of sewerage facilities is a critical factor in urban development, and especially in residential development, Section 201 greatly strengthens the impact of land use planning.

2-2.2 Areawide Waste Treatment Management Planning

(Section 208, PL 92-500). One of the guiding principles behind the 208 program is that water quality problems are closely linked to misuse of natural resources (Ref. 2-58), and that reducing pollution will often require changes in natural resource management. Land use is one of the most basic and vulnerable areas of resource management. The ultimate outcome of 208 planning is to be a comprehensive management program for collection and treatment of wastes and control of pollution from all point and non-point sources. It is expected that necessary measures for controlling these pollution sources would include traditional, structural measures, land use or land management practices, and regulatory programs.

While the 208 program emphasizes the management of land-generated water pollutants and the more general relationship between land use and water quality, the program also relates significantly to air quality planning, planning for solid waste disposal, and other more specialized resource problems, such as coastal zone management planning. It is one of the most comprehensive programs ever attempted in the area of natural resource management.

As noted briefly above, land use considerations are considered important in 208 planning for two reasons: first, they can serve as the basis from which point and non-point source controls can be developed and evaluated, and second, adjustments in the future development patterns can be explored as a means of minimizing or controlling pollution sources and/or minimizing control costs.

Section 208 (b) (2) (F-H) gives the designated agency the authority to consider land use requirements; i.e., land use controls and land management regula-

TABLE 2-1 Generalized Flow Chart for 201 Planning Process Showing Entry of Land Use Considerations

Step 1: List effluent limitations applicable to facility being planned.

Step 2: Assess current situation.
- Describe existing land use pattern.

Step 3: Assess future situation.
- Coordinate facility with applicable State, local, and regional land use management regulations, policies, and plans.
- Use projected land use patterns and densities as basis for determining optimum capacity and location of facilities.
- Give careful consideration to areas subject to flooding; coordinate facility plan with State and/or local programs for floodplain management.
- Coordinate projections of land use with those related to air quality objectives, coastal zone management studies, and other environmental programs.

Step 4: Develop and evaluate alternatives.
- Analyze possible regional solutions by:
  a. Examining effects of interceptor location on land use within and between urban areas, particularly where land is undeveloped;
  b. Examining possible limitation on future expansion due to unavailability of land.
- Examine implications of "no action" alternative with respect to potential effect on land use limitations.
- Consider primary options (including land use-specific) for flow and water reduction, treatment and disposal of effluent, and sludge disposal. (Options should include land application of treated and/or untreated sewage effluents.)
- Examine primary and secondary environmental impacts of alternatives including:
  a. Destruction of sensitive lands.
  b. Violation during construction or operation of Federal, State, or local environmental and/or land use statutes, regulations, or plans.
  c. Changes in rate, density, or type of development of land areas induced by the alternatives.

Step 5: Select Plan
- Public should be provided with information on environmental impacts of proposed alternatives and given opportunity to express views and opinions. These views and opinions should be used in evaluating alternatives and selecting a plan.
- Special attention should be given to the extent to which the proposed plan forecloses future land use options or narrows the range of potential land and water uses.

Step 6: Preliminary design of treatment works.

Step 7: Arrangements for implementation (institutional and financial arrangements).

Source: U.S. Environmental Protection Agency (Revised May, 1975, pp. 5–17.)

tions: "Any plan prepared under such process shall include, but not be limited to—. . . (F) a process to 1) identify, if appropriate, agriculturally and silviculturally related nonpoint sources of pollution, including runoff from manure disposal areas, and from land used for livestock and crop production, and

2) *set forth procedures and methods (including land use requirements)* to control to the extent feasible such sources;. . ." (emphasis added). Similar provisions are made for mine-related and construction activity-related sources of pollution.

It is essential that the 208 planning agency coordinate its efforts with the agency or agencies which have the legal authority over land use planning and control. Federal guidelines for 208 planning do not expect the planning agency to have the authority for enforcement, but they do require it to include in its planning some means for enforcement. In other words, a suitable technical report alone is *not* a sufficient 208 product.

As in the case of 201 planning, a key provision of the 208 guidelines is that projected waste loads (and ultimately treatment facilities) are based on projected land use. Therefore, the starting point for all waste treatment planning and water quality impact is the projected land use plan.

TABLE 2-2 Generalized Flow Chart for 208 Planning Process Showing Entry of Land Use Considerations

Step 1: Identify water quality problems and standards.

Step 2: Construct an inventory of discharges (including point and non-point sources).
- Many non-point sources are land use-specific or, as in the case of stormwater runoff, strongly affected by land use.
- Land use pattern determines geographic distribution of point sources within a study area.

Step 3: Prioritize planning and choose design conditions.
- Selection of design conditions affected by land use pattern in study area.

Step 4: Data collection.
- Gather existing data on land use throughout study area and conduct surveillance as necessary.
- Land use pattern may strongly affect type and level of detail necessary for geophysical, biological, and socioeconomic data.

Step 5: Calibration and verification.

Step 6: Projections.
- Project population, employment, and land use over planning period.
- Determine projected waste loads over planning period; these projections are based on projected land use.

Step 7: Segment analysis.

Step 8: Determine alternative subplans of treatment, control, and flow reduction consistent with eligible waste load reductions.
- Control and flow reduction methods are land use-specific.

Step 9: Screen subplans to select leading alternative subplans consistent with eligible waste load reductions.

Step 10: Combine subplans into alternative areawide plans consistent with eligible waste load reductions.

Source: U.S. Environmental Protection Agency, August, 1975, pp. 3-1 to 3-27.

2-2.3 Continuing Planning Process

(Section 303(e), PL 92-500). "The broad goals of the continuing planning process are to assure that the necessary institutional arrangements and management programs are established to make and implement coordinated decisions designed to achieve water quality goals and standards; to develop a statewide (State and areawide) water quality assessment; and to establish water quality goals and State water quality standards which take into account overall State and local policies and programs, *including those for management of land and other natural resources*; and to develop the strategic guidance for preparing the annual State program plan required under Section 106 of the Act." (Ref. 2-13).

Guidelines originally prepared for this program (Ref. 2-12) have been amended to incorporate 208 requirements for both State and designated areawide planning agencies into a single set of regulations (Ref. 2-13). Hence, 303(e) planning goals are basically the same as 208 goals, except that 303(e) carries on at a State level efforts that were initiated either by designated 208 areawide agencies or by State 208 planners.

It is expected that information would flow both ways, between 303(e) planning efforts and those of 201 and 208 programs. That is, statewide review of water quality problems would enhance 201 and 208 planning, and 201 and 208 plans would provide the central input for 303(e) plans.

As shown in Table 2-3, an essential element of 303 (e) planning is the classi-

TABLE 2-3 Generalized Flow Chart for 303(e) Planning Process Showing Entry of Land Use Considerations

Step 1: Inventory of municipal and industrial sources and analysis of significant pollutant dischargers.

Step 2: Ranking of basin segments in order of abatement priority.

Step 3: Schedule of compliance with water quality regulations.

Step 4: Assessment of municipal needs.

- Based on population served and projected population growth over 20-year period following scheduled date of installation of needed facilities. This assessment is to be based on approved Section 201 and 208 plans where appropriate.

Step 5: Determination of total maximum daily loads.

- Include considerations for seasonal variations.
- Provide for margin of safety in view of uncertainty in estimating non-point sources of pollution and uncertainty of relationship between effluent limitations and water quality.
- Estimates shall be made of total daily thermal load delivered to each river segment.

Step 6: Establish individual point source load allocation. (The sum of all individual loads should not be greater than that estimated in Step 5 for each water quality segment.)

Step 7: Assess individual non-point sources.

Source: U.S. EPA, 1974, 39 pp. with two appendices.

fication of all rivers by segments, as to whether they are water quality-limited or effluent-limited. "Segment" is defined to include land areas that contribute or may contribute to water quality problems, whether of physical, biological, or chemical origin (Ref. 2-13).

2-2.4 Criticisms of Land Use Planning as Required by Federal Water Quality Planning Guidelines

Land use planning requirements, in the context of water quality planning, raise many controversial issues regarding the nature and direction of urban growth, land use controls, etc. (see Ref. 2-21 for fuller discussion). As demonstrated in the Brandywine case (Ref. 2-27), private land owners may view land use planning to preserve water quality as an infringement on individual property rights. The planning will not succeed unless citizens affected clearly understand its objectives and appreciate the need for enforcement where necessary.

Federal vs. Local Control of Planning Process. Since 208 planning must be done for all surface areas of the United States either by areawide or state planning agencies (Ref. 2-13), the need for compatible planning area boundaries and requirements is clear. However, to what extent should EPA control or influence the planning process in more specific matters, such as arriving at an acceptable projected land use pattern or determining the probable impact of one or more non-point sources of pollution? While EPA does not have the power to apply legal sanctions (withholding construction grants and/or NPDES permits), it is required to review and approve plans submitted by states; and lack of approval can result in significant delays in grants and/or permits. If a designated areawide planning agency receives state approval for a management plan which calls for extensive development in land areas which the Regional EPA Administrator considers ecologically sensitive or likely to generate significant levels of non-point pollution, should EPA approval be delayed until the Agency modifies the plan? This question raises the possibility of direct conflict between the planning autonomy of local agencies and the power of the federal government to enforce PL 92-500.

Single-purpose Orientation of Guidelines. Current guidance defines the water quality management plan (Sections 208 and 303(e)) as ". . . a management document which identifies the *water quality problems* of a particular approved State planning area or designated areawide planning area and sets forth an effective management program to *alleviate these problems* and to achieve and *preserve water quality* for all intended uses." (Ref. 2-13, Part 131.1) (Emphasis added) While guidance does require "coordination" of water quality management plans "with affected State, local, and Federal programs, and with other applicable resource and developmental planning . . ." (Ref. 2-13, Part 130.34), the emphasis on water quality is very strong and demands almost exclusive attention by available areawide and State planning personnel. Coordination of water quality planning with transportation, energy, air quality, and economic development planning efforts requires time and mutual education. The November 1, 1978 deadline for plan submission (Ref. 2-13, Part 131), together with minimal funding levels available for planning, will not allow the degree of coordination that would be required to develop integrated *resource* planning.

Possible EPA Control of Community Growth Patterns. Since EPA reviews areawide water quality management plans and funds facilities construction, several issues are raised with respect to control of community growth. For example, if two urban fringe communities each anticipate large population growth, but EPA approves expanded facilities for only one, growth in that community would be favored. Similarly, facility design capacity levels can effectively limit the degree of community growth. Another community growth aspect arises from the location of sewer interceptor lines connecting communities in a regional treatment system. In many cases, development close to the line would be favored over that in more remote communities. Still another problem arises with the choice of upgrading old treatment systems vs. building new ones. Answers to questions of this sort have significant effects on urban sprawl and associated problems of transportation, air quality, etc.

Evaluation and Summary of Criticisms. In a recent statement of policy, the EPA referred to seeking a "truly mutual partnership" between Federal and State in program responsibility (Ref. 2-57). If this aim can be realized, many of the problems and criticisms directed toward the water management planning programs can be dealt with satisfactorily. In states or areas where that partnership cannot be realized, however, the danger of fruitless confrontation followed by resolution in the Courts is real. The concept of property rights in this country is complex, but deeply rooted and firmly held by citizens. Land planners and government regulatory agencies together need above all to reconcile individual and local goals with long-range societal goals.

2-3. IMPACTS OF LAND USE ON QUALITY AND QUANTITY OF WATER RESOURCES

2-3.1 Some Basic Principles

Before discussing specific information pertaining to the physical interaction of land use and water quality, we summarize below some of the basic principles on which this interaction rests.

1. Water, through precipitation and natural drainage flows, is a strong determinant of land surface form; land surface form determines the direction and extent of surface drainage flows.
2. Water is exceedingly abundant in the earth-atmosphere system. On the average, each acre of land in the United States receives approximately one million gallons annually (Ref. 2-11).
3. Water retains its liquid phase over a wide temperature range, which strongly overlaps the range of temperatures commonly encountered near the earth's surface.
4. Liquid water is a widely effective solvent, and, therefore, carries a large and varied load of solutes in all but exceptional flows.
5. Large volumes of water are stored in inland lakes and reservoirs which have relatively slow turnover times.
6. Developed land uses (i.e., excluding managed forests and natural open spaces) significantly alter natural drainage patterns, solute load of drainage flows, and groundwater interactions.
7. Water is a rapidly and extensively recycled compound in the earth-atmosphere system.
8. All life forms are intimately related to, and dependent on, water, and are often sensitive to type and concentration of solutes.
9. Land development invariably requires intensive water use.

2-3.2 Water Quality Impact of Open Space Land Uses

Agriculture. There are approximately 920 million hectares (2.26 billion acres) of land area in the United States (Ref. 2-52). Approximately one-half of this area is classified as farmland; i.e., used for grazing or crops, whereas about seven percent of the land area is urbanized (Ref. 2-52). Consequently, the potential impact of runoff from agricultural and other open space land uses on water quality is very high indeed.

Runoff Volume. Runoff to surface waters is the major mechanism by which agricultural land uses pollute water, although substantial quantities of dissolved pollutants can be carried to groundwater by infiltration. (A third mechanism is by wind transport of particles to surface waters.) Runoff occurs whenever rates of precipitation and/or snowmelt exceed(s) the combined rates of retention and infiltration of surface water to the ground. The rate at which runoff is generated and the total volume generated in any given storm event depend on intensity and duration of rainfall and/or snowmelt; the size, shape, slope, vegetation, and surface configuration of the watershed; and the physical characteristics (particularly degree of perviousness to water) and moisture condition of the surface and subsurface earth materials. Of these parameters, rainfall and/or snowmelt intensity and duration, vegetation, and moisture condition of the surface and subsurface materials can be highly variable in time; therefore, accurate predictions of runoff rate and volume are rarely possible. However, reasonable estimates can be obtained through simulation programs such as "STORM" (Ref. 2-25) and "SWMM" Runoff Block (Ref. 2-24).

In a sensitivity analysis of runoff volume per storm to selected parameters of the "SWMM" model Graham, Costello, and Mallon (Ref. 2-19) found that the interdependent parameters characterizing land use and watershed imperviousness caused the greatest impact. Pervious area infiltration rates, watershed length/width ratio, and overland flow slope were the next three most sensitive parameters.

For agricultural land uses, watershed imperviousness may vary between 5 and 40 percent depending on soil type, vegetation, time of year, and soil moisture condition. Highest degrees of imperviousness are associated with cultivated but unvegetated surfaces, over which a hard-pan crust forms. Vegetated surfaces with frequent shallow depressions and slopes less than 10 percent would have lowest degrees of imperviousness, and consequently generate the least runoff.

Crop type strongly influences runoff. Row crops, such as corn and soybeans, allow the greatest runoff, particularly where rows follow the direction of the slope. Fully seeded crops (not in rows) allow somewhat less runoff. Permanent cover, as in a meadow, allows the least runoff.

Sediments. One of the dominant water pollutants originating on agricultural land is suspended soil particles. Approximately one-half of the sediment reaching lakes and streams originates on cropland (Ref. 2-52). While averages are somewhat meaningless in this context, the overall figures are approximately two billion tons of sediment per year from 412 million acres of cropland, or approximately 5 tons/acre/year (Ref. 2-52). Ten to 30 tons/acre/year may be lost from poorly protected lands (Ref. 2-26).

The erosion process is initiated when forces applied to a soil aggregate are sufficient to loosen the aggregate and release small particles. A thunderstorm lasting 30 minutes may deliver as much as 100 tons of dead weight to each acre (Ref. 2-61). The impact of raindrops on unprotected ground surfaces results in extensive particle loosening near the surface. These small loosened particles are carried with whatever surface flows exist until the flow velocities are reduced sufficiently to allow settling.

This means that there are two fundamental ways of minimizing erosion and the resulting water pollution. One is to reduce raindrop velocities before they strike the soil surface. The other is to increase resistance to flow over the surface and minimize flow velocities over erodible surfaces. In a study quoted by Amemiya (Ref. 2-1), application of 4 tons/acre of wheat straw to a freshly plowed silt loam surface on five percent slope reduced runoff and erosion to zero when 6.25 inches of water was applied at a constant rate of 2.5 inches per hour, whereas no mulch cover allowed 2.83 inches of runoff and 12.42 tons/acre of soil loss.

While croplands in general do not have the highest rates of erosion per acre relative to all other land uses (see Table 2-4), the overall impact of cropland erosion is high because of the high total land area in cultivation. Furthermore, materials eroded from croplands often carry a large load of nutrients and pesticides with them, resulting in a compounded water quality impact.

Nutrients. The major water quality problems caused by high nutrient levels (particularly nitrates and phosphates) is first that they stimulate growth of algae and thereby accelerate the eutrophication process in streams, lakes, and reservoirs, and second, that certain nutrients (eg. nitrites, ammonia) can be toxic to man and other organisms at or above critical concentrations. While conditions conducive to eutrophication vary strongly from one setting to another, roughly 0.3 mg/l inorganic nitrogen and 0.01 mg/l inorganic phosphorus in water is sufficient to support undesirable levels of algal growth (Ref. 2-32). Highly turbid water, caused by suspended solids, supports much less algal growth than more transparent water at comparable nutrient levels, because the turbidity prevents light from reaching the photosynthetic (light-requiring) organisms. Heavily over-

TABLE 2-4 Erosion Rates from Different Land Use Types.

	Metric Tons/ sq km/year	Tons/sq mi/year	Tons/acre/ year	Relative to Forest = 1
Forest	8.5	24	0.04	1
Grassland	85	240	0.4	10
Abandoned surface mines	850	2,400	3.8	100
Cropland	1,700	4,800	7.5	200
Harvested forest	4,250	12,000	18.8	500
Active surface mines	17,000	48,000	75.0	2,000
Construction	17,000	48,000	75.0	2,000

(Relative Total Erosion Losses Nationwide from Different Land Use Types)

Commercial forests	1
Abandoned surface mines	<1
Active surface mines	2
Construction	6
Harvested forests	11
Grassland	11
Cropland	168

Source: U.S.E.P.A. October, 1973, "Methods for Identifying and Evaluating the Nature and Extent of Nonpoint Sources of Pollutants," pp. 6–7.

cast conditions or low water temperatures may also limit algal growth in certain instances.

As shown in Table 2-5, runoff from agricultural land nearly always contains nutrient concentrations much greater than that required for algal growth. Therefore, unless these flows are substantially diluted, streams and lakes receiving them are continuously threatened with eutrophication.

The overall loss of nutrients from agricultural lands in this country is staggering. (See Table 2-5.) In a study of a 5,490 square mile watershed draining into the Des Moines River north of Des Moines, Iowa, Bauman and Kelman (Ref. 2-7) found that 68 percent of the nitrogen input to the river from the watershed came from agricultural losses, and that about 82 percent of the total nitrogen in the river was in the form of nitrate. In contrast, only 11 percent of the phosphorus (dissolved) input was attributed to agricultural losses. A similar study on the Big Muddy River in Illinois showed that about 30 percent of the phosphorus at the mouth of the river could be attributed to agricultural sources (Ref. 2-59). While the percentage contributions from agricultural vs. other sources are highly variable from one situation to another, it is abundantly clear that nutrient runoff from agricultural lands presents a major water quality problem.

Nitrate nitrogen is highly soluble in water and, therefore, free to move with the soil water, whereas phosphorus is less soluble but more strongly bound to soil particles, especially clays. For this reason, phosphates can be expected to enter drainage water less readily than nitrates, but agricultural sediments (acting as a "buffer") may continue to release phosphates to the water above them long after the incoming flows are bringing water with little or no phosphorus.

TABLE 2-5 Estimated Nutrient Runoff from Various Sources

	Nitrogen		Phosphorus	
Source	Pounds/Year (millions)	Usual Concentration in Discharge (mg/l)	Pounds/Year (millions)	Usual Concentration in Discharge (mg/l)
Domestic waste	1,100–1,600	18–20	200–500	3.5–9.0
Industrial waste	>1,000	0–10,000	a	a
Rural runoff				
Agricultural land	1,500–15,000	1–70	120–1,200	0.05–1.1
Nonagricultural land	400–1,900	0.1–0.5	150–750	0.04–0.2
Farm animal waste	>1,000	a	a	a
Urban runoff	110–1,100	1–10	11–170	0.1–1.5
Rainfall[b]	30–590	0.1–2.0	3–9	0.01–0.03

[a] Insufficient data available to make estimate.
[b] Considers rainfall contributed directly to water surface.
Source: Task Group 2610-P Report. 1967. *J. Am. Water Works Assoc.* 59: 344–366.

In 1972, approximately 37 million metric tons (41 million tons) of commercial fertilizer were used in this country, 75 percent of which was used by farmers (Ref. 2-52). This fertilizer contained roughly 20 percent nitrogen, 5.2 percent phosphorus, and 8.8 percent potassium, although these percentages may vary considerably from one application to another (Ref. 2-52). It is estimated that 10 to 15 percent (about one million metric tons), of this fertilizer is lost to waterways in comparison with a total nitrogen loss to waterways of approximately 5.5 million metric tons (Ref. 2-37). While fertilizers contribute less than 20 percent of the total nitrogen losses, fertilizer use will, no doubt, continue to increase over the next several decades, during which time urban demands for water from agricultural watersheds will also increase substantially. From the water quality standpoint, it is essential that nutrient losses from agricultural lands are minimized.

Pesticides. The use of pesticides in agricultural management has increased dramatically during the last two decades. While the total quantity of pesticides used each year is not precisely known, estimates are in the range of 500 million kg. (one billion pounds), of which about 70 percent was used in agriculture and 30 percent by government agencies and the public (Ref. 2-41). Approximately 58 million hectares (143 million acres) of agricultural land was treated for pest control (Ref. 2-39).

While there may be 700 or more pesticidal compounds that have been synthesized and used over the years, about 50 make up a large percentage of total pesticide usage (Ref. 2-39). However, even this number of compounds presents a wide range of physical and chemical properties of concern to water quality management. The fate and distribution of pesticide compounds and their degradation products in the environment is a complex problem.

The pesticide characteristics of most concern from a water quality management standpoint are toxicity to target and other organisms (this influences dosage required); solubility in water; persistence (longevity) in the environment; and the potential for biomagnification. The most troublesome pesticides are those with a high level of toxicity, which dissolve effectively in water, are not degraded rapidly, and can accumulate in plant and animal tissue, thereby allowing biomagnification of the compound. As a group, the chlorinated hydrocarbons (aldrin, chlordane, DDT, dieldrin, endrin, heptachlor, lindane, methoxychlor, mirex, etc.) match these characteristics. The organophosphorus pesticides, on the other hand, are more easily degraded, are not susceptible to biomagnification, and are only slightly more soluble than chlorinated hydrocarbons (Ref. 2-4). Pesticides having the least overall environmental quality hazard include paraquat, acephate, methyl parathion, carboryl, dalapon, and 2, 4-D (Ref. 2-4).

Runoff is the most common mode of water resource contamination by pesticides (Ref. 2-38). It moves either in solution or attached to soil particles. Clearly, the greatest threat to water quality exists when heavy rainfall immediately follows application of pesticides to the land surface. A stunning example of this occurred in 1950, when cotton pesticides washed into 14 tributaries of the Tennessee River in northern Alabama causing extensive fish kills (Ref. 2-62).

The impact of pesticides on water quality can be mitigated primarily by dilution and by rapid chemical and/or microbial degradation. Pesticides which are attached to soil particles and become deposited on the channel or basin floor can cause extensive disruption of the aquatic ecosystem by killing or disabling bottom-dwelling organisms responsible for organic decomposition and mineral recycling, or by affecting developmental stages of fish and other higher aquatic organisms which spend part of their life cycle in the bottom zones.

Aerial transport and careless disposal of pesticides may also cause significant transfer from agricultural land areas to water bodies (Ref. 2-39). For this reason, application, handling, and storage methods are crucial to the control and reduction of pesticide-related water pollution.

Organic Loads. Agricultural land uses generate animal wastes, crop residues, and wastewater from farm buildings which is generally disposed of through septic systems with outfalls to subsurface drainage beds. The primary water quality impact of these organic loads is that they generate large oxygen demands when they enter waterways through surface runoff and/or movement of groundwater.

Approximately 1.8 billion metric tons (two billion tons) of livestock wastes were produced in this country in 1972, or about 10 times the amount produced by the human population (Ref. 2-52). Roughly one-half of these livestock wastes were produced by animals in confined feedlots; the other half were produced by grazing animals (Ref. 2-52). As shown in Table 2-6, oxygen demand of animal wastes is many times that of untreated domestic sewage.

The actual water quality impact of runoff from feedlots and grazing lands is highly variable. In fertile agricultural soils, well separated from the water table (at least five feet) with moderately slow to slowly permeable materials, polluted runoff can be substantially renovated before joining the groundwater; i.e., BOD can be reduced, solids filtered out, bacterial levels reduced to acceptable levels, and, in some cases, nutrient levels substantially reduced. Similarly, grazing areas on moderate slopes and with good, persistent vegetation, can assimilate

TABLE 2-6 Oxygen Demand of Various Wastes.

Source	BOD (mg/l)	COD (mg/l)
Untreated domestic sewage	100–300	400–600
Dairy cattle manure (Dale and Day, 1967)	25,600	
Swine manure	27,000–33,000	70,000
Chicken droppings	24,000	172,000

Source: Miner, J. R. and T. L. Willrich, 1970.

much of the pollutant load of animal wastes before they enter surface water bodies. It is obvious, however, that many forms of mismanagement exist that can result in highly polluted runoff reaching surface water and/or groundwater. In particular, these include overgrazing, particularly on steep slopes (12 percent); spreading animal manures over frozen ground or ground with little vegetation; locating feedlots on floodplain areas, over surficial aquifers, or close to streams; using high density feedlots without roofs; or placing feedlots near abandoned wells, springs, sinkholes, or their openings in the surface.

A large organic load source associated with agricultural land uses is crop residues. Under conditions of good management, however, these residues are stabilized and worked into the upper soil layers, thereby increasing the moisture holding capacity of the soil and adding fertility as it slowly decomposes. The oxygen demand on surface water draining agricultural lands due to crop residues should not be a major problem under most conditions.

Wastewater from farm buildings is a problem in local instances where adequate precautions have not been taken to protect wells, streams, and small impoundments from wastewater drainage beds.

Pathogens. As more and more agricultural animals are raised in confined lots, there is an understandable increase in concern for transmission of disease by organisms released from these land use areas. At present, animal pathogens are seldom the cause of known disease in man or animals, but this is not because of careful planning to minimize disease transmission from agricultural sources (Ref. 2-16). What little information exists on the subject is restricted to bacterial diseases, while virus roles are almost completely unstudied. There is a great deal that needs to be understood about the ecology of disease organisms associated with the agricultural industry.

Silviculture. Over one-third of the land area in the United States is covered with forests, of which approximately two-thirds is classified as commercial forest land (Ref. 2-52).

Runoff Volume. Forest lands can be the least runoff-generating of all land uses, because so much of the incident rainfall is intercepted and deprived of its kinetic energy before striking the ground, and because the resistance to overland flow is so great. In fact, during a rainfall event, a significant fraction of the precipitation can be retained on the tree surfaces themselves and never reach the ground before evaporating. Variable amounts of precipitation reaching the ground do so by running down the branch and trunk surfaces. It is for these reasons that runoff and erosion volumes for all other land uses are compared to those from forest lands as a reference standard. (See Table 2-4.)

Sediments. For reasons just discussed, the sediment production from undisturbed forest lands can be very minimal indeed. Surface erosion and sediment production is accelerated by construction of logging and other roads, by logging, by fire, by disease, and by grazing. Machines used for road construction and logging can compact the upper soil layers enough to reduce infiltration and lead to surface flows. After a forest road is constructed, its impervious surface can cause erosion. Similarly, skid trails for cut logs and fire lanes can lead to significant erosion.

Studies in central Idaho showed that the major source of stream sediments originated with the construction of logging roads, and that 84 percent of all the sediment produced because of the construction came during the first year, nine percent in the second year, and two percent in each of the following four years (Ref. 2-34). Rice and Wallis (Ref. 2-42) showed that road construction in a 10 square kilometer (four square mile) watershed increased the average sediment concentration fivefold in the first year (64 to 303 ppm) and only twice the norm in the second year.

Harvesting methods strongly influence the increase in erosion. Percentages of soil exposed in a logging area varied between 6 and 29 percent depending on the method used (Ref. 2-43). Clearcutting itself may not be as damaging to the stream water quality as the construction of roads giving access to the logging sites (Ref. 2-30). Required harvesting methods may vary from site to site, but in every case steps should be taken to protect the soil surface continuously with materials to intercept and/or slow rainfall and to minimize the loosening or compaction of surface soil layers through machine impact and roads.

Grazing is damaging to forest systems which have poor to moderate ground vegetation, are in areas of occasional drought, or are on steep slopes. These conditions are commonly (not necessarily simultaneously) met in forest lands in the western part of our country.

Nutrients. A lake fed by an undisturbed forest may receive as much as one-half of its total phosphorus input from direct precipitation and more than half of its nitrogen input through precipitation, whereas in managed watersheds, less than five percent of phosphorus input and less than 20 percent of nitrogen inputs come via precipitation (Ref. 2-31).

Since 1963, forest fertilization has started to increase very substantially, particularly in the Pacific Northwest and the South (Ref. 2-52). In the early 1970's, approximately 100,000 hectares (250,000 acres) of forest lands were fertilized, primarily with nitrogen (pelletized urea), and at rates up to 200 lb/acre (Ref. 2-52). Growth enhancement in the vicinity of five percent per year can be realized.

Regardless of the form in which nitrogen fertilizer is applied, the greatest loss of nitrogen from the forest watershed will be as nitrate. Where applications are made from aircraft, direct release to water surfaces is difficult to avoid and will account for a majority of the input to the water. Loss of applied nitrogen to stream channels according to studies carried out to date range between zero and three percent (Ref. 2-52). In most cases, the greatest susceptibility to appreciable loss of added nutrients comes when a significant rainfall hits the fertilized forest area within the first five days after application.

Pesticides. In 1970 about 140,000 kg (310,000 lb) of insecticides and fumigants were used to control insects (Ref. 2-52). The largest target organisms were the gypsy moth and spruce budworm. Herbicides are used primarily to control undesirable weeds in newly planted areas. Herbicides are used more frequently than insecticides, but still cover a small portion of the total forest land area in any given year (Ref. 2-52). In 1969, approximately 108,300 hectares (268,666 acres) were treated for weed control.

Pesticides can enter waterways either by spray drift that reaches the water surface directly, or through surface runoff when rainfall strikes the sprayed area shortly after application. While measurements of pesticide residue concentrations in water ways have been made frequently, there is relatively little conclusive information on the water quality impact of forest land pesticide applications.

Fire Retardants. The two most commonly used fire retardants are Firetrol® which contains ammonium sulfate, magnesium silicate, sodium dichromate, and ferric oxide; and Phoschek® which contains diammonium phosphate, sodium carboxymethylcellulose, sodium orthophenylphenate, and sodium fluorosilicate. The impact of the use of these compounds on water quality is not known (Ref. 2-52).

Recreational Areas. While the concept of "recreational carrying capacity" is in justifiable dispute among managers of recreational lands (Ref. 2-60), excessive use and/or misuse of recreational lands can lead to negative water quality impacts on a local scale. Minimizing this and other negative impacts is not so much a problem of how many visitors should be allowed on a recreational area in a given time period, as how the entire land-water unit and visitor use should be managed.

Runoff Volume. Most open-space recreational land areas have a good complement of natural vegetation cover, modified to varying degrees for visitor access and enjoyment. Increases in runoff volume from a site is directly related to the extent to which land surfaces are made less permeable to water infiltration. Heavily used foot paths without protective mulch or gravel, paved parking lots and other hardsurfaced areas, and large buildings, all increase runoff potential and require planning to maximize infiltration on adjoining land areas. Clearly, runoff potential increases with land slope; therefore, steeply sloped areas should not be used for paths, roads, or buildings.

Sediments. Sediment release to streams and lakes receiving runoff from recreational land areas is a problem only to the extent that native (or planted) vegetation is disturbed by construction activities, misuse, or natural disasters such as fire, floods, and tree disease.

Nutrients. Nutrient losses to receiving waters can come about by runoff from heavily fertilized areas (such as golf courses and other lawn sites), from sediments washed from fertilized areas, and from sewage treatment plant effluents.

Pesticides. Insecticide treatment is common in recreational areas, but no published information on use level or impact on water quality was available to the authors. Surface runoff and spray drift are the two most common modes by which pesticides would affect water quality.

Organic Loads. Sewage disposal presents the greatest potential threat to water quality in recreational areas, but few, if any, generalizations can be made in this area. Difficulty in sewage disposal in certain areas has led to restrictions on number of visitors allowed (for example, see Ref. 2-9). Recreational areas with large visitor numbers generally require a treatment plant of their own, with a capacity set by expected peak daily visitor numbers. When inflow is highly sporadic (corresponding to heavy summer and weekend use, for example), special maintenance procedures may be required for efficient treatment. It may be necessary, for example, to supplement the system with organic matter during low-flow conditions in order to maintain adequate levels of bacterial activity. Effluent quality, therefore, is even more dependent on careful operation and maintenance than in comparable systems with more uniform flow.

Pathogens. The primary threat to water quality in this regard is associated with the operation and maintenance difficulties of sewage treatment plants receiving highly sporadic flows. Effluents should be tested frequently for presence of pathogens which threaten health of humans and animals in recreation areas.

Mining Areas. Mining in this country has affected approximately 5.3 million hectares (13 million acres) of land, including three million hectares undercut by mines and more than 1.2 million surface-mined (Ref. 2-52). Slightly over one million hectares are used to contain mining-related mineral wastes. By the year 2000, the Department estimates that 12 million hectares will be affected by mining operations (Ref. 2-52). A breakdown of total lands affected by mines for states and commodities is given by Grim and Hill (Ref. 2-20).

Mining may produce massive water quality problems for two basic reasons. The first is that surface and sub-surface patterns of natural drainage are often completely disrupted; the second is that mining invariably results in exposure of water-soluble materials to water and oxygen to a much greater degree than had existed prior to mining operations.

Acid Drainage. Perhaps the most serious water quality impact of mining is that generated by the oxidation of pyritic materials (iron disulphide, FeS_2) in the presence of water to form sulfuric acid, as well as iron and other salts. This drainage can arise from both surface and deep mines, and from the mining of coal and many metal mining operations. While it is commonly associated with coal mining, not all coal mining areas release it. It can also be released without mining. The chemical impacts of this drainage include reaction of the acid with clays to release aluminum in concentrations high enough to kill fish, reaction with limestone to yield very hard water, and reactions with soil and minerals to extract heavy metals (manganese, copper, zinc) which are toxic to aquatic organisms. A very extensive study of acid mine drainage in Appalachia has been made (Ref. 2-3).

Sediments. A second major problem with all mining areas is that of sediment. The two biggest sources are soils disturbed by surface mining and vast piles of tailings which are accumulated in mineral processing.

Sediment production by surface coal mining is a rapidly growing problem. In 1920, 8.5 million tons of coal were mined from the surface; in 1947, 149 million tons were mined; and in 1971, 235 million tons, or approximately one-half of the total coal produced, were mined (Ref. 2-52). Materials moved to expose coal seams are often highly erodible; and in hilly or mountainous terrain, the handling of this overburden leads to acute siltation of streams, lakes, and reservoirs. In nearly all cases, specially designed settling basins are required to prevent this siltation.

In ore processing, water is often used as a transport medium. Discharge of wastewater from such operations always requires treatment to remove solids; it may also require treatment to remove toxic chemicals.

Groundwater Pollution. The sinking of deep shafts or the disruption of subsurface drainage patterns may cause serious disruptions of aquifers and groundwater flows. These disruptions also allow polluted surface water to enter groundwater reserves, including those which supply rural (or recreational) users through wells and springs.

Ore extraction processes are sometimes carried out in large shallow basins ("leach mining"). Seepage from these basins can also enter groundwater if they are not adequately sealed to contain the leach liquor (Ref. 2-5-2).

Leachate from mine tailings or spoil banks can also be a serious problem during and after rainfall. Compared with pre-mined conditions, they generally represent more concentrated and more erodible sources of chemicals which cause serious degradation of the streams and lakes they enter. The problem is sometimes compounded by the fact that large tailing piles are saved for future economic extraction and remain there for 20 years or more. In this time extensive erosion can occur, seals between the holding area and groundwater may deteriorate, etc.

Radioactivity. Entry of radioactive compounds into ground and surface waters from mining land areas is a serious concern. A monitoring network has been established which, to date, shows no excessive levels of radioactivity (Ref. 2-52); but mine-related radioactivity does exist in water, and as mining activity increases in relation to power and ore needs, the threat of contamination is real and requires extreme vigilance. An added difficulty in this area is that the effects of long-term, low-level exposure to radioactivity on man and other organisms is very poorly understood.

Nutrients and Pesticides. Fertilizers and pest control chemicals are sometimes used extensively in the reclamation of mine spoil banks. While there is no evidence to date as to identified, harmful effects on water quality, careless management of these compounds could lead to locally severe water quality problems.

2-3.3 Water Quality Impact of Urban Development

The two major sources of urban water pollution are domestic and industrial wastewater (sewage) and storm water runoff. Each has been studied extensively from the standpoint of pollutant contents and treatment methods (Ref. 2-2; 2-5; 2-6; 2-14; 2-28; 2-51; 2-55; and many others). Management models for the storm water component have also been developed, and are now being applied (for example, Ref. 2-24 and Ref. 2-25). While it would be fair to conclude that major urban point sources of pollution are gradually coming under control, urban nonpoint sources are not; in many cases they pose a water threat comparable to that of untreated sewage.

Point Sources of Pollution

Domestic Wastewater. Each person in this country uses an average of approximately 75 gallons of water per day (Ref. 2-49), a large fraction of which is returned to municipal sewage treatment systems. Depending on the type of system used and its overall condition, this flow is joined by variable rates of infiltration and runoff flows. Domestic wastewater contains high levels of solids and nutrients, generates high levels of bacteria (particularly coliforms), and makes high oxygen demands. (See Table 2-7.) The required secondary treatment (PL 92-500, Sec. 301 (b)1(B)) of municipal sewage will reduce effluent levels of suspended solids and biochemical oxygen demand to 30 mg/liter each. No fecal coliform limitations apply at the present time. While a variable fraction of inorganic nutrients are removed from domestic sewage with the removed solids, the effluents from secondary treatment plants remain relatively rich in nutrients. The impact of these nutrient levels on receiving stream water quality is a function of discharge rate relative to stream flow rate pattern, the nutrient concentration of the receiving stream water, the time of year, and other conditions affecting the nature and activity level of organisms in the stream. In some instances, land irrigation of effluents is a preferred solution (Ref. 2-44; Ref. 2-47).

TABLE 2-7 Typical Composition of Domestic Sewage. (All units in mg/1)

Constituent	Concentration		
	Strong	Medium	Weak
Solids, total	1,200	700	350
Dissolved, total	850	500	250
Fixed	525	300	145
Volatile	325	200	105
Suspended, total	350	200	100
Fixed	75	50	30
Volatile	275	150	70
Settleable solids	20	10	5
Biochemical oxygen demand, 5-day, 20°C (BOD_5-20°)	300	200	100
Total organic carbon (TOC)	300	200	100
Chemical oxygen demand (COD)	1,000	500	250
Nitrogen (total as N)	85	40	20
Organic	35	15	8
Free ammonia	50	25	12
Nitrites	0	0	0
Nitrates	0	0	0
Phosphorus (total as P)	20	10	6
Organic	5	3	2
Inorganic	15	7	4
Chlorides[a]	100	50	30
Alkalinity (as $CaCO_3$)[a]	200	100	50
Grease	150	100	50

[a]Values should be increased by amount in carriage water.

Source: Metcalf and Eddy, Inc., 1972, p. 231.

Industrial Wastewater. Estimates of total industrial wastewater generated are difficult to make, but they are in the range of 300 gallons per capita per day (Ref. 2-49). Water consumption by selected industry types varies almost 100-fold. (See Table 2-8.) Treatment requirements for industrial wastewater are highly variable, depending on its load characteristics (see, for example, Table 2-9) and methods of effluent disposal. In most cases, a unique plan is required for each major industrial discharger, in view of the variables mentioned and the overall wastewater management planning strategy adopted for the region in which the industry is located. Clearly, the potential impact on water quality of mismanagement of industrial wastes is staggering, and the recovery from mismanagement in the past may require decades in some instances.

TABLE 2-8 Water Consumption by Selected Industry Types

Process	Consumption
Cannery	
Green beans, gal/ton	20,000
Peaches and pears, gal/ton	5,300
Other fruits and vegetables, gal/ton	2,000–10,000
Chemical industries	
Ammonia, gal/ton	37,500
Carbon dioxide, gal/ton	24,500
Gasoline, gal/1,000 gal	7,000–34,000
Lactose, gal/ton	235,000
Sulfur, gal/ton	3,000
Food and beverage industries	
Beer, gal/1,000 gal	15,000
Bread, gal/ton	600–1,200
Meat packing, gal/ton live weight	5,000
Milk products, gal/ton	4,000–5,000
Whiskey, gal/1,000 gal	80,000
Pulp and paper	
Pulp, gal/ton	82,000–230,000
Paper, gal/ton	47,000
Textiles	
Bleaching, gal/ton cotton	72,000–96,000
Dyeing, gal/ton cotton	9,500–19,000

Source: Metcalf and Eddy, 1972, p. 32.

TABLE 2-9 BOD of Wastes from Selected Industries

Source of Waste	BOD_5, 20 deg C, mg/l
Brewery	500–1,200
Beer slop	11,500
Cannery	300–4,000
Grain distilling	15,000–20,000
Molasses distilling	20,000–30,000
Laundry	300–1,000
Milk processing	300–2,000
Meat packing	600–2,000
Pulp and paper	
Sulfite	20
Sulfite-cooker	16,000–25,000
Tannery	500–5,000
Textiles	
Cotton processing	50–1,750
Wool scouring	200–10,000

Source: McGauhey, P. H., 1968. *Engineering Management of Water Quality*, p. 49.

Non-Point Sources of Pollution

Runoff. One of the most critical consequences of urban development is the increase in impervious surface area, (see Table 2-10), and the resulting increases in runoff volume. At the same time, these impervious surfaces accumulate high levels of dust and dirt generated by urban activities. (See Table 2-11.) With each rainfall, the accumulation is removed to the extent determined by the intensity and duration of the rainfall. (See Table 2-12.) In a study of runoff from a watershed in Durham, North Carolina, downstream water quality was judged to be primarily governed by urban runoff approximately 20 percent of the time (Ref.

TABLE 2-10 Typical Runoff Coefficients for Different Land Use Types.

Land Use Type	Runoff Coefficient
Business	
Downtown	0.70–0.95
Neighborhood	0.50–0.70
Residential	
Single-family	0.30–0.50
Multiunits, detached	0.40–0.60
Multiunits, attached	0.60–0.75
Residential suburban	0.25–0.40
Apartment	0.50–0.70
Industrial	
Light	0.50–0.80
Heavy	0.60–0.90
Parks, cemeteries	0.10–0.25
Playgrounds	0.20–0.35
Railroad yard	0.20–0.35
Unimproved	0.10–0.30
Character of Surface	
Pavement	
Asphalt and concrete	0.70–0.95
Brick	0.70–0.85
Roofs	0.75–0.95
Lawns, sandy soil	
Flat, up to 2% grade	0.05–0.10
Average, 2–7% grade	0.10–0.15
Steep, over 7%	0.15–0.20
Lawns, heavy soil	
Flat, up to 2% grade	0.13–0.17
Average, 2–7% grade	0.18–0.22
Steep, over 7%	0.25–0.35

Source: Hjelmfelt and Cassidy, 1975. *Hydrology for Engineers and Planners*, p. 114.

TABLE 2-11 Dust and Dirt Accumulation in Urban Runoff.

Type	Land Use	Filth per 100-ft-curb	Filth per Acre
1	Single-family residential	0.7	3.2[a]
2	Multi-family residential	2.3	10.5[a]
3	Commercial	3.3	15.1[a]
4	Cropland	3.5[a]	16.0
5	Undeveloped or Park	1.5	6.9[a]

[a]Cropland calculations are per acre; the others are per 100 ft-curb: 460 ft-curb per acre is used for comparison.

(Pollution Parameters in Urban Runoff are shown below—units in mg/gram of dust and dirt)

Parameters	Land Use Type 1	2	3	4[a]	5
Suspended solids (mg/1)	1,000	1,000	1,000	1,000	1,000
BOD (mg/l)	5	3.6	7.7	50	5
COD (mg/l)	40	40	39	550	20
F-Coliform (10^6 MPN/l)[a]	0.65	1.3	0.85	2	0
Settleable solids (mg/l)	100	100	100	100	100
N (mg/l)	0.48	0.61	0.41	30.0	0.05
PO_4 (mg/l)	0.05	0.05	0.07	5.0	0.01
NH_3-N (mg/l)[a]	0.01	0.02	0.01	2.0	0.0

[a]Values for Cropland, F-Coliform, and NH_3-N were arrived at through a process of research, calibration against a severe storm expected impact on the Des Moines River by the Camp Creek Basin, and best judgment.

Source: Revised from Huber, *et al.* Storm Water Management Model User's Manual, Version II. U.S. EPA, Cincinnati, Ohio, pp. 57–58.

2-14). Based on tests made in numerous cities, the URS Research Company estimated the characteristics of runoff water and compared it with treated and untreated sanitary sewage. (See Table 2-13.) While the frequency of rainfall events and total annual rainfall received varies from one region to another, the water quality impact of urban runoff is compensated for to a considerable extent because, with less frequent rainfall, more dust and dirt accumulate between rainfall events which, therefore, cause higher concentrations of pollution in the runoff. At the present time, approximately 30 percent of the total sewered population is served by a combined sewer system (Ref. 2-28). Many combined systems overflow two or more times per month, when rainfall causes total flows to exceed system capacity. Under these conditions, untreated sanitary sewage, plus polluted stormwater, flow directly to the receiving stream.

TABLE 2-12 Percent of Contaminants Removed from Street Surfaces by Runoff Rate/Duration.

Runoff Rate (in./hr)	Runoff Duration (hr)							
	0.25	0.5	1.0	2.0	3.0	4.0	5.0	6.0
0.1	10.9	20.5	36.9	60.1	74.8	84.1	90.0	>90.0
0.2	20.5	36.9	60.1	84.1	>90.0	>90.0	>90.0	
0.3	29.1	49.8	74.8	>90.0				
0.4	36.9	60.1	84.1					
0.5	43.7	68.3	90.0					
0.6	49.8	74.8	>90.0					
0.7	55.3	80.0						
0.8	60.1	84.1						
0.9	64.5	87.4						
1.0	68.3	90.0						

Source: URS Research Co., 1974. p. I-21.

It is clear from the foregoing discussion that a very important management variable affecting water quality in urban areas is the frequency and efficiency of street sweeping. In general, only about one-half of the dust and dirt accumulated on urban streets is now removed in sweeping (Ref. 2-51). Street sweepers tend to be most efficient in picking up large particles (500 microns), whereas street dirt consists predominantly of smaller particles.

Another very significant consequence of urbanization is the speed and intensity of flood flows that are generated in heavy rains. Brater and Sherrill (Ref. 2-8), for example, found that urbanization accompanying a population density change from 100 to 13,000 persons per square mile caused a ten-fold increase in peak runoff rate, while the time to peak flow decreased ten-fold. This means that downstream flooding of receiving streams increases very dramatically with urbanization, particularly when the urbanized land is in the floodplain itself, thereby removing flood-holding capacity while rapidly generating the greatly increased runoff. About 16 percent of the nation's urban areas are located in the natural 100-year floodplain, whereas seven percent of the nation's total land area is in the floodplain (Ref. 2-18). Brater and Sherrill (Ref. 2-8) also found that "hydrologically significant impermeable area" increased linearly from about one percent for 1,000 persons/square mile, to about 10 percent for 7,500 persons/square mile.

Construction Activities. According to projections of the Department of Housing and Urban Development, increases in the U.S. population through the year 2000 will require the *daily* development of about 1,600 hectares (about 4,000 acres) of land for housing, transportation, utilities, communication, and sewer and wastewater treatment systems (Ref. 2-52). While much of the construction

TABLE 2-13 Calculated Quantities of Pollutants which Would Enter Receiving Waters in a Hypothetical City.*

	Street Surface Runoff (Following 1-hr Storm) (lb/hr)	Raw Sanitary Sewage (lb/hr)	Secondary Plant Effluent (lb/hr)
Settleable plus suspended solids	560,000	1,300	130
BOD_5	5,600	1,100	110
COD	13,000	1,200	120
Kjeldahl nitrogen	880	210	20
Phosphates	440	50	2.5
Total coliform bacteria (org/hr)	$4{,}000 \times 10^{10}$	$460{,}000 \times 10^{10}$	4.6×10^{10}

Note: Since the above calculations were based only on a 5-day accumulation of street litter, the above discharge of contaminated runoff could conceivably occur many times in a year.

*The hypothetical city has the following characteristics:
- Population–100,000
- Total land area–14,000 acres
- Land-use distribution:
  - Residential–75%
  - Commercial–5%
  - Industrial–20%
- Streets (tributary to receiving waters)–400 curb miles
- Sanitary sewage–12 mgd

Source: URS Research Co., 1974, p. V-8.

takes place in urbanized areas, major projects also affect rural land areas as well (energy networks, highways, dams, and reservoirs).

Construction-related water pollution is highly site-specific. The significant variables include size and location of the project, rainfall frequency and intensity, pest control measures utilized, resistance of the soil to wind and water erosion, climatic factors affecting rate and type of revegetation possible, physical and chemical properties of sub-surface earth materials, distance to waterways, nature and volume of construction materials which need to be stored on the site, and the requirements for in-water construction operations.

2-3.4 Modeling Applications

In cooperation with the Central Iowa Regional Association of Local Governments (CIRALG), TenEch Environmental Consultants, Inc. and Harland Bartholomew and Associates used the runoff block of the Storm Water Management Model (SWMM) to investigate land use-water quality interactions in the Des Moines, Iowa metropolitan area.

Approximately six percent of the study area, or 30,000 acres, is urban residential, with a population of slightly more than 290,000. Domestic sanitary sewage generated by this population is approximately 30 million gallons per day, compared with the total sewage generation in the study area of about 42 million gallons per day by 315,000 people (Ref. 2-10). About one-third of the urban residential land area (excluding streets) is covered by impervious surfaces, and an additional 5,000 acres consists of paved streets. Therefore, about 43 percent of the total urban residential land area, including streets, is covered with completely impervious surfaces, the remainder having varying degrees of perviousness.

Agricultural land uses occupy approximately 80 percent of the study area, including 350,000 acres of cropland. Urban non-residential land uses and non-urban streets occupy about two percent each; permanent open space (recreation areas, cemeteries, etc.) covers nearly four percent, and water surfaces one percent. Approximately five percent of the land is either forested or vacant (Ref. 2-10).

Hypothetical Test Plots. Based on information gathered in earlier phases of the Des Moines 208 study, and with advice from CIRALG, typical land use and surface characteristics were selected as input to the SWMM model. (See Table 2-14.) Pollutant generation by each of five land use types was based on other studies (see Table 2-11), and modified to include cropland factors.

Based on the SWMM computation output, the five land use types were ranked in order of increasing runoff and pollutants generated for each pollution parameter. (See Table 2-15.) Pollutants varying most strongly with land use are COD, settleable solids, and phosphates. Those varying least with land use are total runoff, total suspended soils, and coliform levels.

Under the simulated conditions, undeveloped and single-family residential land uses produced the least runoff and pollutants, while commercial/industrial and cropland produced the most. While croplands predictably generated nutrients (nitrates, phosphates, ammonia) at the highest rate, commercial/industrial areas generated almost as much, according to these computations.

Soil erosion contributes very strongly to the load of suspended solids and therefore, it is understandable that croplands lead in this category. (See Table 2-15.) However, multi-family and commercial/industrial areas produce two to five times the rate of settleable and suspended solids without erosion as compared to croplands. Urban activities generate large quantities of solids which enter runoff flows readily, because of extensive smooth, paved surfaces in commercial/industrial and multi-family areas.

TABLE 2-14 Conditions Used in SWMM Runoff Model for Analyses of 10 Acre Land Use Tracts

Land slope: 5.0%
Crop management factor: 0.4 cropland
0.1 all others
Soil erodibility factor: 0.32
Erodible distance: 209 ft
Impervious factor: Residential–medium density–.25
Pasture, forest and vacant–.15
Cropland–.30
Residential–high density–.55
Non-residential–low intensity–.75
Erosion control practice factor: 1.0
Urban catch basins per acre: 10
Rural catch basins per acre: 1
Infiltration rate: 0.52, 3.0 (min., max.)
Horton's infiltration decay rate (1/sec): 0.00115
Catch basin storage volume (cu ft): 15
Concentration of BOD in catch basin (mg/l): 100
Resistance of impervious area: 0.013
Resistance of pervious area: 0.25
Retention of impervious area (in.): 0.062
Retention of pervious area (in.): 0.184

Rainfall Pattern During Simulated Storm

15-Min. Interval No.	Rainfall Intensity
1	0
2-5	0.2 in./hr
6-7	0.6 " "
8	0.2 " "
9-10	0

Impact of Development. The SWMM runoff model was used to estimate the water quality impact of projected urban development in a 15,000 acre sub-area in the Des Moines 208 region. The baseline for comparison was the simulated runoff quality under existing conditions. Future conditions were those established in the preliminary Intensity Development Pattern by CIRALG, with assistance from Harland Bartholomew and Associates (Ref. 2-10). A third set of conditions was simulated, namely that of future development modified in the following ways to protect water quality: 1) provide continuous open space corridors along water courses; 2) reduce impervious factors in selected instances through project design modifications (pervious pavements, etc.) and additional project open space; 3) improve overall agricultural cropping and management

TABLE 2-15 Rank Order of Land Use Types Based on Amount of Flow and Pollutants Generated Over a Hypothetical 10 Acre Tract for Two and One-half Hours of a Specified Storm.[a]

	Least 1	2	3	4	Most 5
Runoff	UD(77,739)	SF(1.66)	Crop(1.99)	MF(3.61)	CI(4.89)
BOD	UD(907)	SF(1.31)	MF(7.85)	CI(18.4)	Crop(19.8)
Suspended solids	SF(181,139)	UD(1.12)	MF(1.18)	CI(1.70)	Crop(3.87)
Coliform bacteria	UD = 0	SF(33,314)	MF(1.07)	CI(1.14)	Crop(1.74)
COD	UD(1,595)	SF(2.11)	MF(7.68)	CI(11.8)	Crop(100.2)
Settleable solids	SF(78)	UD(1.04)	Crop(6.76)	MF(13.5)	CI(36.2)
Nitrates	UD(444)	SF(1.07)	MF(11.9)	CI(26.2)	Crop(26.4)
Phosphates	UD(45)	SF(1.06)	MF(11.7)	CI(26.2)	Crop(39.0)
Ammonia	UD(44)	SF(1.04)	MF(11.7)	Crop(26.0)	CI(26.0)
Suspended solids minus erosion	UD(9,729)	SF(1.03)	Crop(6.82)	MF(11.6)	CI(25.9)

[a]The full set of conditions used to make these calculations appears in Tables.

Note: (1) Legend: UD–Undeveloped or park
SF–Single-family residential
MF–Multi-family residential
CI–Nonresidental–low density commercial/industrial
Crop–Cropland

(2) Relative values of runoff/pollutants generated are shown with each land use type; large numbers in parenthesis are absolute values.

practices; 4) limit construction on land areas with nine percent slopes or greater with highly erodible soils.

Land use data for each of the three conditions just described is as follows:

	(1) 1975 Existing	(2) Future High Degree of Development	(3) Future High Degree of Development with Improved Water Quality Practices
Residential, medium density (acres)	2,941	5,214	4,995
Residential, high density (acres)	102	357	334
Non-residential (acres)	274	984	654
Cropland (acres)	8,307	6,116	5,894
Pasture, forest and vacant (acres)	2,594	1,255	2,463
Population	16,700	50,100	43,500

Model output showed that projected future development would approximately double the pollution level compared with existing conditions in all parameters, except total runoff, suspended solids, and settleable solids. (See Table 2-16).

TABLE 2-16 Pollutants Generated in 15,000 Acre Subarea Under Existing and Hypothetical Development Conditions[a]

	(1) Existing (1975)	(2) Future High Level of Development	(3) Future High Level of Development with Modifications to Improve Water Quality
Runoff (liters)	1.09	1.08	1 (=1.45 X 10^8)
BOD (gm)	1 (=5.11 X 10^6)	1.87	1.71
Suspended solids (gm)	3.17	2.96	1 (=8.48 X 10^8)
Coliform bacteria (MPN)	1 (=1.59 X 10^{11})	1.92	1.78
COD (gm)	1 (=4.22 X 10^7)	1.97	1.81
Settleable solids (gm)	1 (=6.87 X 10^5)	1.32	1.20
Nitrates (gm)	1 (=3.29 X 10^6)	1.79	1.64
Phosphates (gm)	1 (=4.80 X 10^5)	1.84	1.69
Ammonia (gm)	1 (=2.54 X 10^5)	1.74	1.59
Suspended solids minus erosion (gm)	1 (=2.28 X 10^7)	1.41	1.28

[a]See text for use and population factors for each of the above.

Note: Relative values of runoff/pollutants generated are shown for each condition; numbers in parenthesis are absolute values.

Modifications made in projected development to minimize water quality impact brought substantial reductions in all parameters, but only in two cases (runoff and suspended solids) was the improvement better than existing conditions. This means that considerable effort will be required in the area of non-point pollution control, simply to maintain the status quo in water quality under the conditions simulated here.

Even though much of the projected development was projected to take place on land now in agriculture, all three nutrient parameters nearly doubled for future development without modifications. This indicates clearly that the control of nutrient losses is the concern of urban, as well as rural, land owners. Pollution by suspended solids decreased with development, and could be reduced very substantially by management practices and by leaving highly erodible soils undeveloped or unfarmed.

Not surprisingly, the parameters increasing most with development are oxygen demand (BOD, COD) and coliform levels. This increase is particularly troublesome in surface waters which are also enriched by nutrients, since nutrient-rich waters will stimulate even more development of oxygen-demanding materials; i.e., aquatic plants and organisms that feed on them.

2-3.5 Summary of Land Use-Water Quality Studies

Land use-water quality information currently available is far too extensive and complex to summarize briefly. However, in order to establish an estimate of relative seriousness among water quality impacts of different land uses, three summary tables have been prepared. (See Tables 2-17, 2-18 and 2-19.)

TABLE 2-17 Impact of Land Use on Water Quality by Point-Source Pollutants

Land Use		Type of Pollutant				
		Organic Wastes	Nutrients	Sediments	Thermal Wastes	Toxic Compounds
Type	Intensity					
Residential	High	●	●			
	Medium	●	●			
	Low	◓	○			
Commercial	High	◓	○			
	Low	○	○			
Industrial	High	●	◓	○	◓	○
	Low	○	○			○
Forest	Mature					
	Immature					
	Recently cut					
Agriculture	Cultivated crops					
	Pasture, orchard					
	Vacant					
Mining	Surface					
	Deep shaft					

Key to Symbols: ● Denotes primary or major impact.
○ Denotes possible or minor impact.
◓ Denotes intermediate impact.

Source: Modified from Shubinski and Nelson, 1975, p. 5.

TABLE 2-18 Impact of Land Use on Water Quality by Non-Point Source Pollutants

Land Use		Type of Pollutant					
Type	Intensity	Organic Waste	Nutrients	Sediments	Thermal Waste	Toxic Compounds	Other
Residential	High	●	●	○	○		◐(1) (2)
	Medium	○	○	○			○(1)
	Low	○	○	○			
Commercial	High	○	○	○	◐		○(2)
	Low						
Industrial	High	◐	◐	○	◐	○	○(2)
	Low	○	○	○	○		
Forest	Mature	○					○(3) (4)
	Immature	○					○(3) (4)
	Recently cut	○	○	○	◐		○(3) (4)
Agriculture	Cultivated crops	○	●	●			○(3) (4)
	Pasture, orchard	◐	○				○(4)
	Vacant	○	○	○	○		
Mining	Surface			●		◐(5)	●(6)
	Deep shaft			◐		○(5)	

Key to Symbols:
● Denotes primary or major impact.
○ Denotes possible or minor impact.
◐ Denotes intermediate impact.
(1) Deicing compounds and lead.
(2) Oil drippings from parked cars.
(3) Herbicides.
(4) Pesticides.
(5) Heavy metals.
(6) Acid drainage.

Source: Modified from Shubinski and Nelson, 1975, p. 5.

TABLE 2-19 Water Quality Impact of Land Development

Land Use Changes	Water Quality Parameters				
	Organic Wastes	Nutrients	Sediments	Thermal Wastes	Toxic Compounds
From forest land					
to residential (high)	●	◓	○	○	
to residential (low)	◓	○	○	○	
to commercial	◓	○	○	◓	
to industrial	●	○	○	◓	○
to cropland	○	●	●	○	
From cropland					
to residential (high)	○	+	+		
to residential (low)	○	+	+		
to commercial	○	+	+		
to industrial	○	+	+	○	○
to forest	+	+	+	+	
From pasture, open field, vacant					
to residential (high)	○	○	0/+		
to residential (low)	○		0/+		
to commercial	○	○	0/+		
to industrial	○	○	0/+	○	○
to cropland		●	●		
to forest		+	+	+	

Key to Symbols: + Indicates positive water quality impact.
0/+ Indicates that impact may be positive or negative.

2-4 LAND USE PLANNING AND WATER QUALITY IN THE FUTURE

2-4.1 Planning Goals and Objectives

As a nation, our land use policy will determine, to a large degree, the environmental quality we enjoy in the decades ahead. The use (or abuse) of land affects air, water, and land quality in persistent, complex ways—many of them poorly understood.

Land use planning, particularly as it affects water quality, should have the following general goals:

1. To guide the safe and productive use of land and water resources by private and public land owners;
2. To protect surface and sub-surface water resources from land use-generated pollutants which threaten the integrity and function of aquatic ecosystems;
3. To develop land use policies and procedures which a) maximize infiltration

of rainfall where it is received; b) minimize alteration of natural surface and sub-surface drainage flows; and c) minimize land use-generated changes in the temperature of water returned to natural flows or reservoirs;

4. Minimize land uses which have unknown but potentially hazardous water quality impacts.

More specific objectives are required for land use planning to meet these goals in a particular setting. In establishing land use planning objectives, the planner needs to understand economic goals and limitations of the study area, recent trends and projections of population growth, the physical characteristics of land resources (soil types, geology of surface and sub-surface materials, topography, drainage patterns and hydrology), the nature of the vegetation, the existing land use pattern, the water quality impact of present land uses, existing laws governing land use and water quality, and existing and potential water quality problems in the area, institutional arrangements (existing and potential) for implementing land use policies, and controls protecting water quality.

The planner must at all stages maintain good communication with the citizens affected by the land use plan under consideration. He will find, in many cases, that he is in the position of reconciling the goals and objectives of individual land owners with the long-range goals and objectives of the community, region, and state in which the study area is located.

2-4.2 Land Use Controls for Cost-Effective Water Quality Protection

The list of land use controls potentially effective in water quality protection is a very long one, and one which every planner should know. However, each setting is somewhat unique geographically, economically, socially, and politically; therefore, land use controls effective in one area may be very different from those which are effective in another.

A thorough review of land use issues, including land use controls, was published by the Council on Environmental Quality (Ref. 2-15). We do not intend to summarize or add to that review in this chapter. We do, however, wish to emphasize that the set of land use controls selected for a particular area must satisfy two basic requirements: that they are effective in meeting water quality goals and objectives; *and* that the citizens affected understand the water quality objectives and appreciate the need for enforcing the controls.

The basis for land use controls in protecting water quality is an understanding of land use-generated water quality impacts in a particular region. Therefore, a selection of land use controls should begin only after a thorough review of all relevant information regarding land use-water quality interactions.

2-4.3 General Conclusions

1. Land use-generated non-point pollution sources present serious threats to water quality nationwide. In particular, agricultural, mining, and urban land uses pollute water significantly.
2. In urbanized areas, increases in degree of impervious cover (paving, rooftops, etc.) are responsible for many downstream water quality problems, as well as increased flooding problems.
3. Land planners at all levels of government and in the private sector bear major responsibility for identifying water polluting land uses and estimating relative water quality impacts of all significant point and non-point pollution sources.
4. Many land use-generated sources of water pollution can be controlled through land management practices which minimize surface runoff and maximize ground infiltration of surface water. In agriculture, reducing runoff also reduces costly losses of soil, nutrients, and pesticides.
5. Where voluntarily applied management practices are not adequate, land use controls may be required to protect water quality from the effects of intense and/or improper land use practices. This may be particularly applicable in urban and mining land uses. Full cooperation is required among local, state, and federal levels of government in order to arrive at locally acceptable sets of land use control policies.
6. The cost for proper assessment of land use-generated water quality problems and their solutions may be very costly in certain locations; therefore, cost-sharing mechanisms among all levels of government may be necessary in this area for many years to come.

REFERENCES

2-1 Amemiya, Minoru, "Land and Water Management for Minimizing Sediment," *In* Willrich, T. L., and G. E. Smith, editors, *Agricultural Practices and Water Quality*, Iowa State University Press., Ames, Iowa, 1970, pp. 35–45.

2-2 American Public Works Association, Water Pollution Aspects of Urban Runoff, U.S. Department of the Interior, Federal Water Pollution Control Administration, Washington, D.C., 1969.

2-3 Appalachian Regional Commission, *Acid Mine Drainage in Appalachia*, Washington, D.C., 1969.

2-4 Ashton, Peter and Richard C. Underwood, editor, Non-Point Sources of Water Pollution, *Proceedings of the Southeastern Regional Conference*

Conducted on May 1 and 2, 1975 in Blacksburg, Virginia at Virginia Polytechnic Institute and State University, Published by Virginia Water Resources Center, Blacksburg, Va., 1975.

2-5 AVCO Economic Systems Corporation, A Multi-phasic Component Study to Predict Storm Water Pollution from Urban Areas, U.S. Department of Interior, Offices of Water Resources Research, Washington, D.C., 1970.

2-6 AVCO Economic Systems Corporation, Storm Water Pollution from Urban Land Activity, U.S. Department of the Interior, Federal Water Quality Administration, Washington, D.C., 1970.

2-7 Bauman, E. Robert and Sheldon Kelman, Effects of Surface Runoff on the Feasibility of Municipal Advanced Waste Treatment, *In* Willrich, T. L. and G. E. Smith, editors, *Agricultural Practices and Water Quality*, Iowa State University Press, Ames, Iowa, 1970, pp. 344–362.

2-8 Brater, Ernest F. and James D. Sherrill, Rainfall-Runoff Relations on Urban and Rural Areas, U.S. Environmental Protection Agency, EPA-670/2-75-046, 1975.

2-9 Bury, Richard L., Recreation Carrying Capacity–Hypothesis of Reality, *Parks and Recreation* **11**(1): 22 ff., 1976.

2-10 Central Iowa Regional Association of Local Governments (CIRALG), *Des Moines 208 Areawide Waste Treatment Management Plan, Phase I Interim Report*, January 1976.

2-11 Clark, John, Coastal Ecosystems, Ecological Considerations for Management of the Coastal Zone, The Conservation Foundation, Washington, D.C., 1974.

2-12 *Code of Federal Regulations: 40 Protection of the Environment*, Parts 130 and 131, Published in the *Federal Register*, **39**, no. 107, Monday, June 3, 1974, pp. 19634–19644.

2-13 *Code of Federal Regulations: 40 Protection of the Environment*, Parts 130 and 131, Published in the *Federal Register*, **40**, no. 230, Friday, November 28, 1975, pp. 55334–55349.

2-14 Colston, Newton V., Jr., Characterization and Treatment of Urban Land Runoff, U.S. Environmental Protection Agency, Cincinnati, Ohio, 1974.

2-15 Council on Environmental Quality, *Environmental Quality, The Fifth Annual Report of the Council on Environmental Quality*, Council on Environmental Quality, Washington, D.C., 1974.

2-16 Diesch, Stanley L., "Disease Transmission of Water-Borne Organisms of Animal Origin," *In* Willrich, T. L. and G. E. Smith, editors, *Agricultural Practices and Water Quality*, Iowa State University Press, Ames, Iowa, 1970, pp. 265–285.

2-17 Dornbush, James N., John R. Andersen, and Leland L. Harms. Quantification of Pollutants in Agricultural Runoff, U.S. Environmental Protection Agency, Washington, D.C., 1974.

2-18 Goddard, James E., *An Evaluation of Urban Floodplains*. U.S. Department of the Interior, Office of Water Resources Research, Washington, D.C., 1973.

2-19 Graham, Philip H., Lawrence S. Costello, and Henry J. Mallon, Estimation of Imperviousness and Specific Curb Length for Forecasting Stormwater Quality and Quantity, *J. Water Pollution Control Federation* **46**(4): 717–725, 1974.

2-20 Grim, Elmore and Ronald D. Hill, Environmental Protection in Surface Mining of Coal, U.S. Environmental Protection Agency, Cincinnati, Ohio, 1974.

2-21 Harbridge House, Inc., Key Land Use Issues Facing EPA, U.S. Environmental Protection Agency, Washington, D.C., 1974.

2-22 Heaney, James P. *et al.*, Urban Stormwater Management Modeling and Decision-Making, U.S. Environmental Protection Agency EPA-670/2-75-022, 1975.

2-23 Hjelmfelt, Allen T., Jr., and John J. Cassidy, *Hydrology for Engineers and Planners*, Iowa State University Press, Ames, Iowa, 1975.

2-24 Huber, Wayne C. *et al.*, Storm Water Management Model User's Manual, Version II, U.S. Environmental Protection Agency, Cincinnati, Ohio, 1975.

2-25 Hydrologic Engineering Center, U.S. Army Corps of Engineers, Urban Storm Water Runoff, "STORM," U.S. Army Corps of Engineers, Davis, Ca., 1975.

2-26 Johnson, H. P. and W. C. Moldenhauer, "Pollution by Sediment: Sources and the Detachment and Transport Process," *In* Willrich, T. L., and G. E. Smith, editors. *Agricultural Practices and Water Quality*, Iowa State University Press, Ames, Iowa, 1970, pp. 3–20.

2-27 Keene, John C. and Ann Louise Strong, The Brandywine Plan, *Am. Institute of Planning J.* **36**(1): 50–58, 1973.

2-28 Lager, John A. and William G. Smith, Urban Stormwater Management and Technology: An Assessment. U.S. Environmental Protection Agency, Cincinnati, Ohio, 1974.

2-29 Landon, Ronald A., "Hydrogeology and Water Quality Management," *J. Hydraul. Div.–Am. Society of Civil Engineers* **101** (HY2): 285–289, 1975.

2-30 Lieberman, J. A. and M. D. Hoover, "The Effect of Uncontrolled Logging on Stream Turbidity," *Water and Sewage Works* **95**: 255–258, 1948.

2-31 Likens, Gene E. and F. Herbert Bormann, "Linkages between Terrestrial and Aquatic Ecosystems," *Bioscience 24*(8): 447–456, 1974.

2-32 Lund, J. W. G., Phytoplankton, *In Eutrophication: Causes, Consequences, Correctives* (Proceedings of a Symposium), National Academy of Sciences. Washington, D.C., 1969, pp. 306–330.

2-33 McGauhey, P. H., *Engineering Management of Water Quality*, McGraw-Hill Book Company, San Francisco, California, 1968.

2-34 Megahan, W. F. and W. J. Kidd, Effect of Logging Roads on Sediment Production Rates in the Idaho Batholith, U.S.D.A. Forest Service, Research Paper INT-123, 1972.

2-35 Metcalf and Eddy, Inc., *Wastewater Engineering: Collection, Treatment, and Disposal*, McGraw-Hill Book Company, 1972.

2-36 Miner, J. R. and T. L. Willrich, "Livestock Operations and Field-Spread Manure and Sources of Pollutants," *In* Willrich, T. L. and G. E. Smith, Editors, *Agricultural Practices and Water Quality*, Iowa State University Press, Ames, Iowa, 1970, pp. 231–240.

2-37 National Academy of Sciences, Accumulation of Nitrate, NAS, Washington, D.C., 1972.

2-38 Nicholson, H. Page, "The Pesticide Burden in Water and its Significance," *In* Willrich, T. L., and G. E. Smith, Editors, *Agricultural Practices and Water Quality*, Iowa State University Press, Ames, Iowa, 1970, pp. 183–193.

2-39 Nicholson, H. Page, "Agricultural Chemicals and Water Quality," *In Water Resources Utilization and Conservation in the Environment*, M. C. Blount, Editor, Taylor County Printing Co., Reynolds, Ga., 1975, pp. 214–227.

2-40 Okun, Daniel A., "Water Management in England: A Regional Model," *Environmental Science and Technology* **9**: 918–923, 1975.

2-41 Pimentel, D., "Pesticides and Pest Control in the Future," *In Agricultural Wastes: Principles and Guidelines for Practical Solutions*, Cornell University. Syracuse, New York, 1974.

2-42 Rice, R. M. and J. R. Wallis, "How a Logging Operation Can Affect Stream Flow," *Forest Industries* **89**: 38–40, 1962.

2-43 Rice, R. M., J. S. Rothacher, and W. F. Megahan, Erosional Consequences of Timber Harvesting: An Appraisal, *Proceedings of a Symposium on Watersheds in Transition*. Fort Collins, Colorado, 1972, pp. 321–329.

2-44 Sanks, Robert L. and Takashi Asano, Editors, *Land Treatment and Disposal of Municipal and Industrial Wastewater*, Ann Arbor Science, Ann Arbor, Michigan, 1976.

2-45 Schmidt, Curtis J. and Ernest V. Clements, III., Demonstrated Technology and Research Needs for Reuse of Municipal Wastewater, U.S. Environmental Protection Agency, EPA-670/2-75-038, 1975.

2-46 Shubinski, Robert P., and Steven N. Nelson, *Effects of Urbanization on Water Quality*, American Society of Civil Engineers, Technical Memorandum No. 26, New York, N.Y., 1975.

2-47 Sopper, William E. and Louis T. Kardos, Editors, Recycling Treated Municipal Wastewater and Sludge through Forest and Cropland, Pennsylvania State Unversity Press, University Park, Pa., 1973.

2-48 Task Group 2610-P, "Sources of Nitrogen and Phosphorus in Water Supplies," *J. Am. Water Works Assoc.* **59**: 344–366, 1967.

2-49 Todd, David Keith, Editor, *The Water Encyclopedia*, Water Information Center, Inc., Port Washington, N.Y., 1970.

2-50 Train, Russell E., "Land Use Planning: A Review," *In Land Use Planning Abstracts*, Volume II, December, 1975, published by Environment Information Center, Inc., 1975, pp. 11–12.

2-51 URS Research Company, Water Quality Management Planning for Urban Runoff, U.S. Environmental Protection Agency, Washington, D.C., 1974.

2-52 U.S. Environmental Protection Agency, Methods for Identifying and Evaluating the Nature and Extent of Non-Point Sources of Pollutants, U.S. E.P.A., Washington, D.C., 1973.

2-53 U.S. Environmental Protection Agency, Guidelines for Preparation of Water Quality Management Plans, U.S. E.P.A., Washington, D.C., 1974.

2-54 U.S. Environmental Protection Agency, Clean Water, Report to Congress—1974, U.S. EPA, Washington, D.C., 1974.

2-55 U.S. Environmental Protection Agency, Water Quality Management Planning for Urban Runoff, U.S. EPA, Washington, D.C., 1974.

2-56 U.S. Environmental Protection Agency (Revised), *Guidance for Preparing a Facility Plan*, U.S. EPA, Washington, D.C., 1975.

2-57 U.S. Environmental Protection Agency, Water Quality Strategy Paper (Third Edition), U.S. EPA, Washington, D.C., 1975.

2-58 U.S. Environmental Protection Agency, Guidelines for Areawide Waste Treatment Management Planning. U.S. EPA, Washington, D.C., 1975.

2-59 Verduin, Jacob, "Significance of Phosphorus in Water Supplies," *In* Willrich, T. L., and G. E. Smith, Editors, *Agricultural Practices and Water Quality*, Iowa State University Press, Ames, Iowa, 1970, pp. 63–71.

2-60 Wagar, J. A., "Recreational Carrying Capacity Reconsidered," *J. Forestry* 72(5): 274–278, 1974.

2-61 Wischmeier, W. H. and D. D. Smith, "Rainfall Energy and its Relationship to Soil Loss," *Trans. Am. Geophys. Union* **39**: 285–291, 1958.

2-62 Young, L. A. and H. P. Nicholson, "Stream Pollution Resulting from the Use of Organic Pesticides," *Progressive Fish Culturist* **13**: 193–8, 1951.

3

Evaluation of wastewater treatment alternatives

Joseph L. Pavoni

TenEch Environmental Consultants, Inc.
Louisville, Kentucky

Jerry R. Perrich

Olin Corporation
Brandenburg, Kentucky

3-1. INTRODUCTION

Many areas throughout the country are currently planning water quality control strategies which will provide adequate wastewater treatment and will insure that the quality of local water resources meet Water Quality Standards. The efficiency of municipal wastewater treatment is a prime factor in the quality of regional water resources and, therefore, must be addressed by water quality control strategies.

The technology of wastewater treatment has expanded rapidly in the last 20 years, so that many alternative treatment techniques are available to the water quality control planner. However, data describing the applications and efficiencies of the various wastewater treatment techniques is scattered throughout the literature, including textbooks, journals, and government agency reports, and is not readily available. Consequently, there is a need for a single reference source outlining process technologies, effectiveness, and costs of currently available wastewater treatment techniques.

This chapter is focused on meeting this need, and is comprised of three complementary presentations. The first is a general description of wastewater treatment techniques. The authors recognize that many of the citizens or planning commissions and water quality planning committees (such as might be established under PL 92-500) do not possess technical backgrounds. Consequently, the first presentation in this chapter describes the nature, origin, and characteristics of domestic wastewaters; the effect of these specific pollutants on the environment; and currently available wastewater treatment techniques. These techniques are described at length to provide insight for planners into the mechanisms, efficiencies, and limitations of each wastewater treatment technique. It is anticipated that this portion of the chapter might be of interest to water quality management planning committee members, as background reading to acquaint them with wastewater treatment technology.

The second section of this chapter is a reference source and starting point for water quality engineering consultants and planners in water quality management planning. This section describes effluent quality classifications and alternative wastewater treatment schemes and also presents cost data for currently available wastewater treatment techniques and schemes.

In this section, alternative effluent qualities are discussed, as well as the factors which enter into decisions regarding desirable and achievable effluent qualities. Specific qualities of effluents, ranging from the minimum acceptable according to federal law to the maximum achievable with current technology, are identified and classified alphabetically according to the degree of treatment they represent.

After the development of specific effluent qualities, which may be considered the goals of wastewater treatment, wastewater treatment schemes, consisting of previously presented treatment techniques, are constructed. These treatment techniques are combined in a logical and consistent manner to yield different treatment schemes for each effluent quality. Generally, more than one treatment scheme is constructed for each effluent quality category.

Although the treatment schemes are developed without regard to cost, water quality management decisions must include economic considerations. Consequently, cost data for various treatment techniques are summarized from the literature and presented in graphical form as a function of wastewater volume treated. The costs of complete treatment plant schemes are also developed. The cost data presented reflect the best estimates available in the current literature, and is keyed to second quarter 1976 costs. A methodology for updating the cost estimates is included in the text.

This second presentation is intended to serve as general background to the nontechnical reader, as well as to delineate alternative treatment schemes and cost estimates for the technical reader.

The third and final section in this chapter is an example of how the information and data in this chapter may be utilized in water quality management planning. A case study is discussed, and cost estimates for upgrading the quality of wastewater treatment in the sample study are constructed for various wastewater treatment alternatives. The alternative water quality control strategies are ordered by effluent quality, and the costs are tabulated for each quality and strategy. The expenditures associated with each degree of effluent quality are thereby clearly delineated.

This final presentation presents a rational methodology for evaluating alternative water quality management plans on the basis of wastewater treatment costs within the given study area.

3-2. WASTEWATER CHARACTERISTICS

This section describes the biological, chemical, and physical characteristics of wastewater, thus providing a starting point for the technical analysis of treatment techniques, schemes, and economics. Included in this section are descriptions of the origin of wastewater, the flow of wastewater to the treatment plant, the various constituents contained in wastewater, the laboratory methodology utilized to measure these pollutants, and the effect of pollutants on the environment.

3-2.1 Water Use and Wastewater Production

Man's consumptive use of water in the environment is depicted in Figure 3-1, which outlines the water consumption cycle. The water source for a typical inland municipality may be groundwater (wells, springs), surface water (rivers, lakes) or rainfall (collection of runoff from specially designed pads or cisterns). The water is treated at a water treatment plant to remove impurities and pathogens, and perhaps to add beneficial constituents (e.g., fluoride). An extensive underground distribution system delivers the treated water to the consumer, where it is used for a wide variety of purposes; e.g., drinking, washing, laundering, toilet waste conveyance, lawn watering, fountains, fire fighting. This water may also be utilized by industries for cooling, processing, steam production, etc.

During some of the above uses the water is returned directly to the environment; e.g., lawn watering. Most often, however, the used water is collected in a sewer system and transported to a wastewater treatment plant. It is generally reported in the literature that from 60 to 80 percent of the water supplied to the consumer is returned as wastewater, the remainder being lost. Wastewater flows from domestic consumers are estimated at 100 gallons per day (gpd) per person for purposes of treatment plant design (Refs. 3-1, 3-2).

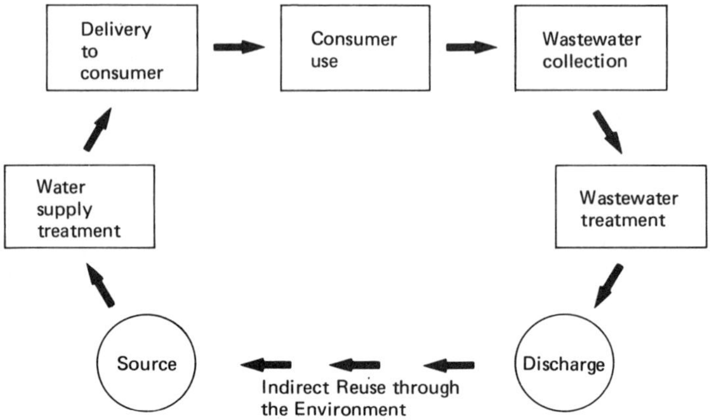

Fig. 3-1. The water consumption cycle.

In addition to wastewater production by domestic consumers, there are two other major potential sources of wastewater flow—groundwater infiltration into sewers and rainfall inflow. The magnitude of the infiltration flow will depend primarily on soil type, groundwater levels, and the physical characteristics and age of the sewers. Groundwater infiltration may range from 1,000 to 40,000 gpd per mile of sewer. Common design practice is to allow for 30,000 gpd per mile for infiltration (Refs. 3-1, 3-2).

In communities where storm sewers and sanitary sewers are combined, the runoff from rainfall is directed into the combined sewer system, and swells the normal flow. The amount of inflow due to rain depends on the frequency, duration, and intensity of storms and the land area which is drained by the sewers. It has been reported that during heavy rains the storm flow rate may reach 100,000 gpd per mile of sewer (Ref. 3-1).

At the step following collection in the water use cycle, the wastewater is treated in a wastewater treatment facility to remove waste constituents and to destroy pathogens before it is discharged to the environment. In some instances, the effluent may be utilized for beneficial, nonconsumptive purposes; e.g., irrigation or recreational facilities. In a very few cases the effluent has been recycled directly. Most frequently, however, the effluent is discharged to a nearby stream and is reused indirectly by a downstream municipality as a water source.

Various constituents in wastewater are potentially harmful to the environment and to human health. In the environment, the wastes may cause eutrophication of water bodies (i.e., accelerated filling and siltation due to an overabundance of nutrients), destruction of animal and plant life, and aesthetic degradation. The wastes also contain pathogenic bacteria and viruses, which, if not removed or inactivated, could adversely affect individual and community health. Thus, the treatment of wastewater has two purposes—to protect human health and preserve environmental quality.

3-2.2 Wastewater Flow

The flow of wastewater into a wastewater treatment plant reflects the pattern of water usage in the community and the hydraulic characteristics of the sewer system. A residential community typically demands water, and produces domestic wastewater at a high rate in the early morning and early evening. This diurnal flow averages about 100 gallons per capita day. Commercial districts produce wastewater at a fairly constant rate during their hours of operation. Typical flow volumes for business establishments are displayed in Table 3-1. Industrial wastewater flow and peak flow periods vary widely and are highly dependent on industry type and location.

The time of occurrence and magnitude of the peak flow at a wastewater treatment plant are also functions of the hydraulic characteristics of the sewer system. An extensive system will tend to delay the time of occurrence and mitigate the magnitude of the peak flow. It is becoming common, in current treatment plant design, to provide an equalization basin at the beginning of the treatment process to even out peak flows and reduce the effects of varying flows on treatment efficiency. For wastewater flows which are predominantly domestic wastewater at a high rate in the early morning and early evening. This which describes the maximum and minimum flows that occur as ratios to the average flow (see Table 3-2).

TABLE 3-1 Typical Commercial Wastewater Flows (Ref. 3-1).

Establishment	Unit	Average Flow gpd/unit
Shopping center	Employee	60
Small business	Employee	20
Restaurant	Meal	7
Airport	Passenger	5
Theater	Seat	5
Motel	Person	50
Hotel	Person	100
Factory	Employee per shift	25
Self service laundry	Machine	350
Bowling alley	Alley	200
Store	25 ft frontage	400
Auto service stations	Set of pumps	500

TABLE 3-2 Domestic Wastewater Flow as Ratios to the Average (Ref. 3-2).

Description of Flow	Ratio to the Average
Maximum daily	2.25 : 1
Maximum hourly	3 : 1
Minimum daily	0.67 : 1
Minimum hourly	0.33 : 1

3-2.3 Composition of Wastewaters

Pure water does not exist in nature because, as a universal solvent, water always carries other substances with it. Solid matter may be in solutions, emulsions, or as a separate layer. Gases may be present in solution or as bubbles in suspension.

The nature of the materials in the water determines the relative quality of the water and the potential use which man can derive from it. Constituents which are undesirable, at least as far as man is concerned, may be labeled "pollutants." It should be noted that the term "polluted" is of necessity a relative description, since any constituent carried in water might be considered a pollutant if it is present in an excessive concentration. "Polluted" and "unpolluted" waters carry many of the same constituents, but in polluted water one or more exist in concentrations which are undesirable. Table 3-3 lists the nature and source of commonly occurring wastewater constituents.

The parameters of water quality routinely used to indicate pollution are: suspended solids, biochemical oxygen demand, chemical oxygen demand, nitrogen, phosphorus, and bacteria. Table 3-4 presents the range of values of those constituents commonly contained in domestic wastewater. Also included are the values of these constituents, selected by EPA to represent typical domestic wastewater (Ref. 3-3). A thorough discussion of these constituents, including standardized test procedures utilized in their measurement and their impact on water bodies, follows.

Suspended Solids. As commonly defined, the suspended solids content of wastewater is that matter which is carried in suspension as opposed to other solid matter in solution or colloidal dispersion. Together, the classifications of "suspended" and "filterable" describe the total solids content of wastewater.

Analytically, the total solids content is that matter which remains as residue after evaporation of the wastewater at 103 or 105°C. Naturally, matter having a significant vapor pressure at these temperatures will evaporate and not be included in the total solids analysis.

Figure 3-2 presents the classifications of solids typically found in medium strength domestic wastewater. The division between suspended and filterable solids is an operational definition, based on the retention of suspended solids in

TABLE 3-3 Wastewater Constituents (Ref. 3-1).

Parameter	Source
Physical	
Solids	Carriage water[a], dom. and industrial waste
Temperature	Dom. and industrial wastes
Color	Dom. and industrial wastes
Odor	Decomposing sewage, industrial wastes
Chemical	
Organic	
Proteins	Domestic and commercial wastes
Carbohydrates	Domestic and commercial wastes
Fats & oils	Domestic, commercial, and industrial waste
Grease	Domestic, commercial, and industrial waste
Surfactants	Domestic and ind. wastes
Inorganic	
pH	Ind. waste
Chlorides	Carriage water, domestic wastes, gd. water inflow
Alkalinity	Dom. wastes, carriage water, gd. water inflow
Nitrogen	Dom. and agricultural wastes
Phosphorus	Dom. and ind. wastes, natural runoff
Gases	
Carbon dioxide	Carriage water, dom. and ind. wastes
Oxygen	Carriage water, surface water inflow
Hydrogen sulfide	Decomposition of domestic wastes
Methane	Decomposition of domestic wastes
Biological	
Bacteria	Domestic wastes
Viruses	Domestic wastes

[a]Carriage water refers to the domestic water supply.

TABLE 3-4 Concentrations of Pollutants in Domestic Wastewater (Ref. 3-1).

Pollutant	Range (mg/l)	Typical (mg/l)
Suspended solids	100–350	230
BOD_5	100–200	200
COD	250–1,000	350
NH_3-N	15–50	18
Total N	20–85	30
Phosphorus	6–20	13
UOD	219–680	383

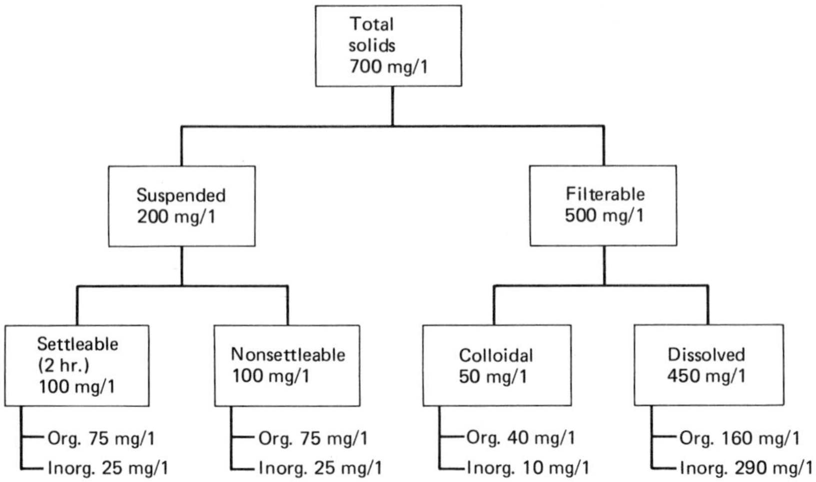

Fig. 3-2. Solids content of medium strength domestic wastewater. (Ref. 3-1)

the surface of a specified filter. The filter size is commonly chosen so that the minimum diameter of the retained suspended solids is 1 micron. Of the suspended solids, the settleable solids are those which settle in a cone shaped container (called an Imhoff cone) in a defined time period. The common settling periods utilized to determine settleable solids content are ½, 1, and 2 hours.

Filterable solids are those able to pass through a specified filter, and can be further defined as colloidal and dissolved on the basis of size (see Figure 3-3) (Ref. 3-1). Colloidal solids range in size from 10^{-3} to 10^{-6} mm, while dissolved solids are less than 10^{-6} mm and exist as molecules and ions in true solution.

Settleable solids can be removed by settling in quiescent conditions. Nonsettleable and colloidal solids can be removed by filtration and coagulation–flocculation processes which create larger settleable particles. Dissolved solids can be removed in precipitation and adsorption processes.

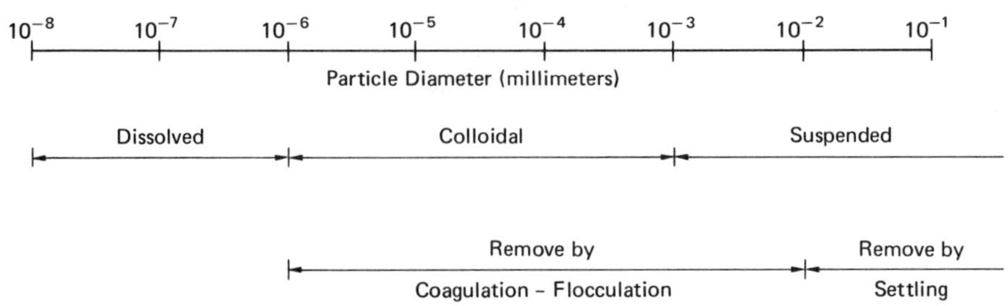

Fig. 3-3. Classifications and sizes of particles in wastewater. (Ref. 3-1)

Each of the solids classifications mentioned may include organic and inorganic (i.e., mineral) compounds. Organic matter may serve as substrate (i.e., a food source) for aquatic bacterial growth, whereas inorganic matter may act as nutrients (i.e., essential compounds required in small amounts).

All the forms of solids may have detrimental effects on water quality if present in excessive concentrations. Dissolved solids levels may make the water unsuitable for certain uses, while colloidal and suspended solids content may increase turbidity, thereby decreasing light penetration into the water and inhibiting aquatic growth. The suspended solids concentration is an easily measured parameter, and is often used as a rough index of water quality.

Oxygen Demanding Constituents. Orgainc and inorganic solids in wastewater, particularly the filterable matter, may be metabolized by bacteria and other microorganisms which, under aerobic conditions, draw on the dissolved oxygen reservoir in the water for their metabolic processes. Dissolved oxygen consumed by microbial activity is replenished through the diffusion of oxygen from the air into the water. The total amount of oxygen which would be consumed if all the oxidizable matter were oxidized is referred to as the Theoretical Oxygen Demand. This demand can be calculated from the stoichiometry of the oxidation equation of the specific compounds in the wastewater.

Various analytical procedures are available to measure the oxygen demand of a wastewater. The two major techniques currently utilized are the Biochemical Oxygen Demand (BOD) and the Chemical Oxygen Demand (COD). The former is a bioassay procedure, while the latter is a chemical digestion process. Both have significant interpretations, though they cannot always be directly related to one another.

The Biochemical Oxygen Demand precedure is described below in an excerpt from an EPA training manual (Ref. 3-7).

> "The BOD test is used for determining the relative oxygen requirements of municipal and industrial wastewaters. Application of the test to organic waste discharges allows calculation of the effect of the discharges on the oxygen resources of the receiving water. Data from BOD tests are used for the development of engineering criteria for the design of wastewater treatment plants. The BOD test is an empirical bioassay-type procedure which measures the dissolved oxygen consumed by microbial life while assimilating and oxidizing the organic matter present. The standard test conditions include dark incubation at 20°C for a specific time period (often 5 days). The actual environmental conditions of temperature, biological population, water movement, sunlight and oxygen concentration cannot be accurately reproduced in the laboratory. Results obtained must take into account the above factors when relating BOD results to stream oxygen demands.

The sample of waste, or an appropriate dilution, is incubated for 5 days, at 20°C in the dark. The reduction in dissolved oxygen concentration during the incubation period yields a measure of the biochemical oxygen demand."

The oxygen demand determined by the previous procedure is referred to as the 5-day BOD (BOD_5). It is commonly estimated that approximately two thirds of of the oxygen demand from the metabolism of carbonaceous matter by microorganisms is exerted in these five days. The complete demand should occur in approximately 20 days (BOD_{20}). In addition to the carbonaceous oxygen demand, there may be an oxygen demand from the metabolism of ammonia nitrogen by nitrifying microorganisms. Since the nitrifying bacteria grow very slowly, it is assumed their oxygen demand is negligible until after the fifth day. The Ultimate Oxygen Demand (UOD) of a waste is the sum of the carbonaceous and nitrogenous oxygen demands, and is defined mathematically as (all quantities in mg/l):

$$UOD = 1.5\ BOD_5 + 4.6\ NH_3 \qquad \text{Eq. 3-1}$$

The EPA has defined another quantity, the effluent strength factor, as the UOD minus the dissolved oxygen in the effluent. This is an attempt to measure the total impact of an effluent discharge on the dissolved oxygen concentration in the receiving stream.

The Biochemical Oxygen Demand analysis is one of the most widely used measures of water quality. It is primarily a measure of the organic material present which can be biologically oxidized, and consequently has particular application in determining efficiencies of various biological unit treatment processes. However, it is inherently nonreproducible, and is often misused as an invariant quantity rather than a complicated measurement which is highly dependent on test conditions.

The Chemical Oxygen Demand is another common measure of the oxygen demand of a wastewater. The COD is a rigorous, high temperature, acidic, chemical digestion process, in which oxygen demanding material is consumed by a powerful oxidizing agent (potassium dichromate). A sample of the wastewater, or an appropriate dilution, is mixed with a precisely measured amount of potassium dichromate solution and concentrated sulfuric acid. The sample is then heated and refluxed for a specified time period (usually two hours). The amount of potassium dichromate remaining after digestion is measured by titration, and is an inverse measure of the oxygen demand of the wastewater. Any organic or inorganic compound present in the wastewater which is capable of being oxidized in the test conditions will be included in the Chemical Oxygen Demand determination.

The BOD to COD ratio is an indication of the degree of biotreatability of the waste. Ratios of 0.8 or higher are indicative of wastes highly amenable to biological treatment, while lower ratios show increasing resistance to biological treatment.

Nitrogen. Nitrogen in domestic wastewaters occurs primarily in the ammonia and organic nitrogen forms. Contributions from inorganic nitrogen forms (nitrite and nitrate nitrogen) are usually negligible. However, there may be a measurable concentration of inorganic nitrogen following biological treatment, when a portion of the ammonia nitrogen is oxidized by the microorganisms to inorganic nitrogen forms. Organic nitrogen normally exists in wastewater as proteins and nucleic acid components arising primarily from excreta and foodstuffs. Urea in urine and the biodegradation of organic nitrogen accounts for the presence of ammonia nitrogen in wastewater.

Inorganic nitrogen is measured by colorimetric test procedures. Ammonia nitrogen is measured by boiling off the ammonia from a slightly alkaline solution of the waste, capturing it in an acid solution, and measuring its concentration by titration or colorimetry. Organic nitrogen is measured by digestion of the waste in a highly acidic solution, which converts the organic nitrogen to ammonia, and then measuring the ammonia by the above procedure. The Total Kjeldahl Nitrogen Procedure measures the organic and ammonia nitrogen forms together. The sum of all the nitrogen forms (ammonia, organic, and inorganic) are known as the Total Nitrogen content of the wastewater.

Nitrogen in treatment plant effluents contributes to eutrophication problems. In addition, organic and ammonia nitrogen may be metabolized by bacteria causing an oxygen demand in the receiving water. Ammonia nitrogen also consumes chlorine during the chlorination process, thereby inhibiting disinfection.

Phosphorus. Phosphorus exists in wastewater as inorganic phosphate ions (PO_4^{3-}, HPO_4^{2-}, $H_2PO_4^-$ and polyphosphate polymers) and as organic phosphates. The organic phosphate is bound in organic molecules, such as RNA, DNA and nucleotides, while the inorganic phosphates exist as ions in solution. At normal wastewater pH (6.8–7.2), the predominant inorganic ions are HPO_4^{2-} and $H_2PO_4^-$ (Ref. 3-8).

The concentration of phosphate ions in water can be measured colorimetrically, by the formation of a blue complex. Polyphosphates and organic phosphates are converted to phosphate ions by chemical digestion, and then measured colorimetrically.

The sources of phosphorus in wastewater are excreta, food wastes, and synthetic detergents. Over half the concentration of phosphorus in domestic wastewater is contributed by detergents, in which phosphates are used as "builders" to enhance detergent action.

The importance of effluent phosphorus levels is their effect on eutrophication of the receiving waters.

Bacteria and Viruses. Bacteria and viruses are ubiquitous in nature, and domestic wastewater typically contains a large and diverse population of these microorganisms, which originate primarily from fecal matter. Bacteria are a normal part of the flora of the intestines of warm blooded animals, and aid in the breakdown and assimilation of foodstuffs in the gut. It has been estimated that up to one fourth of the weight of fecal matter is bacteria cells (Ref. 3-9). Viruses are not normal inhabitants of the gut but may be present through ingestion and replication. An individual may or may not excrete viral particles, depending on whether or not the individual has been recently infected.

In addition to the normal nonpathogenic bacteria in the gut, pathogenic bacteria may be ingested, for which the gut is an ideal environment for growth and reproduction. They may produce clinical disease symptoms in the host, or may exist subclinically. Eventually they are excreted by the host into the wastewater. In a few cases they may be carried in the host permanently, although no disease symptoms are present, and constantly shed in the excreta. In a similar manner pathogenic and nonpathogenic viruses are ingested, may reproduce in the gut, may produce disease in the host, and are excreted.

Pathogenic bacteria strains have been identified in the following genera: ESCHERICHIA, SALMONELLA, SHIGELLA, KLEBSIELLA, and PROTEUS. Diseases such as gastroenteristis, bacterial dysentery, typhoid fever, and cholera are caused by some of these bacteria. The latter two are the most important in the view of public health as they are readily transmitted by water and case fatality rates may range up to 50 percent if untreated (Ref. 3-10).

Pathogenic viruses that have been identified include the groups, Enterovirus, Reovirus, Adenovirus, and Hepatitis. They are responsible for diseases such as poliomyelitis, meningitis, respiratory diseases, hepatitis, and others.

The most common transmission route for these pathogens is through water. Pathogens may survive ineffective wastewater treatment and be passed into natural waters, where they are a danger to persons drinking untreated water or inadvertently ingesting the water while swimming, boating, etc. Treatment plant effluents are also often used as a water source by downstream municipalities. Consequently, the destruction of pathogenic organisms in wastewater effluents is of the utmost importance. The major danger in the release of pathogens to the environment is the possibility of completing the fecal-oral transmission route.

Unfortunately, it is very difficult to quantitatively measure pathogenic organisms in water. However, there are relatively simple and routine methods for measuring the concentration of coliform bacteria in water. Coliforms are defined by Standard Methods to be "gram negative non spore forming, rod shaped bacteria which ferment lactose with gas formation within 24 hours" (Ref. 3-5). The ESCHERICHIA and AEROGENES (also known as ENTEROBACTER) genera are the major coliform species, although other species may occasionally appear as coliforms in the test procedure. It is believed that some, but not a majority, of the coliform species are pathogenic. Since ESCHERICHIA species occur only in the gut of warm blooded animals, their presence in water is an indication of fecal contamination. AEROGENES species live in the gut, but also in soils and other habitats, so their presence in water is not a strong indication of fecal contamination. As the coliform test is commonly run, ESCHERICHIA and AEROGENES are not discriminated, so the presence of coliform bacteria in water must be regarded as an indication, but not a verification, of possible fecal contamination, and the attendant danger of infection from pathogenic organisms (Ref. 3-11). Thus, coliform bacteria are used as an indication of biological water quality in natural and drinking waters, and as a measure of the efficiencies of disinfection in wastewater treatment.

It is estimated that domestic wastewater contains 100 million coliforms per 100 ml, and 7000 viral particles per 100 ml (Refs. 3-1, 3-9, 3-12). Most are removed from the water in the conventional secondary biological treatment, with chlorination being the final step to destroy the remaining microorganisms. Unfortunately, viruses, being more resistant to disinfection than bacteria, may persist in water after disinfection processes which destroy all pathogenic bacteria. Furthermore, virologists consider ingestion of just one viron to be sufficient to cause disease in a host. Since there are no routine methods available to quantify viral concentrations in water, there can be no absolute verification of the absence of viruses in drinking water. For these reasons, the EPA currently recommends that wastewater not be directly reused as a drinking water source (Refs. 3-12, 3-13).

3-2.4 Eutrophication

Organic growth in water bodies is dependent on the water system's temperature, oxygen tension, pH, concentration of nutrients, and amount of light (for photosynthetic activity). Autotrophic organisms, such as algae, utilize light, inorganic nutrients from the water, and carbon dioxide to produce cell mass. When they die, they are decomposed by heterotrophic organisms, such as bacteria, which utilize the organic substrate and dissolved oxygen to produce cell mass and energy. The next higher orders in the food chain consume algae and bacteria and in turn are consumed themselves by the next order of life. Essentially, all the organic life in a water body is dependent on the autotrophs to produce organic material.

In natural water bodies, the limiting factor in autotrophic growth is usually the amount of inorganic nutrients available. In a few cases, the amount of light may be the limiting factor because of high turbidity levels in water.

The principal inorganic nutrients required for growth are nitrogen and phosphorus. Nitrogen in any form can be utilized by algae, although nitrate nitrogen is the principal nutrient form (Ref. 3-14). In the absence of sufficient nitrogen, certain blue-green algae are able to fix nitrogen from the air and so provide a nitrogen source in the water.

Phosphorus is utilized as phosphate. Organic forms of phosphate are degraded to inorganic phosphate, in which form it is assimilated by algae and bacteria and so passed up the food chain. Other nutrients are also required, but in very small quantities. These trace nutrients are almost always present in natural waters.

Since nitrogen can be fixed from the atmosphere if not present in sufficient concentration in water, phosphate levels are usually the limiting factor in algal growth, and subsequent biologic activity throughout the food chain. If an excessive amount of phosphate is present in a water system, algal blooms may occur where an algal species exhibits sudden and rapid growth only to die out as the microenvironment changes slightly. Their decomposition by bacteria depletes the water's oxygen supply, thereby promoting anaerobic conditions.

An eutrophic water body is one rich in nutrients and organic matter; consequently, it sustains a high rate of biosynthesis and nutrient turnover. The quality of an eutrophic water system is often aesthetically displeasing and unsuitable for man's uses.

Water bodies progress naturally from an oligotrophic state, in which the concentration of organic matter is low and algal growth is limited, to an eutrophic state. This eutrophication or natural aging process, usually occurs over a period of many centuries. However, man can hasten the process by adding nutrients such as phosphorus and nitrogen, and organic matter to water. The predominant sources of added nutrients are wastewater effluents and agricultural runoff. Various types of water systems may be classified according to their phosphorus concentrations as follows: (Ref. 3-15)

1. Oligotrophic lake–less than 0.01 mg/l PO_4-P.
2. Eutrophic lake–more than 0.1 mg/l PO_4-P.
3. Polluted river–0.6 mg/l PO_4-P.
4. Domestic wastewater–9 to 10 mg/l PO_4-P.

Thus, the phosphorus and nitrogen contained in wastewater effluents, if not removed, may contribute significantly to the eutrophication potential of a water body and the subsequent degradation of water quality.

3-3. WASTEWATER TREATMENT TECHNIQUES

There currently exist a large variety of treatment techniques designed to remove pollutants from wastewater. These techniques can be combined to form process trains; i.e., treatment schemes, which will accept wastewater as the influent and "produce" a specified degree of treated water as the effluent. Currently available treatment techniques are described in this section for planners and/or community members as background reading to acquaint them with the mechanisms, efficiencies, and limitations of each wastewater treatment technique.

3-3.1 Conventional Treatment Processes

The objective of wastewater treatment is to separate wastes from water. The separated wastes are ultimately disposed, and the effluent is discharged to a receiving body of water. In one sense, all wastewater treatment processes can be considered separation processes.

There are physical, chemical, and biological separation processes. Sedimentation, screening, and distillation are examples of physical processes. Coagulation, ion exchange, and pH adjustment are typical chemical processes, while various forms of biological digestion are biological processes. In the biological processes organic wastes are metabolized by living organisms, while in the physical and chemical processes physical and chemical properties are utilized for waste separation.

As might be expected, actual wastewater treatment techniques are usually combinations of all three processes, although one removal process may predominate. For example, in the activated sludge process, most of the organic waste is metabolized by the sludge biomass—a biological process. Then this biomass is separated from wastewater by sedimentation—a physical process. Suspended solids in wastewater may become entrapped and coagulated by exocellular polymers produced by the sludge biomass and so are removed from the wastewater—a chemical process. Another example is carbon adsorption, a physical process in which organic adsorption is caused by chemical forces. However, a biological slime may also develop on the carbon particles and biologically remove a portion of the organic wastes.

In the literature, treatment processes are described by the predominant removal action—physical, chemical, or biological. Since many physical and chemical processes overlap and tend to occur together, it has become common to classify these processes together as physical-chemical, or p-chem. Thus, biological treatment techniques rely primarily on biological means of waste removal, and p-chem treatment techniques rely on physical and chemical means of removal.

Historically, biological techniques have been widely utilized since they are generally cheaper to build and operate than p-chem techniques. Also, p-chem technology has not been available on a full-scale operational level until very recently. This situation is changing as p-chem treatment technology advances, and many new installations have been built which utilize p-chem treatment exclusively. Table 3-5 compares biological and p-chem techniques.

Conventional wastewater treatment consists of pretreatment, primary sedimentation, secondary biological treatment, chlorination, and discharge. As wastewater initially enters a treatment plant, it is screened to remove gross objects (rags, sticks, etc.) and then flows through a grit chamber where pebbles and grit are allowed to settle out. These two steps together are termed pretreatment. Some small plants do not pretreat the wastewater. Instead, the gross objects may be ground in a comminutor to smaller particle size, and the wastewater fed directly to the primary or secondary treatment process.

TABLE 3-5 A Comparison of Biological and Physical Chemical Treatment (Ref. 3-14).

Physical-Chemical vs Biological	Biological vs Physical-Chemical
1. Less land area 1/4 – 1/2	1. Cheaper to build and operate
2. Greater flexibility in design and operation	2. Operators need less training
3. Potential for superior pollutant removals	
4. Not affected by toxic substances	
5. Lower sensitivity to wastewater variations	

Primary treatment consists of settling in a primary sedimentation tank where settleable solids are removed as sludge from the bottom of the tank, and floating material is skimmed off the top. Properly designed, primary sedimentation tanks remove up to 90 percent of the settleable solids, from 50 to 65 percent of the suspended solids, and from 25 to 40 percent of the BOD_5. Detention times range from one to three hours, depending on the settling characteristics of the wastewater.

Following primary treatment, the wastewater may be processed in a secondary treatment scheme. The most common secondary biological treatment techniques include trickling filtration and activated sludge digestion. Historically, trickling filters have been most frequently utilized because of their ease of operation. Currently, activated sludge is more popular, and is likely to remain so, because of its reliability and control, its suitability for large volumes of wastewater, and its higher degree of treatment. The activated sludge treatment technique and its variations are described in section 3-3.2.

Following biological secondary treatment, the wastewater in a typical conventional treatment plant is chlorinated in a chlorine contact basin and then discharged.

P-chem treatment techniques are much less standard, and it is difficult to apply the terms "primary" and "secondary" with any meaning. For the purposes of this discussion, and as is generally understood in the literature, the term "conventional secondary effluent" refers to the effluent from a wastewater treatment plant utilizing biological secondary treatment.

Conventional secondary effluent will normally contain the following pollutants:

1. Suspended and colloidal solids;
2. Organic materials which are resistant to or have escaped from biological treatment, or which are by-products of bacterial metabolism;

3. Nutrients–primarily phosphorus and nitrogen;
4. Dissolved solids such as chlorides and other mineral salts, which are usually present in the original water supply, but which are increased during water usage;
5. Bacteria and viruses.

Tertiary and advanced treatment techniques are generally understood to be techniques which are applied after employment of conventional secondary treatment, in order to affect higher removals of pollutants than is possible with conventional secondary treatment.

3-3.2 Activated Sludge

The heart of this treatment technique is the aeration tank, in which a bacterial biomass (i.e., activated sludge) consumes organic matter to produce cell mass and energy. Oxygen is supplied by diffusers and/or mixers in the aeration tank. The effluent from the aeration tank flows to a settling tank where the activated sludge settles to the bottom. The clarified effluent from the settling tank is disinfected and discharged. The sludge from the bottom of the clarifier is withdrawn and split into two streams. A portion of the sludge is recycled to the aeration tank, while the remainder of the sludge receives additional treatment and ultimate disposal.

The conventional activated sludge technique described above is illustrated in Figure 3-4. There are several variations on this technique which permit varying degrees of control and specific pollutant removals in the activated sludge process.

Operational Requirements. In order for the activated sludge to function properly, certain environmental constraints must be met. Bacteria in the activated sludge mass, treating typical domestic wastewater, are mesophilic (prefer temperature between 25–40°C) and neutrophilic (pH 6.5–7.5), because domestic wastewater usually varies within those ranges. The principal nutrient requirements for the biomass are nitrogen and phosphorus. Nitrogen and phosphorus require-

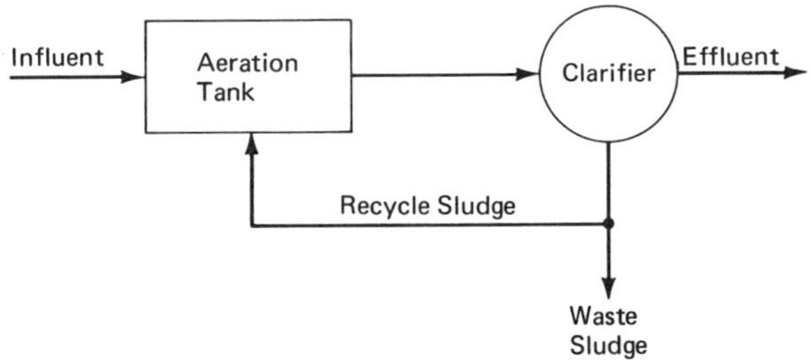

Fig. 3-4. The conventional activated sludge treatment technique.

ments for cellular growth are about 12 and 2 percent of the cell mass. In addition, other inorganic nutrients are required in trace amounts. It should be noted that domestic wastewater usually contains a sufficient supply of all nutrients (Ref. 3-1).

The final requirement for cell metabolism is oxygen. Air is mixed into the aeration tank either by mechanical mixers or by air diffusers. The air supply must be sufficient to satisfy the oxidation requirements of the waste and to provide adequate mixing. Metcalf and Eddy (Ref. 3-1) state that in conventional activated sludge processes 500–900 cu ft of air per lb BOD removed is required for cell metabolism.

In diffused air systems, a minimum of 3 cfm of air/ft of tank length is necessary to insure adequate mixing and prevent the deposition of solids. Fair, *et al.* (Ref. 3-9) further define operational requirements of the activated sludge process:

> "Plants of this kind are normally designed to effect about 90% removal of the BOD from the incoming wastewater or the primary-tank effluent reaching secondary biological-treatment units. Common design values for plants without reaeration units are (1) a BOD loading of 50 lb or less per 100 lb of mixed liquor suspended solids (MLSS); (2) air requirements of 500 to 700 cu ft per lb of BOD removed when the BOD loading does not exceed 50 lb per 100 lb MLSS; (3) the return of activated sludge in sufficient volume to maintain an MLSS concentration of 2000 to 2500 mg per l; and a returned-sludge concentration of not more than 100 to 120 relative to the sludge-volume index. These interlinked specifications generally imply the provision of 4 to 8 hrs of aeration with 25% sludge return, an air supply of ½ to 2 cu ft per gal of wastewater, and a sludge wasteage of 5000 to 10,000 gal per mg of wastewater."

Modifications of the Activated Sludge Process. There are many modifications of the conventional activated sludge process. Table 3-6 lists the characteristics of the most common process modifications, and Figure 3-5 presents the schematic flow charts and organic removals associated with these process modifications.

Complete Mix. Completely mixed activated sludge is physically similar to the conventional plug flow process, except that the influent wastewater and the recycled sludge are fed into the tank along the entire length of the aeration basin instead of just at the head end. However, there is a net flow of water and sludge towards the effluent end of the tank in the completely mixed system. The mixed liquor suspended solids (MLSS) in the completely mixed aeration tank is about twice that of conventional plug flow reactors. Conventional, completely mixed activated sludge is less liable to upset by shock loads than the plug flow process and is suited to varying applications.

TABLE 3-6 Characteristics of Activated Sludge Treatment Modifications (Refs. 3-1, 3-2).

Process	Aeration System	Loading		MLSS (mg/l)	Aeration Period (hr)	BOD eff. (%)	Application
		lb BOD/lb MLSS	lb BOD/1000 ft^3				
Conventional plug flow	Diffused air, mechanical	0.2–0.4	20–40	1500–3000	4–8	85–95	domestic wastes, susceptible to shock loads
Conventional complete mix	Diffused air, mechanical	0.2–0.6	50–120	3000–6000	4–8	85–95	gen. application, resistant to shock loads
Step aeration	Diffused air	0.2–0.4	40–60	2000–3500	5–7	85–95	gen. application, wide range of wastes
Contact stabilization	Diffused air, mechanical	0.2–0.6	60–75	1000–3000[a] 4000–10,000[b]	½–2[a] 3–6[b]	80–90	flexible, expansion of existing system
Extended aeration	Diffused air, mechanical	0.05–0.15	10–25	3000–6000	20–30	75–95	flexible, small flows, package plants
Oxygen aeration	Mechanical	0.25–1.0	100–250	6000–8000	3–5	85–95	gen. application, need economical oxygen source

[a]Contact unit.
[b]Solids stabilization unit.

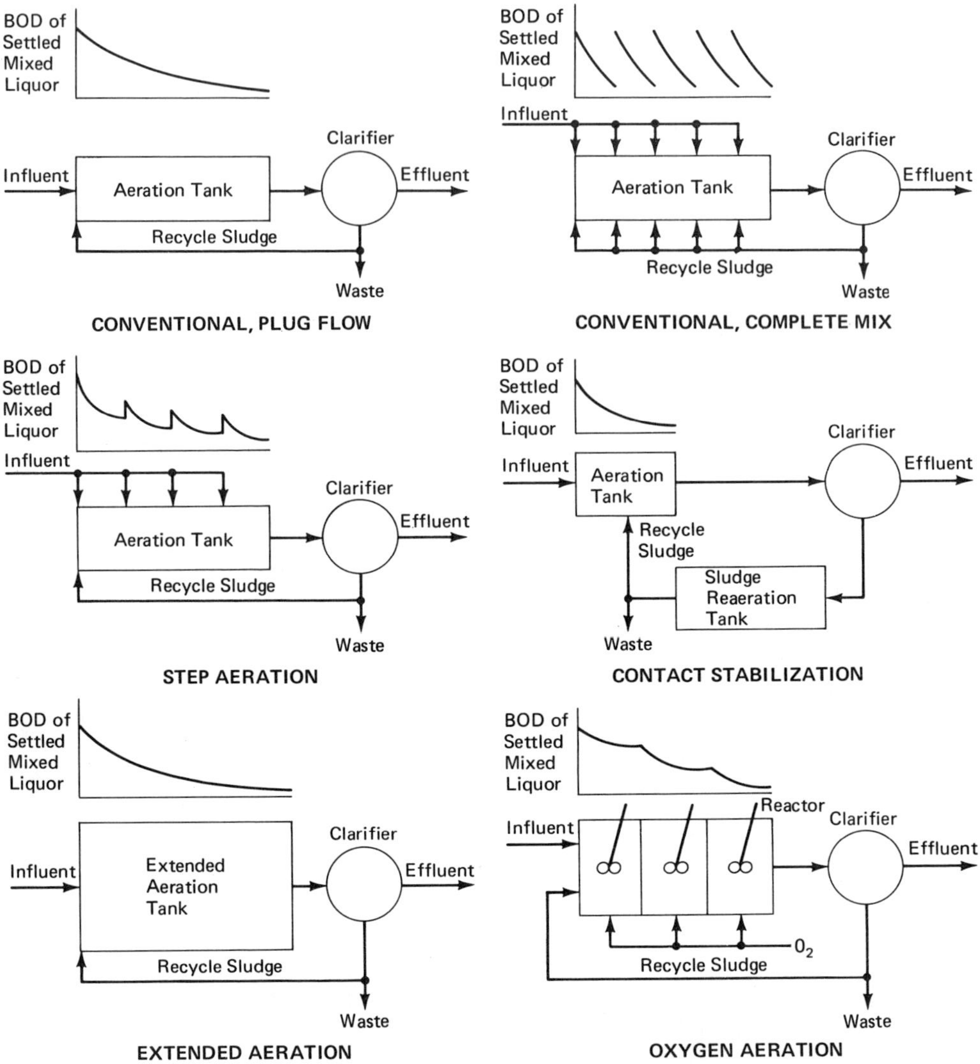

Fig. 3-5. Activated sludge treatment modifications. (Refs. 3-1, 3-9)

Step Aeration. In the step aeration activated sludge process, the waste is introduced at points along the side of the aeration tank; however, the recycled sludge enters the head end of the tank. Thus, the oxygen demand in the step aeration process is more evenly distributed throughout the basin than in the conventional plug flow reactor, in which a high oxygen demand exists at the head end of the aeration tank and a lesser demand toward the effluent end. The flow regime in the step aeration process is plug flow, and the MLSS concentration is usually about 50 percent higher than in a conventional plug flow system.

Contact Stabilization. The contact stabilization modification of the activated sludge process allows the rapid assimilation of organic matter from wastewater by the activated sludge mass, and then separates the sludge for stabilization in a separate aerator. The wastewater detention time in the aeration chamber is very short, often less than one hour. The sludge is aerated in the stabilization basin for three to six hours to allow the microbes to consume the organic matter that was removed from the wastewater in the aeration tank. The stabilized sludge is then recycled to the aeration tank. Soluble organics, which would have been removed in the conventional activated sludge process, are not effectively removed during the contact stabilization process. Therefore, contact stabilization is not as consistently efficient as conventional activated sludge.

Extended Aeration. The extended aeration activated sludge process differs from the conventional process primarily because of the extremely long aeration period utilized—up to 30 hours. The long detention time inherent to this process allows the sludge to proceed well into the endogenous respiration phase so that a high level of organic removal is achieved and optimal sludge settling characteristics are developed. At one time, it was thought that no excess sludge would be produced in extended aeration systems, since the sludge was allowed to proceed far into endogeny; however, it is now recognized that solids must be wasted to prevent solids buildup in the system. Since a large aeration tank must be built to provide a long detention time, extended aeration is usually applied to small wastewater flows.

One benefit of the extended aeration process not found in other activated sludge systems is the nitrification which occurs during the process. Ammonia nitrogen in wastewater is metabolized to nitrate by nitrogen bacteria in the activated sludge biomass. These bacteria are slower growing than other bacterial species in the activated sludge mass, and consequently do not develop sufficient mass to permit effective nitrification during the mean cell residence time (i.e., sludge age) of the conventional activated sludge process (up to 5 days). However, in the longer residence time of the extended aeration modification (10-30 days), these bacteria can accumulate to a significant level so that extensive nitrification is possible. Operational problems associated with small extended aeration plants, which are usually built as package plants and left unattended, often preclude achieving the high organic removals that are possible with this process.

Pure Oxygen. The pure oxygen activated sludge process is a relatively new modification to conventional biological treatment (Ref. 3-19). In this process, pure oxygen is supplied to the mixed liquor in a series of closed, completely mixed reactors, and unused oxygen is recycled for greater efficiency. Up to four times the normal amount of dissolved oxygen can be supplied to the mixed liquor during this process, thereby promoting increased bacterial activity, decreased sludge volume, and improved sludge settleability. Additionally, aeration tank volume requirements are reduced for the pure oxygen process. The cost of oxygen is a pertinent economic factor in the consideration of this process. Metcalf and Eddy (Ref. 3-1) delineate the benefits of the pure oxygen system as:

> "Quantitatively, the benefits to be derived from the use of pure oxygen can be illustrated by considering its application in (1) proposed and existing complete mix systems and (2) existing conventional plug-flow aeration tanks.
>
> In the first case, the benefits can be evaluated by considering the complete-mix activated-sludge process design example presented previously. If the mixed-liquor solids concentration in the reactor could be increased to about 5,000 mg/liter by using pure oxygen instead of air then the reactor volume could be reduced by about 31 percent. Thus, in complete-mix systems, pure oxygen could be used to reduce the required aeration tank volume in new designs or to extend the useful life of existing systems. Economic comparisons must include the cost of pure oxygen or an oxygen plant and its operation.
>
> In the case of a conventional plug-flow activated-sludge system that is overloaded, conversion to pure-oxygen aeration could be used to extend the useful life of the system and might eliminate the need for additional treatment facilities. In the case of a conventional plug-flow system where sufficient oxygen is being supplied, conversion to pure oxygen will not significantly improve the performance."

Efficiencies of Activated Sludge Processes. The activated sludge processes, including pretreatment and primary settling, are efficient in removing organic constituents and suspended solids from wastewaters. Table 3-7 presents typical primary and secondary effluent characteristics.

TABLE 3-7 Wastewater Characteristics[a] (Refs. 3-1, 3-3).

	Raw Wastewater (Range)	Typical Wastewater	Primary Effluent	Secondary Effluent
Suspended solids	100–350	230	85	20
BOD_5	100–300	200	135	20
COD	250–1,000	350	225	50
NH_3-N	15–50	18	16	15
Total N	20–85	30	25	20
Phosphorus	6–20	13	11	10
UOD	219–680	383	276	99

[a]Based on average treatment efficiencies. These values will vary from plant to plant, and within any one plant with time.

Some organic materials are removed in the primary settling tank and a large proportion of the remainder are metabolized by the activated sludge. Those present in the secondary effluent are termed refractory and are presumed to be nonbiodegradable by the activated sludge process.

Most of the suspended solids are settled in the primary settling tank, and a majority of the remainder are removed by the activated sludge. The primary removal of suspended solids by the activated sludge is not through assimilation by the microbes, but rather by flocculation of the sludge in the final clarifier, which enmeshes the suspended solids.

During secondary treatment, the total nitrogen content of wastewater is reduced by approximately one-third. The activated sludge mass synthesizes a portion of the organic nitrogen to cell matter and excretes a portion as ammonia.

Ammonia nitrogen contained in wastewater may be reduced slightly during secondary treatment. A portion of the influent ammonia content is oxidized by nitrifying bacteria. This ammonia decrease is partially balanced by the production of ammonia by bacteria as a metabolic end product from organic nitrogen. In the extended aeration process, nitrification usually converts most of the ammonia nitrogen to nitrate nitrogen.

Phosphorus will settle out and will be taken up as a nutrient during biological secondary treatment. However, the amounts removed are small, usually less than 30 percent, so that most of the phosphorus remains unaffected by activated sludge treatment.

The Ultimate Oxygen Demand is reduced about 75 percent in the conventional activated sludge process. Most of this reduction occurs in the removal of organic carbonaceous material. The UOD of secondary effluent is attributable mainly to ammonia nitrogen.

The extended aeration modification might be expected to reduce the ammonia in the effluent by 90 percent (Ref. 3-1). Assuming all other removal efficiencies are the same as for conventional, activated sludge, this would raise the removal efficiency of UOD to about 90 percent.

3-3.3 Trickling Filters

Trickling filters consist of an open structure containing media, through which wastewater flows (trickles). The structure is commonly a circular concrete basin 6 to 12 feet deep, although other shapes and sizes are utilized. Trickling filter media is typically rough rock or slag and from 2 to 4 inches in diameter; however, trickling filters have also been constructed with wood slats and synthetic media. Removal of dissolved organic material from the wastewater stream takes place as the wastewater trickles over the media on which a bacterial slime growth is established. Periodically, portions of the slime become detached

TABLE 3-8 Trickling Filter Classifications (Ref. 3-81).

	Low Rate	Intermediate Rate	High Rate	Super Rate
Organic loading, lb $BOD_5/1000\ ft^3/day$	5–20	15–30	30–60	50–100
Hydraulic loading, mgad	1–4	4–10	10–30[a]	30–90[a]
Recirculation	Seldom	Usually	Always	Always
Application	Complete treatment	Complete or roughing	Complete or roughing	Roughing

[a]Includes recirculation.

(slough off) and are carried out of the filter by the wastewater. A clarifier following the trickling filter removes the sloughed bacterial slime and some suspended solids.

There are two major differences between the conventional trickling filter treatment technique and the conventional activated sludge process. First, the bacterial growth is fixed on the media in the filter, whereas in the activated sludge process it is suspended in the wastewater as a dispersed floc. Second, all the solids from the secondary clarifier following the trickling filter are wasted, as opposed to being partially recycled as in the activated sludge technique.

The design and operation of a trickling filter treatment plant is similar to that of an activated sludge treatment plant except for those elements directly related to the trickling filter. Pretreatment and primary treatment precede the biological treatment portion of the plant. Primary effluent is applied to the trickling filter, and flows through the media in a matter of minutes. The secondary clarifier removes the sloughed slime and the wastewater flows to a chlorine contact chamber for chlorination and subsequent discharge.

A common modification of this typical trickling filter technique entails the recirculation of a portion of the effluent from the secondary clarifier back to the trickling filter. This affords additional treatment to the wastewater, and helps equalize hydraulic loading variations.

Trickling filters may be classified according to the hydraulic and organic load supplied, as displayed in Table 3-8. Low rate filters are utilized as the complete biological portion of a secondary treatment process, while super rate filters are useful as roughing filters to prepare a wastewater for further biological treatment. Intermediate and high rate filters have been utilized for both complete treatment and roughing applications.

Trickling Filter Effluent Quality. Under normal loading conditions, the effluent quality which might be expected to be produced by a completely operated trickling filter treating typical domestic wastewater is presented in Table 3-9. Data are only available for low and high rate filters, although intermediate filters might be expected to produce an effluent quality somewhere between low and high rate filter effluent quality. As the data in Table 3-9 indicates, low rate trickling filters are slightly less efficient than the typical activated

TABLE 3-9 Trickling Filter Effluent Quality Ranges (Ref. 3-81).

	Low Rate	High Rate
BOD_5 (mg/l)	10–35	30–75
Suspended solids (mg/l)	15–40	35–80
Effluent nitrified	frequently in summer	no

sludge process with regard to both BOD_5 and suspended solids removal; and high rate filters are less efficient than low rate filters.

Low wastewater flow rates through the filter (as found in low rate filters) and warm wastewater temperatures characteristic of summer, allow the establishment of nitrifying bacteria in the bacterial slime. The ammonia nitrogen in the wastewater may then be nitrified, substantially reducing the UOD of the effluent. However, since nitrifying bacteria are strongly temperature dependent, seasonal temperature variations often preclude the possibility of consistent nitrification of wastewater effluents.

From the data in Table 3-9, it may be concluded that a well designed and carefully operated low rate trickling filter plant may meet the secondary treatment standards required by the 1972 Amendments. However, an acceptable effluent quality would require optimal operation–a requirement that cannot be met in most applications. Therefore, most trickling filter installations will require some additional alternate treatment technique to routinely meet secondary treatment standards.

3-3.4 Package Plants

A typical package wastewater treatment plant consists of the extended aeration activated sludge treatment process without pretreatment or primary treatment. Domestic wastewater is passed through a comminutor, which grinds the large solids, and then flows into the extended aeration tank. The detention time in the aeration basin is approximately 24 to 30 hours (typical for extended aeration processes), and consequently the organic loading on the activated sludge biomass is very low. Following clarification, the wastewater is disinfected (chlorination) and discharged. The entire sludge mass from the clarifier is recycled to the aeration tank; however, occasionally, small portions of the settled sludge are pumped to a digestor or holding tank.

A holding tank is a covered tank which serves only to contain the waste sludge. No solids treatment is deliberately encouraged. When the holding tank is full, contents are pumped into a tanker truck and hauled to landfills for disposal or to large treatment facilities for treatment.

More recently, package plant manufacturers have installed aerobic sludge digestors. The waste sludge is aerated in the digestor so that aerobic digestion is encouraged, which reduces the solids content of the sludge. However, since some of the waste sludge is nondegradable, the aerobic digestor must periodically be pumped out and the solids trucked to disposal or further treatment.

Package plants are commonly constructed of steel, and may be fabricated at the vendor plant or on site. Site preparation is usually minimal, with the plant often installed above grade. Package plants range in size up to one mgd and occasionally larger, but frequently are less than 0.5 mgd.

Operation of package plants is automated, so that constant supervision by trained personnel is not required. An individual, usually with only limited technical training, visits the package plant daily to perform routine maintenance and check on the operation of the plant.

Application and Utilization of Package Plants. The primary purpose of package plants is to provide local wastewater treatment facilities to areas which do not have wastewater treatment because of a lack of sewers to a regional treatment facility. Typical package plant installations include: subdivisions not yet served by municipal sewers; small communities with a wastewater flow too low for a custom built plant; mobile home parks; and highway rest stations. According to one manufacturer, a single package plant can be successively installed in several locations to provide temporary service until extension of municipal sewers is completed (Ref. 3-82).

The utilization of package plants has a substantial impact on land use. Land areas not sewered to a normal treatment facility may be developed if package plant installation is authorized. Land development is thereby hastened in areas where it would have been otherwise delayed.

Package Treatment Plant Efficiencies. Package plants are potentially capable of achieving pollutant removals typical of the extended aeration activated sludge process. In addition, under favorable conditions, the extended aeration process is capable of extensive nitrification. However, there are several interfering factors, and consequently, package plant performance is usually less than predicted.

First, since most package plants allow no provision for solids disposal, grit and "ash" from biological oxidation increases in the aeration tank, plant operation consequently suffers, and the solids level in the effluent rises.

Second, the daily visit for operation and maintenance work by the operator is usually insufficient to keep the biological process operating at optimal efficiency.

Finally, the typical operator does not possess technical training or skills which qualify him to operate wastewater treatment plants. Therefore, he is often incapable of recognizing potential and/or developing problems. As long as the plant is in good mechanical operating condition, it is assumed, sometimes erroneously, to be adequately treating the wastewater.

Seven package treatment plants in the Louisville Kentucky metropolitan area were monitored for wastewater treatment efficiency in 1974 in conjunction with development of a water quality management plan (Ref. 3-83). Data from these typical package plants revealed that BOD_5 treatment efficiency averaged 77 percent. Plants equipped with lagoons as a tertiary treatment technique averaged 83 percent BOD_5 reductions. Neither of these removal efficiencies are sufficient to meet secondary treatment standards (30 mg/l BOD_5; i.e., 85 percent efficiency in treating typical domestic wastewaters).

Since conventionally operated package plants cannot be expected to routinely meet required secondary treatment standards, some upgrading technique is required for these facilities.

3-3.5 Suspended Solids Removal

As shown in Table 3-7, the suspended solids concentration in secondary effluent is approximately 20 mg/l. These suspended solids consist of colloidal and non-settleable solids which escape enmeshment in the activated sludge as it settles in the clarifier, and some of the small floc particles of sludge which fail to settle.

A common rule of thumb is that 1 mg of suspended solids in the secondary effluent equals $^1/_2$-$^3/_4$ mg of BOD_5. Clearly, the 20 mg/l suspended solids represents a substantial portion of the effluent BOD_5. In addition, the suspended solids contained in the effluent may increase turbidity levels in the receiving stream.

Removal of these residual suspended solids is a solid-liquid separation process. The three processes most commonly utilized in this application are microstraining, mixed media filtration, and chemical coagulation (Ref. 3-20).

Microstraining. Microstraining utilizes a drum-type filter to screen suspended particles from the effluent stream. Figure 3-6 shows the essential construction of

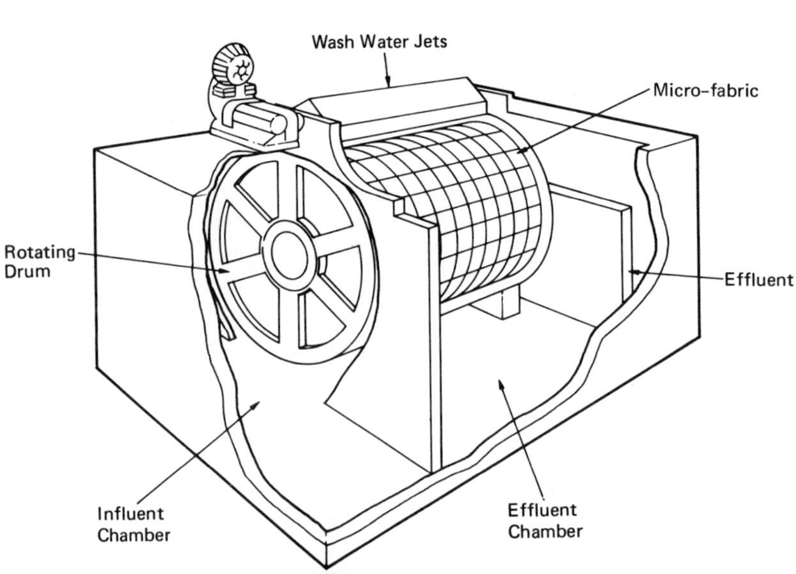

Fig. 3-6. Typical microstrainer. (Ref. 3-14)

the microstrainer. The water enters the open end of the revolving drum and flows out the sides, which consist of fine woven mesh screens and supporting rib structure. In this application, the screen (on which the solids are retained) commonly has a mesh of 23 or 35 microns. At the highest point of the drum, high pressure jets wash the solids from the screen into a trough, which recycles the solids to the sedimentation tank. Ultraviolet lights are sometimes mounted above the screen to prevent biological growth from binding the screen.

The microstraining unit operates with a high flow rate and low hydraulic characteristics. Typical filter flow rates are 6–10 gpm per sq ft at a head differential of six inches (Ref. 3-14).

A major disadvantage of microstraining is its inability to handle fluctuations in solids content and shock loads. For an increase in solids loading, it is possible to reduce the rotational speed of the drum, but this reduces the capacity of the unit. Furthermore, microstraining has no effect on colloidal solids which pass through the unit unhindered.

Mixed Media Filtration. Sand filters have been used since the 19th century in treating surface waters to produce municipal drinking water. The slow sand filter (English filter) was developed in England in the 19th century, and it is characterized by flow rates of 0.05–0.10 gpm per sq ft. Slow sand filter beds are cleaned by scraping off the top layer of sand. The rapid sand filter was developed for use in America, where surface waters were not amenable to slow filtration. Flow rates from 2–5 gpm per sq ft are used and the filter is cleaned by backwashing. More recently, the rapid sand filter has been utilized to filter wastewater effluents following secondary biological treatment.

A sand filter is basically a basin of graded and selected sands through which water is filtered. Flow may be either up or down through the filter, and may be induced by gravity or pressure. In the latter case, the filter is covered and the water is forced through the bed under higher pressures and flow rates than normally can be achieved in gravity filters.

The bed of a typical downflow filter consists of 18 to 30 inches of sand, supported by 6 to 18 inches of gravel. An underdrain system supports the gravel and sand, collects the filtered water, and distributes the backwash water. The filter is cleaned by forcing water up through the sand and gravel, thereby fluidizing and expanding the bed. The particles scour one another and dislodge the filtered particles, which are carried out of the filter by the wash water. Additional energy may be imparted to the filter grains by water jets or air scouring to enhance the mechanical self-scrubbing action.

Many theories have been advanced to explain filtration. Table 3-10 presents the accepted concepts of removal mechanisms. The physical or mechanical mechanisms are functions of physical parameters such as grain size, porosity, and bed depth. The chemical mechanisms are functions of the chemical and surface characteristics of the filter bed and the suspended matter. Coagulation-

TABLE 3-10 Particle Removal Mechanisms in Filtration (Ref. 3-1).

Physical	Chemical	Other
1. Straining	1. Chemical adsorption	1. Coagulation–flocculation
2. Sedimentation	2. Physical adsorption (e.g. electrostatic forces)	2. Biological growth
3. Inertial impaction	3. Adhesion	
4. Interception		

flocculation may bring about the formation of larger particles, which allow removal by one of the preceding mechanisms. Bacterial growth may decrease the pore volume, thereby increasing filter efficiency.

After backwashing, a single medium filter of uniform density will grade itself naturally as it settles with the largest particles on the bottom and the finest on the top. Consequently, the majority of the filtering occurs in the top 2 inches of the filter, where the pore volume is the smallest. Ideally, filtration should proceed with the coarsest medium first and the finest medium last. More efficient filtration takes place in upflow filters where a majority of the entire filter bed depth is used instead of only the first few inches. However, with upflow filtration, there is the problem of inadvertantly expanding the bed and destroying the filtering action.

Dual media and multimedia filters (i.e., mixed media filters) are an attempt to more optimally utilize the filter bed depth and retain the advantages of the downflow configuration. Two or more media of different densities and grain sizes are placed in the same filter. When particle settling occurs after backwashing, the larger and less dense medium remains on top of the filter, while the smaller and denser medium settles to the bottom. Thus, the larger medium acts as a rough filter and the smaller medium "polishes" the water. Typical materials employed in mixed media filter construction include synthetic resin, anthracite coal, sand, and garnet. Particle sizes of these materials are specified in filter design to produce the desired filter bed composition.

Chemical Coagulation–Flocculation. By far the most efficient means of removing residual solids in wastewater is through chemical addition (Ref. 3-20). Properly performed, chemical addition to promote coagulation and flocculation can remove both suspended and colloidal solids. Neither microstraining nor filtration is capable of effectively removing colloids. Suspended and colloidal particles in water acquire a net negative or positive charge due to ionization of molecules on the particle surface and/or preferential adsorption of charged ions from solution. Solids in wastewater effluents are, characteristically, negatively charged. They tend to remain in suspension due to mutual repulsion from their similar, acquired charges.

Coagulation and flocculation are mechanisms by which the dispersed particles can be made to agglomerate into larger particles, which then may be removed by standard liquid-solid separation techniques. In recent years, there has been some confusion as to the use of the terms coagulation and flocculation. Common and useful definitions of the terms are that "coagulation" refers to the agglomeration of particles by charge reduction and "flocculation" refers to agglomeration by inter-particle bridging. The statements which follow offer abbreviated versions of the current theories of coagulation-flocculation. A thorough treatment of coagulation-flocculation may be found in Stumm and O'Melia (Ref. 3-21); Black, Birkner, and Morgan (Ref. 3-22); and Birkner and Morgan (Ref. 3-23).

Dispersed particles may be induced to coagulate by added counter ions (i.e., opposite in charge to the net charge of the suspended particles) to the solution. The counter ions tend to reduce the net charge on the particles so that the particles may agglomerate as a result of collisions which occur in the suspending liquid medium. Flocculation is brought about by the addition of natural or synthetic polymers. Typically, these polymers exhibit a net charge opposite to the suspended particles; although similarly charged polymers have been observed to promote flocculation. A portion of the polymer bonds to a particle and the remainder extends out into solution. It may bond to another particle and so bridge between the two and tie them together. Eventually, a "floc" of entangled particles and polymer is produced which is large enough to be separated from solution. The particles in a coagulation suspension appear as tight packets which settle rapidly, while a flocculated suspension has a loose, fluffly floc that settles slowly.

Iron salts, aluminum salts, lime, and synthetic organic polymers are commonly used to promote coagulation-flocculation. Table 3-11 lists the forms normally employed. Polymers are often added with coagulants as aids to enhance coagulation.

TABLE 3-11 Chemicals Utilized for Coagulation-Flocculation

Chemical	Formula
Aluminum sulfate (Alum)	$Al_2(SO_4)_3 \cdot 18H_2O$[a]
Sodium aluminate	$NaAlO_2$
Ferric sulfate	$Fe_2(SO_4)_3$
Ferric chloride	$FeCl_3$
Ferrous sulfate (Copperas)	$FeSO_4 \cdot 7H_2O$
Calcium oxide (lime)	CaO
Calcium hydroxide (lime hydrate)	$Ca(OH)_2$
Sodium carbonate (soda ash)	Na_2CO_3
Polymers	varies

[a]Waters of hydration vary from 13 to 18.

Suspended and colloidal particles may also be removed by enmeshment in precipitates. If metal salts, such as alum, $Al_2(SO_4)_3$, and ferric chloride, $FeCL_3$, or metal oxides and hydroxides such as lime, CaO and $Ca(OH)_2$, are overdosed in the water to produce supersaturation, a precipitate of metal hydroxide or metal carbonate will form (e.g., $Al(OH)_3$, $Fe(OH)_3$, $Mg(OH)_2$, $CaCO_3$) which will enmesh the solids in the precipitating floc, thereby removing them from the water (Ref. 3-24).

Chemicals for the removal of suspended solids may be added to the primary treatment tank, to the aeration tank, or after the aeration tank (as tertiary treatment). The advantages and disadvantages of addition at these locations are listed in Table 3-15. Kugelmann (Ref. 3-14) suggests that chemical addition after the activated sludge tank is the most advantageous location for suspended solids removal, and presents three techniques which can be utilized. In the simplest method, chemical addition to the activated sludge mass occurs as it leaves the aeration tank and enters the final clarifier. The chemical enhances sludge settleability as well as coagulating suspended and colloidal solids. A more comprehensive method combines the above scheme and mixed media filtration of the effluent from the clarifier to remove the small floc particles which escape the clarifier. The most complex system includes chemical addition, sedimentation, and mixed media filtration following the biological clarifier. In these systems, microstraining is not included because it is ineffective with a chemical floc.

Removal Efficiencies. Microstraining is the least versatile suspended solids removal system, and is most susceptible to shock loads. Mixed media filtration is reliable and slightly more efficient than microstraining. Neither microstraining nor mixed media filtration can remove colloidal solids. Chemical addition is the most versatile and reliable of the suspended solids removal systems, and is capable of removing colloidal matter. Table 3-12 lists wastewater constituents

TABLE 3-12 Efficiencies of Suspended Solids Removal Techniques[a] (Ref. 3-1)

	Microstraining or Multimedia Filtration	Chemical Addition
Suspended solids	75–90%	60–80%
BOD_5	45–70%	75–90%
COD	35–60%	60–70%
NH_3	–	5–15%
Org N	15–40%	30–50%
PO_4	–	90–95%[b]

[a]From conventional secondary effluent.
[b]If a phosphate precipitating chemical is used.

removal efficiencies which might be expected with these three suspended solids removal techniques.

3-3.6 Dissolved Solids Removal

Dissolved inorganic and organic constituents in wastewater may be measured by an evaporation procedure in order to yield a value known as the total dissolved solids (TDS) in solution. The TDS of wastewater effluents commonly ranges between 100 and 500 mg/l (Ref. 3-1). Since the Public Health Service Drinking Water Standards of 1962 set a limit of 500 mg/l TDS in potable water, the TDS content in effluents is usually not excessive and has no deleterious effect on the receiving water (with respect to excessive dissolved solid concentrations). However, inorganic phosphorus and nitrogen forms are included in the TDS measurement, and their importance in eutrophication of water bodies has already been noted. Consequently, treatment techniques which remove TDS will also reduce the concentrations of phosphorus and nitrogen in effluents, thereby reducing eutrophication potential.

Dissolved organic constituents contained in treatment plant effluents originate from three sources: influent organics which escape treatment; influent refractory compounds which cannot be biologically degraded; and refractory metabolic by-products of the activated sludge biomass. This dissolved organic matter represents an oxygen demand in the receiving water if utilized by microorganism as substrate. It may also be noxious or toxic.

Four treatment techniques are currently utilized for removal of dissolved constituents in wastewater effluents. The first two, activated carbon and ion exchange, are surface sorption phenomena, while the second two, electrodialysis and reverse osmosis, are membrane phenomena.

Sorption and Membrane Phenomena. Sorption occurs when two phases come into contact at a common surface. At liquid-solid interfaces both solvent and solute molecules interact with the solid surface. The affinity of the solvent molecules is shown by the degree to which the solvent "wets" the surface.

The degree of attraction of the solute for the solid surface is the major factor in the transfer of the solute from the solvent to the solid surface. This attraction is a function of two forces; 1) the solute-solvent affinity, and 2) the solute-solid affinity.

Solutes with a low affinity for their solvent (lyophobic) are easier to remove from solution by sorption than solutes with a high affinity for their solvent (lyophilic). Thus, water constituents which are hydrophobic (water hating) are easier to remove from solution by sorption than hydrophilic (water liking) solutes (Ref. 3-24).

The second force, affinity of the solute for the solid, is a surface phenomenon which depends on the character of the solid surface. Current concepts of sorption account for this surface phenomenon as a function of electrical attraction, van der Waal attraction (i.e., physical, ideal adsorption), and chemical bonding (i.e., activated, chemisorption). It is believed that most sorption processes include all three factors, although one factor may predominate.

Activated carbon sorption is principally due to chemical bonding and, therefore, is very effective in removing organic compounds from effluents. Ion exchange is a function of the electrical attraction between solute ions and the solid and thus, dissolved ions are most effectively removed from solution.

The major factors which effect sorption processes include surface area for phase transfer, pH, effect of mixed solute interactions, temperature, and physical and chemical natures of the adsorbate (solute) and absorbant (solid).

A membrane can be generally defined as a phase separating two other phases which acts as a barrier to flow between the two phases. A separation membrane will allow selective transport of molecular or ionic species between the two phases; i.e., it has a different permeability to different constituents ("perm–selective").

Reverse osmosis membranes are usually constructed of cellulose acetate, and are manufactured to give specific properties; in particular they are highly permeable to water, but tend to reject solutes of all kinds. At present, the most satisfactory explanation for this rejection phenomena is the difference in the diffusion rates of the solutes and of water across the membrane (Ref. 3-24).

Electrodialysis membranes are manufactured from a variety of materials, but share in the ability to discriminate in permeability based on ionic charge. Cationic membranes are permeable to cations but not to anions, and anionic membranes are permeable to anions but not cations. Both type membranes have very low permeability to nonionized species and to water.

Current theories postulate that electrodialysis membranes possess fixed and mobile ions (e.g. all cations are fixed and anions are mobile or vice versa). Ion transport across the membrane will occur for solute ions which possess a charge similar to the mobile ions, but not for the solute ions of opposite charge (Ref. 3-24).

Activated Carbon Sorption. Activated carbon is prepared in a two step process–carbonization and activation. First, an organic material is subjected to high temperature pyrolysis in order to produce a carbon char. In the activation step, the char is partially oxidized at high temperature in an oxygen poor atmosphere, which results in oxidation along bedded planes in the carbon, so that macro- and micropores are formed, thereby vastly increasing the surface area of the carbon particles. Up to 50 percent of the carbon may be lost in the activation step. The activated carbon is then divided by size into powdered (less than 0.07 mm) and granular (larger than 0.1 mm) (Refs. 3-1, 3-24).

In the activated carbon treatment technique, wastewater is contacted with the carbon for periods from 15 to 60 minutes. The contact technique, varies with the type of carbon employed. Powdered carbon is flash mixed with the wastewater and allowed to settle in a sedimentation tank where coagulants may be added to induce settling. Granular carbon may be utilized in beds, or in columns. Flow configurations may be upflow, expanded, or downflow, and either gravity or pressure may be utilized as the driving force. Countercurrent operation of activated carbon filters enhances adsorption (Ref. 3-25).

In addition to sorption, granular carbon columns and beds will also filter the wastewater, thereby removing suspended solids. Although filtration has a beneficial effect, the solids removed may clog the column or bed. Therefore, it is necessary to provide backwashing to keep solids from clogging the carbon contactor, thereby eliminating the necessity for backwashing.

Biological growth often occurs on the carbon particles, increasing the organic removal efficiency, but also increasing the solids content in the effluent from the carbon contactor. Therefore, a vigorous and frequent (once a day) backwashing of activated carbon columns may be required to control biological growth.

When the adsorptive capacity of the activated carbon is exhausted, the carbon is removed from the contacting unit and regenerated in a furnace, where high temperatures reactivate the carbon by desorbing and destroying the organic matter. Carbon losses in the furnace range from 5 percent to 15 percent, and must be made up with new carbon. The economics of the adsorption process require the spent carbon to be regenerated in all but the smallest applications.

Both powdered and granular carbon systems are capable of 70 to 80 percent organic removal. Field applications have demonstrated capacities ranging from 0.25 to 0.87 lb COD/lb carbon (Ref. 3-14).

Kugelman (Ref. 3-14) has compared the relative merits of granular and powdered carbon and concluded that granular carbon is the most useful form for wastewater applications:

> "Both granular and carbon and powdered carbon contact are excellent processes for organics removal from wastewater. Powdered carbon processes have advantages over granular carbon in that: powdered carbon is cheaper than granular carbon [10¢/lb vs 30¢/lb] ; powdered carbon plants require a much smaller carbon inventory and can be dosed to meet the demand. However, granular carbon exhibits higher practical adsorption capacity, does not require complex dewatering procedures prior to regeneration and most important has been successfully regenerated at full scale... Thus, an engineer designing a treatment plant using carbon adsorption would most likely choose granular carbon."

Ion Exchange. Ion exchange is a process in which ions are exchanged between a liquid and a solid surface phase. The ion exchange driving force is defined as the relative affinities of the ions for the solid surface. Ion exchange has widespread usage in removing hardness, iron, and manganese from drinking water supplies. In industrial wastewaters, ion exchange may be utilized to recover valuable waste materials and by-products contained in effluent streams.

Ion exchange is an equilibrium process, as shown in Equations 3-2 and 3-3:

Catonic exchange:

$$B_1^+ + R^-B_2^+ \rightleftharpoons R^-B_1^+ + B_2^+ \qquad \text{Eq. 3-2}$$

Anionic exchange:

$$A_1^- + R^+A_2^- \rightleftharpoons R^+A_1^- + A_2^- \qquad \text{Eq. 3-3}$$

In the equations above, R is a solid phase and B and A are cations and anions, respectively. It is possible to use both natural materials and synthetic resins as the solid ion exchange surface, although synthetic resins are usually more advantageous, since they can be tailored for particular applications. The exchanging ions B_2^+ and A_2^- are often H^+ and OH^-, though other ions may also be used (e.g. Na^+, Cl^-).

When an ion exchange resin is exhausted, it is flooded with a concentrated solution of the exchanging ion, which drives Equations 3-2 and 3-3 to the left and regenerates the resin. The concentrated brine solution produced during the regeneration process, which containing the exchanged ions requires ultimate disposal.

The resin affinity for ions is affected by several factors: valence, ion radius, physical characteristics of the resin, etc. In general, the following rules of thumb describe exchange preferences:

1. At low concentrations and ordinary temperatures, exchange preference increases with the valence of ions.
2. In similar conditions, but at high concentrations, the effect of different valences decreases.
3. At low concentrations, ordinary temperatures, and constant valence, exchange preference increases with increasing atomic number.
4. At high concentrations or high temperatures, the effect of increasing atomic number decreases.
5. Organic molecules of high molecular weight exhibit high exchange potentials.

Suspended solids contained in wastewater will clog ion exchange contactors and cause operating problems. Consequently, filtration is usually practiced prior to ion exchange in order to accomplish suspended solids removal.

A major problem involved in the ion exchange process is the disposal of the concentrated regenerative brines. In effect, the ion exchange process concentrates the undesirable ions into a small waste stream, and it is necessary to dispose of this brine waste.

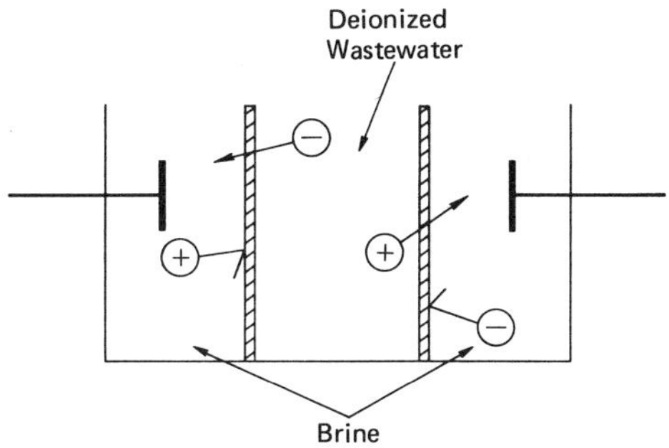

Fig. 3-7. Electrodialysis.

Electrodialysis. Perm-selective membranes and an electric potential driving force are combined in the electrodialysis process to extract ions from solution. Figure 3-7 shows a simple electrodialysis cell in which wastewater may be deionized.

In the cell, two perm-selective membranes (one cationic and one anionic) separate the wastewater from the brine solution and electrodes. As an electric potential is applied across the cell, the cations in the wastewater migrate through the cationic permeable membrane towards the cathode and concentrate in the right hand portion of the cell. Anions are prevented from migrating from the right side of the cell. The net result is the removal of ions from the wastewater stream and concentration in the brine. In actual applications, the simple cell is expanded to a row of alternating cation and anion permeable membranes. The brine must ultimately be disposed of by some appropriate scheme (Ref. 3-15).

There are several major operational problems which have led the EPA to reject electodialysis as a viable wastewater treatment technique at this time (Ref. 3-14). These problems include scale formation, leakage of electrolyte and current, current inefficiency, stability, poisoning of the membrane, and fouling. Extensive pretreatment of the wastewater is required in order to control these problems.

Reverse Osmosis. Whereas electric potential is the driving force in electrodialysis reactions, pressure is utilized in the reverse osmosis process to force a solvent through a semipermeable membrane. The passage of solutes is retarded by the membrane, so that a purified solvent is obtained. Figure 3-8 presents the principle of operation of the reverse osmosis unit. There are several different spatial configurations which have been developed for reverse osmosis in field applications including tubular, spiral wound, and hollow fin fibers.

Full-scale operational reverse osmosis processes are relatively new, and have found extensive application in the desalting of brackish water to produce drinking water. More recently, this process has been applied to the demineralization of secondary treated wastewater effluents (Ref. 3-26).

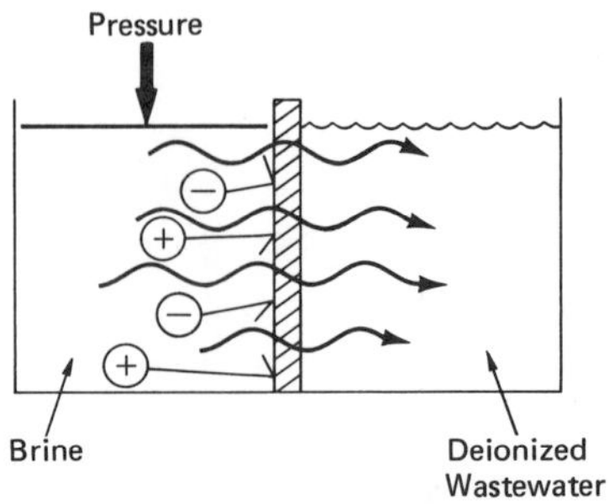

Fig. 3-8. Reverse osmosis.

The major problem experienced in applying the reverse osmosis process to wastewater effluents entails membrane fouling by dissolved, suspended, and colloidal matter. These may clog the membrane and significantly reduce the throughput of the reverse osmosis unit. Effluent pretreatment to remove the suspended and colloidal solids is required to control membrane fouling. In addition, chemical cleaning procedures may be necessary to maintain the efficacy of the reverse osmosis membrane.

Removal Efficiencies. The pollutant removal efficiencies for activated carbon adsorption, reverse osmosis, electrodialysis, and ion exchange are listed in Table 3-13. It appears from data contained in this table that reverse osmosis is by far the most efficient technique for removal of both dissolved organic and inorganic matter. However, this data does not include the effect of reduced throughput due to membrane fouling. Assuming that fouling can be controlled by appro-

TABLE 3-13 Efficiencies of Removal of Dissolved Solids in Wastewater after Conventional Biological Secondary Treatment (Ref. 3-1).

	Activated Carbon Adsorption	Ion Exchange[a]	Electrodialysis[b]	Reverse Osmosis[a]
Suspended Solids	80–90%	–	–	95–98%
BOD_5	70–90%	40–60%	–	95–99%
COD	60–75%	30–50%	–	90–95%
NH_3	–	85–98%	30–50%	95–99%
Org N	50–90%	80–85%	–	95–99%
PO_4	–	85–98%	30–50%	95–99%

[a]Assuming a prefiltration step to remove suspended solids.
[b]Assuming prefiltration and carbon adsorption to remove suspended solids and dissolved organics.

priate means (i.e., activated carbon adsorption and filtration), reverse osmosis is an excellent technique for the removal of dissolved matter from wastewaters.

The electrodialysis technique requires extensive wastewater pretreatment, possesses numerous operating problems, and is effective only in removing inorganic matter. The efficiencies listed in Table 3-13 for electrodialysis are typical efficiences, although much higher efficiencies could be obtained with extremely complex operating equipment and procedures.

The ion exchange technique requires prefiltration to remove suspended material which would otherwise clog the contactor. Ion exchange is moderately efficient in removing organic matter, but very efficient in removing inorganic matter.

Activated carbon adsorption achieves significant dissolved organic material removal, although dissolved inorganic constituent removal is negligible. This process can be designed to filter suspended solids, or the suspended solids can be removed in a pretreatment filtration step.

The concentrated brines form the reverse osmosis, electrodialysis, and ion exchange techniques are commonly deposited in evaporation ponds. Thus, the ultimate disposal of the dissolved matter removed from the wastewater stream is to the land. The material removed in the activated carbon technique is oxidized in the regeneration furnace to carbon dioxide and water, with ultimate disposal to the atmosphere.

The activated carbon technique has been accepted as a viable wastewater treatment. However, reverse osmosis, electrodialysis, and ion exhange have not been widely applied to wastewater treatment, in spite of their high treatment effiency. Operational problems arising from treatment of wastewater effluents have limited their usefulness.

3-3.7 Nitrogen Removal

This Section describes the currently available techniques for removing nitrogen forms from conventional biological secondary effluent. Section 3-2.3 defined total nitrogen as including the organic, ammonia, and inorganic (nitrite and nitrate) nitrogen forms.

All forms of nitrogen in wastewater effluents are potentially detrimental. The inorganic forms are capable of being utilized by plants as a nutrient source and may, therefore, contribute to the eutrophication of receiving waters. Ammonia nitrogen can be utilized by some bacteria as an energy source, with the resultant depletion of dissolved oxygen in the receiving water. Ammonia nitrogen also reacts with chlorine in the chlorination step of wastewater treatment to produce chloramines, which are less effective in disinfection than free chlorine (see Section 3-3.12). Organic nitrogen represents an oxygen demand on the receiving water through metabolism by microorganisms. Typical wastewater effluents contain very little organic nitrogen, as it is almost completely metabolized in the biological treatment process.

Where eutrophication is not a consideration, it is only necessary to consider the effects of the organic nitrogen forms on the environment. In such instances, nitrification of the ammonia nitrogen to its nitrate form may be sufficient to alleviate the potential detrimental effects of effluent nitrogen levels on receiving waters.

Currently Available Nitrogen Removal Techniques. There are a variety of techniques which have been investigated in this country and have been shown to either nitrify effluent streams or remove nitrogen completely from conventional secondary biological effluent. Table 3-14 presents currently available nitrogen removal techniques.

In physical-chemical treatment techniques, nitrogen either as ammonia, ammonium ion, or nitrate, is removed entirely from the wastewater, as opposed to being converted to another nitrogen form. Biological processes may accomplish nitrification, in which case ammonia nitrogen is converted to nitrate nitrogen, or they may be extended to denitrification, in which case the nitrate nitrogen is converted to gaseous nitrogen. The following discussion briefly summarizes the potential nitrogen treatment techniques presented in Table 3-14.

In the ammonia stripping process the wastewater pH is elevated to 9.5 or more, at which point the ammonia nitrogen in solution exists primarily as NH_3. Although ammonia is very soluble in water, it can be removed by exposure of the wastewater to sufficiently large volumes of air—as in a stripping tower. The

TABLE 3-14 Nitrogen Removal Techniques (Ref. 3-28).

	Technique	Equation
Chemical	Ammonia stripping	$NH_3\,(\text{solution}) \xrightarrow{\text{air}} NH_3\uparrow\text{gas}$
	Ion exchange (ammonia)	$NH_4^+ \rightarrow \text{Clinoptilolite} \cdot NH_4$
	Ion exchange (nitrate)	$NO_3^- \longrightarrow \text{resin} \cdot NO_3$
	Reverse osmosis	
	Electrodialysis	
Physical	Distillation	
	Breakpoint chlorination	$2NH_3 + 3\,Cl_2 \longrightarrow N_2\uparrow + 6HCL$
	Catalytic reduction	$NO_3^- + Fe^{2+} \xrightarrow{Cu} N_2\uparrow + Fe^{3+}$
Biological	Nitrification	$NH_4 \xrightarrow{\text{organisms}} NO_3^-$
	Denitrification	$NO_3^- \xrightarrow{\text{organisms}} N_2\uparrow$

ammonia then escapes during the stripping operation into the atmosphere. An ammonia stripping process has been operated at the South Tahoe Advanced Waste Treatment Plant in California since 1968, during which time it has been shown that scaling and poor efficiency at low temperatures are the major operational problems associated with this process (Ref. 3-27).

The naturally occurring mineral clinoptilolite has the capacity to selectively remove ammonia in wastewater via an ion exchange process. EPA studies on large scale pilot plants have shown ion exchange with clinoptilolite is possible, and further studies are being conducted to provide a long-term evaluation of this process. The major problems associated with this process at the present time include the varying exchange capacities of different lots of clinoptilolite; the poisoning of clinoptilolite, by trace metals such as magnesium, over a longer period of time; and the disposal of the concentrated brine solution containing ammonia produced from the regeneration of the clinoptilolite. The efficiency of ammonia nitrogen removal by this process has been reported to be in excess of 95 percent (Ref. 3-14). Operational data is not available on application of this process on a full scale basis (Ref. 3-28).

Ion exchange with synthetic resins for nitrate removal has been studied by the EPA (Ref. 3-28). Study results indicate that ion exchange with synthetic resins is not a viable nitrate removal technique since resins with a high selectivity for the nitrate ion have not been developed to date, and the regeneration of such resins is currently inefficient. In addition, there is the problem of ultimate disposal of the highly concentrated, nitrogen-rich, regeneration brine which is produced during the process.

It is possible to remove ammonia, nitrite, and nitrate ions from wastewater by deionization and demineralization techniques such as reverse osmosis, electrodialysis, and distillation. However, these techniques are not selective and require the removal of 85 to 90 percent of all the ions in solution in order to remove 85 to 90 percent of the nitrogen ions. Distillation is particularly problematic since ammonia distills with water unless an acid pH is maintained during the process. However, at depressed pH levels nitrous acid may distill over into the condensed water. All deionization techniques produce a concentrated brine which must receive ultimate disposal. The EPA has concluded that deionization and demineralization techniques are applicable only for wastewater which is destined for reuse, since they are too expensive for wastewater treatment schemes in which the resulting effluent is discharged to the environment (Ref. 3-28).

Breakpoint chlorination is a technique by which ammonia is converted to nitrogen gas through chlorine oxidation. The nitrogen gas formed during oxidation escapes into the atmosphere. Although the actual reactions by which chlorine in water oxidizes ammonia to nitrogen are complicated, it is generally agreed that the equation in Table 3-14 adequately describes the overall process. There are also some side reactions during breakpoint chlorination in which nitric and nitrous oxides may be formed. However, the quantities of these compounds formed are insignificant compared to the formation of nitrogen gas (Refs. 3-1, 3-24). While the stoichiometric ratio of $Cl_2:NH_3$ is 3:2 (molar) and 7.6:1 (weight), the actual ratio in practice has been observed to be slightly greater. Nitrogen removal efficiencies in breakpoint chlorination operations have been reported to be greater than 95 percent with final effluent concentrations of 0.1 mg/l ammonia N (Refs. 3-14, 3-28).

The major problems associated with the breakpoint chlorination process include the close monitoring requirements to avoid effluent overdosing with chlorine, which is uneconomical and may have undesirable effects on water quality, and increased mineralization (8 to 9 mg/l chloride ion for each mg/l ammonia originally present in the effluent). The potentially adverse water quality effect associated with chlorine overdosing can be eliminated by passing the wastewater, after breakpoint chlorination, through an activated carbon column. The activated carbon catalytically converts the free and combined chlorine remaining in the effluent to chloride ion (the mechanism is not known) (Ref. 3-29). The EPA concludes that breakpoint chlorination should be limited to applications where the receiving water can accept effluents with increased mineralization, or where it is used as a polishing technique after the bulk of the effluent ammonia nitrogen has been removed (Refs. 3-14, 3-28).

Another physical chemical nitrogen removal technique involves the reduction of nitrate to nitrogen gas with ferrous ion in the presence of copper. Although technically possible, this process is not considered viable for general application since careful pH control is required; phosphate and carbonate ions, which are ubiquitous in wastewater, interfere with the reactions; and the ferrous-ferric hydroxide sludge which is produced is difficult to dewater and to ultimately dispose (Ref. 3-28).

The physical-chemical nitrogen removal techniques are currently beset with operational problems which limit their effectiveness. At the present time, the ammonia stripping process appears to be the most viable p-chem technique, providing that scaling can be controlled and low air temperatures are avoided.

The biological processes for nitrogen conversion or removal utilize naturally occurring bacteria to convert most of the nitrogen in wastewater to nitrate nitrogen (nitrification) and then to gaseous nitrogen (denitrification). The biological approach to nitrogen control in effluents has been tested in pilot plant studies and has received broad support in the literature. It is generally regarded as an extremely efficient and viable treatment technique.

Ammonia Stripping. Ammonia is readily soluble in water (900 grams per liter in cold water to 74 g/l in hot water) and undergoes hydrolysis as shown in Equation 3-4, to form ammonium ion.

$$NH_3 + H_2O \rightleftharpoons NH_4^+ + OH^- \quad K = 2 \times 10^{+9} \text{ @ } 20°C \qquad \text{Eq. 3-4}$$

The 50 percent equilibrium for this hydrolysis occurs at about pH 9.3. Therefore, it is necessary to raise the pH of wastewater above 9.3 (preferably much higher) in order to insure that most of the ammonia nitrogen exists as NH_3 species, and therefore can be stripped out of the wastewater.

Since ammonia is highly soluble in water, it is necessary that large volumes of air be contacted with the wastewater to force the ammonia out of solution into the gas phase. The temperature of the air is also critical since cold air is less efficient than warm air, and therefore, the stripping efficiency is dependent on the ambient temperature. O'Farrell, *et al.* (Ref. 3-31) have reported that a decrease in air temperature from 80°F to 43°F reduces the efficiency of ammonia transfer by 30 percent.

An operational ammonia stripping process is represented in Figure 3-9 (Refs. 3-31, 3-32). The process is a staged, crosscurrent, gas-liquid separation in which air is pulled through the stages by one blower located at the end of the train. The water cascades through towers designed to provide a high surface area to optimize liquid-air interphase transfer. Ammonia stripping process technology has been well developed in the chemical industries for some time.

The ammonia stripping process is particularly well suited to follow a lime coagulation process in which the wastewater pH has been adjusted to about 9.5. However, further addition of lime is required to elevate the pH to a level which will allow an efficient ammonia stripping operation.

A major problem of the stripping process is the formation of calcium carbonate scale inside the towers, which reduces the surface area available for transfer and eventually clogs the tower. The formation of scale is a function of the reactions of lime in water (lime is the cheapest chemical available for the adjustment of pH). At normal pH, lime added to wastewater reacts with the bicarbonate according to Equation 3-5 to produce calcium carbonate and hydroxyl ions.

$$Ca(OH)_2 + HCO_3^- \longrightarrow CaCO_3\downarrow + H_2O + OH^- \qquad \text{Eq. 3-5}$$

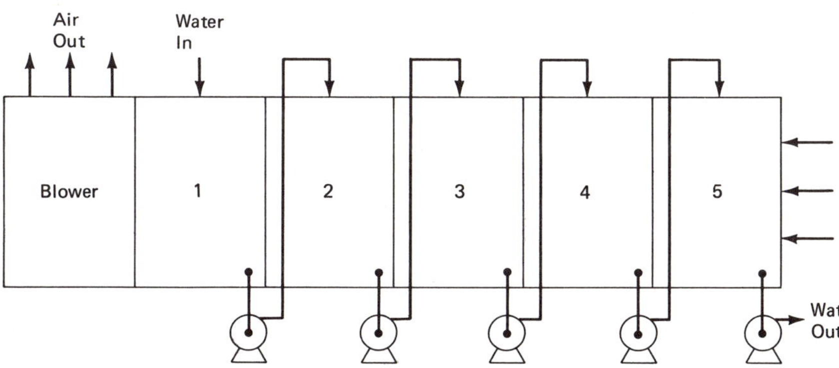

Fig. 3-9. Ammonia stripping. (Refs. 3-31, 3-32)

As lime is added, the pH rises to about 9, and all the bicarbonate will have been consumed and precipitated as calcium carbonate. Further addition of lime will raise the pH above 9 and put calcium ions into solution. When the wastewater is contacted with large volumes of air to strip out ammonia, carbon dioxide in the air dissolves into the wastewater, hydrolyzes to carbonate ion, and precipitates out as calcium carbonate to form scale inside the stripping towers. The scale may be very hard and difficult to remove, or soft and easily washed away. Both types of scale have been encountered, but there is no explanation in the literature for the preferential formation of either type of scale (Ref. 3-30). In addition, no studies have been reported in which attempts were made to use other chemicals (e.g., sodium hydroxide) to raise pH, thereby eliminating calcium ion addition to solution.

Scale formation can be controlled by continuous maintenance, which consists of washing and scraping the towers to remove scale. Scaling rates in stripping towers have been reported as high as 125 mg/l as $CaCO_3$ at pH 11.5 and 16 mg/l as $CaCO_3$ at pH 10.5 (Ref. 3-31). The ammonia removal efficiencies at these pH levels were 86 percent and 80 percent, respectively.

Where applicable, the ammonia stripping technique is advantageous for the following reasons: it can easily by associated with lime coagulation; it is simple in concept and relies on a well-known technology; it is the most economic system for nitrogen removal; and it is efficient. Its major disadvantages include: its efficiency is highly dependent on air temperatures; freezing may occur in the towers at low temperatures; scaling may occur; and the ammonia transferred into the air may be redissolved by rain and washed into the receiving water, thus negating the ammonia removal from the environment's water system (Ref. 3-14).

Biological Nitrification and Denitrification. Biological nitrification is performed by aerobic chemoautotrophic bacteria ubiquitous in domestic wastewaters. These bacteria (NITROSOMONA, NITROSOCOCCUS, NITROBACTER, and others) catabolize ammonia and nitrite and produce nitrate according to equations 3-6 and 3-7 (Ref. 3-33).

$$2NH_4^+ + 3O_2 \longrightarrow 2NO_2^- + 4H^+ + 2H_2O \qquad \text{(Eq. 3-6)}$$

$$2NO_2^- + O_2 \longrightarrow 2NO_3^- \qquad \text{(Eq. 3-7)}$$

Investigations have shown that nitrifying bacteria are very sensitive to pH. These organisms metabolize optimally at pH 8.4 (90 percent of optimum between pH 7.8 and 8.9). Outside of this pH range their activity drops markedly (Ref. 3-34). These bacteria also respond to temperature changes, though not so markedly, and the temperatures of normal domestic wastewaters are usually sufficiently high to allow an acceptable rate of metabolic activity.

The nitrifying bacteria do not grow as rapidly as normal bacteria in an activated sludge biomass, and their growth in a typical activated sludge process is consequently inhibited. Therefore, the effluent from conventional activated sludge shows little or no nitrification.

Currently there are two biological techniques which encourage the growth of nitrifying bacteria. The first is the extended aeration modification of the activated sludge process. In the extended aeration process the sludge age (10-30 days) is sufficiently long to allow the establishment of a nitrifying bacteria level that is a significant portion of the activated sludge biomass. Thus, the effluent from the extended aeration process is characteristically well nitrified.

The second technique currently utilized for nitrification was first reported by Barth, *et al.* in 1968 (Ref. 3-35). Since then, other investigators have developed similar processes, and extensive pilot plant data has been accumulated in the literature. As part of a three stage nitrification–denitrification process, nitrifying bacteria are established in a second aeration tank, which receives effluent from a conventional activated sludge process. The nitrifying bacteria are the predominant bacteria type in the second sludge aeration tank sludge biomass, and nitrification is the primary function of this aeration chamber, although further reduction of the carbonaceous oxygen demand also occurs. The advantages of this technique over extended aeration are that a smaller aeration tank is required and more control over the nitrification process is possible.

EPA design criteria for a nitrification aeration tank include: pH control in a range from 7.6 to 7.8, a detention time of 3 to 6 hours, MLVSS of 1500-3000 mg/l, and sufficient oxygen to meet the stoichiometric nitrogenous and carbonaceous demand with a minimum of 1 mg/l dissolved oxygen at all times. Nitrification sludge is less dense than activated sludge and does not compact well. Recycle of up to 100 percent should be provided in the nitrification aeration basin system (Ref. 3-28).

Biological denitrification is accomplished by facultative chemoheterotrophic bacteria which, in the absence of oxygen will utilize nitrogen in the nitrate ion as a terminal electron receptor producing nitrogen gas and small amounts of nitrous oxide and nitric oxide. Equation 3-8 is the overall equation depicting the denitrification process (Ref. 3-16).

$$2\,NO_3^- + 10\,e^- + 12\,H^+ \longrightarrow N_2 + 6\,H_2O \qquad \text{(Eq. 3-8)}$$

The biological nitrification-denitrification process is presented schematically in Figure 3-10. The effluent from the activated sludge process flows into the nitrification aeration basin where bacteria convert ammonia to nitrate and further reduce the organic carbon in the effluent. The effluent then flows to the denitrification chamber where the denitrifying bacteria convert the nitrate to nitrogen gas. Since the effluent from the nitrification tank is very low in organic carbon,

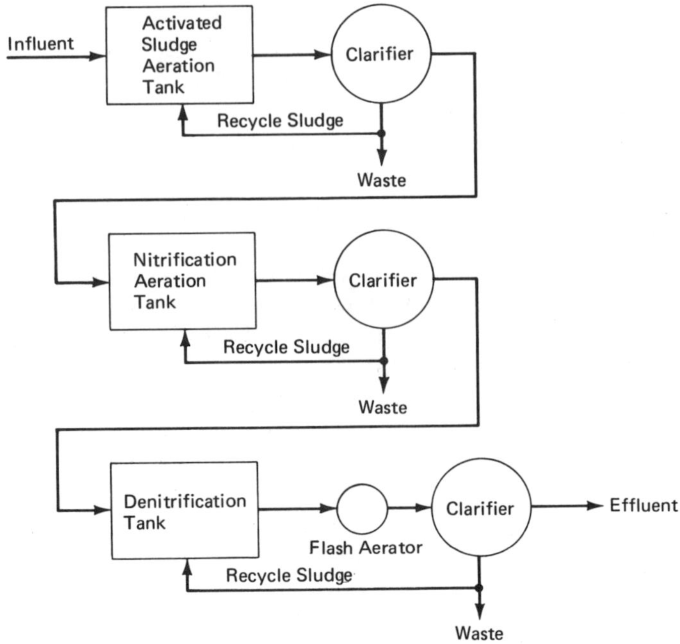

Fig. 3-10. Biological nitrification–denitrification.

an organic substrate must be added for the denitrifying organisms. Methanol is recommended as the substrate additive because of its price and availability. Equation 3-9 is the basic equation for denitrification, using methanol as the organic carbon source.

$$5\ CH_3OH + 6\ HNO_3 \longrightarrow 5\ CO_2 + 3\ N_2\uparrow + 13\ H_2O \qquad \text{(Eq. 3-9)}$$

The equation indicates that 5 moles of methanol must be supplied for each 6 moles of nitrate nitrogen in the waste stream or approximately 2 mg/l methanol to 1 mg/l nitrate nitrogen on a weight basis. However, pilot plant operation has indicated that a portion of the methanol is converted to cell mass while some is consumed aerobically, with a resulting process requirement of 3 to 4 mg/l methanol per mg/l nitrate.

The optimum pH range for the denitrification process is 6.5 to 7.5. However, no hydrogen or hydroxyl ions are consumed or produced in the metabolic processes and, therefore, pH control is not explicitly required.

Since anaerobic conditions are required for nitrate reduction in the denitrification process, there must be some means of preventing oxygen in the atmosphere from dissolving in the denitrification tank. Air tight enclosures are not recommended because the nitrogen gas and carbon dioxide produced in the metabolic processes must be allowed to escape. It is believed by several investigators that the denitrifying tank will seal itself with a floating scum cover. Mixing the denitrification chamber must, therefore, be sufficiently gentle to avoid damaging this scum layer but adequate to maintain the bacteria in suspension (Refs. 3-28, 3-34).

EPA design criteria for denitrification tanks include: a pH range from 6.5 to 7.5 (the pH of the effluent from the nitrifying tank may have to be adjusted downward; a detention time of 1 to 3 hours (longer if low winter temperatures are expected); and a MLVSS of 1300 to 1900 mg/l (Ref. 3-28).

The effluent from the denitrification tank flows through a flash aerator, allowing the release of nitrogen and carbon dioxide gases produced during the process. Flash aeration also provides an acceptable dissolved oxygen level in the treatment systems effluent. If this flash aeration is omitted, small nitrogen and carbon dioxide bubbles may form in the clarifier and prevent settling of the sludge. Properly aerated, the denitrification sludge will settle well in the clarifier, very similarly to typical activated sludge. Return sludge capabilities of 100 percent are recommended for the denitrification process.

In addition to suspended growth denitrification, columnar denitrification has been shown to be a viable means to achieve nitrate reduction. Columnar reactors have been designed similar to deep bed filters with both fine (2 to 4 mm) and large (15 to 25 mm) grain media. The fine grained media columns also act as filters to remove suspended solids. Periodic sloughing of the biological growth requires that a filtration step be included to insure a high quality finished water.

Properly operated, the biological nitrification and denitrification processes have been shown to convert or remove nitrogen in an extremely efficient manner. Efficiencies up to 100 percent have been reported in the removal of ammonia, nitrite, and nitrate from wastewater (Refs. 3-14, 3-17, 3-28, 3-33, 3-34, 3-36, 3-37, 3-38, 3-39). Biological nitrogen treatment is therefore generally regarded as an extremely efficient and viable treatment technique.

3-3.8 Phosphorus Removal

Phosphorus is present in typical domestic wastewater as inorganic and organic phosphate. Both forms can eventually serve as nutrients for biological growth and play a key role in the eutrophication of waters receiving wastewater effluents (see Section 3-2.4).

Since phosphates are required for biological growth, it might be expected that some phosphate removal would occur in the activated sludge process. This, in fact, does occur. However, the phosphate concentration of wastewater is two to five times higher than that required for the growth of activated sludge, so that only 20 to 50 percent of the influent phosphorus is removed in the activated sludge process. The remaining phosphorus portion is discharged in the secondary effluent.

It is possible, under special operating conditions, to force the activated sludge to assimilate more phosphate from solution than is actually required for growth. This "luxury uptake" has been a source of controversy both in the elucidation of the phenomena and in the description of the basic mechanism involved. However, luxury uptake has not been shown to be effective in removing significant quantities of phosphate; and, consequently, biological removal of phosphorus cannot be considered a viable technique.

Since most phosphorus in domestic wastewater exists as inorganic phosphate ions, demineralization techniques such as reverse osmosis, electrodialysis, and ion exchange will effectively remove it from solution (Refs. 3-15, 3-26). However, these processes are not selective for the phosphate ion and consequently, to achieve a high degree of phosphate removal, the wastewater must be essentially demineralized of all ions in solution. One exception is the use of alumina minerals in an ion exchange process to remove phosphate from wastewaters. In general, demineralization techniques are applicable to wastewater renovation and reuse processes, but are usually considered to be too costly for wastewaters which are discharged to the environment.

Ion Exchange with Alumina. In 1969, Ames (Ref. 3-40) summarized the work on phosphate removal from wastewater and demonstrated the possibility of an ion exchange process using alumina to remove phosphates from water. Alumina (i.e., oxides and hydroxides of aluminum) occurs naturally in bauxite ore and is purified commercially by the aluminum industry. Alumina has a preference for phosphate ions second only to hydroxide ions in anion exchange properties. Consequently, it is possible to selectively remove phosphate from wastewater with alumina ion exchange, and then regenerate the alumina with a high concentration of hydroxide; e.g., sodium hydroxide.

Ames demonstrated this process on a bench scale using one-liter columns. An interesting feature of this process is that the regenerate sodium hydroxide brine can itself be regenerated by lime addition which precipitates phosphate as hydroxyapatite. Consequently, the regenerate brine can be reused.

Although technically feasible, ion exchange has not seen widespread application since chemical application is simpler in concept and use and less expensive.

Chemical Addition. Probably the most efficient method of phosphorus removal is through chemical addition (Ref. 3-41). Phosphate can be precipitated from solution by a wide variety of chemicals. Economic factors usually reduce the choice of chemicals to iron salts, aluminum salts, and lime.

Both ferrous and ferric salts of iron can be utilized to precipitate insoluble phosphate products. The basic reactions are shown in Equations 3-10 and 3-11.

$$Fe^{3+} + PO_4{}^{3-} \longrightarrow FePO_4 \qquad \text{(Eq. 3-10)}$$

$$Fe^{3+} + 3\,H_2O \longrightarrow Fe(OH)_3 + 3\,H^+ \qquad \text{(Eq. 3-11)}$$

Equation 3-10 describes the formation of insoluble ferric phosphate, while Equation 3-11 outlines a typical hydrolysis reaction of the ferric ion to form insoluble ferric hydroxide. Other hydrolysis reactions of the ferric ion also occur and form complicated aquocomplexes. The formation of these hydrolysis products compete with the formation of the ferric phosphate, and therefore, an excess of iron salt must be added to insure the precipitation of all the phosphate present in the water. However, the excess metal is not wasted since the insoluble ferric hydroxides aid in coagulation and flocculation of the phosphate precipitate, as well as destabilizing colloidal dispersions and precipitating suspended solids which may be present in the wastewater. It should also be noted that the hydrolysis reactions liberate hydrogen ions which will consume alkalinity and lower the pH.

The stoichiometric mole ratio of ferric ion to phosphorus is 1 : 1, while the weight ratio is 1.8 : 1 (see Equation 3-10). However, the actual iron dose required to precipitate all the phosphate in a wastewater stream is a function of the phosphate concentration, the alkalinity, and the suspended solids concentration. Consequently, bench tests are required for proper design of phosphorus removal processes.

Ferrous iron also reacts with phosphate in water to form insoluble precipitates. The reactions and mechanisms by which ferrous iron reacts to form phosphate salts and hydrolysis products are very complicated and poorly understood. One anomaly is that although ferrous and ferric ions have different valences, and different stoichiometry of reactions with phosphate in water, the overall consumption of ferrous ion in the precipitation of phosphate is the same as for ferric ion.

The optimum pH range for precipitation of phosphate with ferric ion is from 4.5 to 5, although significant removals can be obtained at higher pH levels. Ferrous ion precipitates phosphate in the pH range from 7 to 8.

Aluminum ions also form insoluble phosphate precepitates and complicated aquacomplexes according to Equations 3-12 and 3-13.

$$Al^{3+} + PO_4^{3-} \longrightarrow AlPO_4 \qquad \text{(Eq. 3-12)}$$

$$Al^{3+} + H_2O \longrightarrow Al(OH)_3 + 3H^+ \qquad \text{(Eq. 3-13)}$$

These equations are similar to those describing phosphate precipitation with iron salts. The aluminum phosphate reaction competes with the formation of various hydroxides, which act as coagulants and can improve the settling of the aluminum phosphate complex.

As with iron salts, aluminum addition produces hydrogen ions which consume alkalinity and lower the pH of the wastewater. The magnitude of this pH change depends on the amount of alkalinity in the wastewater—the higher the alkalinity, the less change in pH for a given dose of alum.

The stoichiometric mole ratio of aluminum to phosphorus is 1 : 1 (weight ratio is 0.87 : 1); however, in practice the ratio is much higher. Black and Veatch (Ref.

3-41) report that aluminum to phosphorus mole ratios of 2.3:1 (weight ratio of 2:1) should be used to achieve 95 percent phosphate reduction. The competing side reactions partially account for the excess of aluminum which must be used. The optimum pH range for phosphate removal with aluminum is 5.5 to 6.5, though significant removal will occur above pH 6.5.

Some of the common iron and aluminum salts utilized for phosphate removal include ferric chloride, ferric sulfate, ferrous chloride, ferrous sulfate, aluminum sulfate, and sodium aluminate. The hydrolysis of sodium aluminate produces hydroxyl ions instead of hydrogen ions and tends to elevate the wastewater's pH level.

Lime also is very effective in the removal of phosphorus. At a high pH, phosphate and calcium react according to Equation 3-14 to form the mineral hydroxyapatite.

$$3\ HPO_4^{2-} + 5\ Ca^{2+} + 4\ OH^- \longrightarrow Ca_5(OH)(PO_4)_3 + 3H_2O \qquad \text{(Eq. 3-14)}$$

This equation is not exact, since all ionic forms of phosphate may enter into the reaction, and hydroxyapatite may be formed with slightly different molecular formulas. The solubility of hydroxyapatite at pH levels greater than 9 is extremely low, so that phosphate removals accomplished with lime addition are very efficient. Normally in wastewater treatment, lime doses are based on the amount of lime needed to raise the pH and precipitate suspended solids. This quantity (up to several hundred mg/l as $CaCO_3$) is an order of magnitude greater than the amount necessary to react with the phosphate in the waste stream. Thus, the calculation of lime dosages necessary to precipitate phosphate is dependent on the pH, alkalinity, and suspended solids of the wastewater, rather than the concentration of phosphorus.

Location of Chemical Addition. Chemical addition for phosphorus removal may be accomplished either in the primary or the secondary portion of conventional biological treatment. As part of the primary treatment, the chemical is added to the influent wastewater and the phosphate precipitated in the primary settling tank. As part of secondary treatment, the chemical may be added either to the aeration tank or to the wastewater as it flows from the aeration tank to the clarifier. In both cases, the phosphate precipitate is removed in the secondary clarifier. Alternatively, phosphate precipitation may be carried out as part of a tertiary treatment scheme following conventional biological treatment.

Table 3-15 lists the advantages and disadvantages of chemical addition for phosphorus removal at primary, secondary, and tertiary locations. Greater than 95 percent phosphate removal can be achieved through the proper choice of chemicals and application location (Refs. 3-14, 3-42). The selection of chemicals and application location for phosphorus removal by chemical addition must be based on the total wastewater treatment scheme selected.

TABLE 3-15 Comparison of Phosphorus Removal by Chemical Addition at Different Locations of a Treatment Plant (Ref. 3-14).

Primary	Secondary	Tertiary
ADVANTAGES		
1. Applicable to all plants	1. Lowest capital	1. Lowest P in effluent
2. Increased BOD and suspended solids removal	2. Lower chemical dose than primary	2. Most efficient metal use
3. Lowest degree of metal leakage	3. Improved stability of activated sludge	3. Lime recovery demonstrated
	4. Polymer not required	
DISADVANTAGES		
1. Least efficient utilization of metal	1. Careful pH control to get P <1 mg/l	1. Highest capital cost
2. Polymer required for flocculation	2. Overdose of metal may cause low pH toxicity	2. Highest metal leakage
3. Sludge more difficult to dewater than primary sludge	3. Cannot use lime because of excessive pH	

3-3.9 Stabilization Ponds

Domestic wastewater can be treated by the natural physical, chemical, and biological processes which occur in a shallow pond. Such structures, designed to receive and hold wastewater for natural treatment, are known as stabilization ponds, oxidation ponds, lagoons, polishing lagoons, and maturation ponds, and have been applied to the treatment of raw domestic wastewater, primary and secondary effluents, and industrial wastes.

The advantages of wastewater treatment by ponding include lower capital and operating costs than an equivalent mechanical plant, the ability to regulate the effluent discharge volume, and the insignificant effect of infiltration and inflow on treatment efficiency. Disadvantages of ponding include the extensive land area required, potential odor problems, and little control over the treatment processes proceeding in the pond (Ref. 3-2).

Design and Operation of Ponds. Stabilization ponds are shallow, flat bottomed ponds surrounded by an earth dike; they can be designed to any shape, but the length usually does not exceed three times the width. Pond depths typically vary from two to twenty feet, depending on the treatment purpose. Depths less than two feet encourage the growth of emergent vegetation.

Ponds may be classified as aerobic, aerobic–anaerobic, or anaerobic. Aerobic ponds are shallow ponds, up to 5 feet deep, in which bacteria and algae stabilize the wastewater under aerobic conditions. Oxygen is supplied by natural diffusion across the pond surface, by algal photosynthetic processes, and, in some cases, by mechanical aerators. Best results are achieved if the pond is mixed periodically. Aerobic ponds typically are used for treatment of secondary effluents or wastes with a high fraction of soluble organics. The soluble organics are metabolized by bacteria, and the nutrients in solution are utilized by algae. Although the efficiency of BOD_5 conversion of soluble organics in aerobic ponds is high, (up to 95 percent), the growth of algae is a source of BOD increase. If the algae are not removed from the effluent, they may exert a BOD load on the receiving water that is acutally higher than that of the wastewater influent to the pond (Ref. 3-1).

Aerobic-anaerobic ponds are designed and operated to produce anaerobic bottom conditions in which anaerobic bacteria digest settleable wastes and dissolved material in the absence of dissolved oxygen. The surface zone of the aerobic-anaerobic pond is aerobic, allowing bacteria and algae to consume soluble organics and nutrients. Between the surface and bottom zones is an intermediate zone, marginally aerobic/anaerobic, in which facultative bacteria decompose organic wastes. The aerobic-anaerobic pond is sometimes referred to as facultative, and is typically used for the treatment of raw, screened, or primary settled wastewaters.

Anaerobic ponds are usually designed to be deeper than aerobic and facultative ponds (up to 20 feet deep), to encourage anaerobic conditions throughout the pond depth. Only an extremely shallow surface zone of the anaerobic pond is naturally aerobic. Waste stabilization occurs by typical anaerobic processes in which organic wastes are converted to carbon dioxide, methane, other gaseous products, organic acids, and bacterial cell tissue. Frequently, odors are a problem with anaerobic ponds, and efficiencies are lower than with aerobic and facultative ponds, though up to 85 percent BOD_5 conversion is possible and 70 percent is routinely achieved. Anaerobic ponds may be utilized for the complete treatment of raw domestic and industrial wastewaters.

Stabilization ponds may be designed in series to affect more complete treatment of wastewaters. A series of ponds may all be of one type, or may be varied. A pond series, such as anaerobic-facultative and anaerobic-facultative-aerobic, allows better process control and higher treatment efficiencies than single ponds or a series of similar ponds.

The design of ponds is extremely complex, since a multitude of processes are occurring in the pond at the same time, including: sedimentation, oxidation, digestion, gas exchange, photosynthesis, mechanical aeration, evaporation, and

TABLE 3-16 Design Parameters for Stabilization Ponds (Ref. 3-1).

Parameter	Aerobic	Aerobic-Anaerobic	Anaerobic
Depth, ft	2–5	2–8	8–20
Detention time, days	10–40	7–30	20–50
pH	6.5–10.5	6.5–9.0	6.8–7.2
Optimum Temperature, °C	20	20	30
BOD_5 loading, lb/acre/day	60–120	15–50	200–500
BOD_5 conversion	80–95	80–95	50–85
Principal conversion products	Algae, CO_2 bacterial cells	Algae, CO_2 bacterial cells CH_4	CO_2, bacterial cells CH_4
Algae concentration, mg/l	80–200	40–160	—

seepage. Fair, *et al.* (Ref. 3-9) state that the rate of BOD conversion in a pond is a function of detention time, depth, BOD loading, temperature, visible light energy input, and efficiency of conversion of light energy into chemical energy. Climatic conditions are also an important consideration, since a pond is much less efficient in cold weather when bacterial metabolism is slowed; also, the possibility of an ice cover exists. Table 3-16 lists the characteristic design parameters for stabilization ponds.

Treatment of Secondary Effluents in Ponds. Aerobic, facultative, or anaerobic ponds may be used singly or in combination in the treatment of domestic wastewater. Aerobic ponds are best suited for the treatment of secondary effluents for two reasons. First, they encourage algal growth, which consumes nutrients; and second, they are extremely efficient in the conversion of the BOD remaining in the secondary effluent.

As a tertiary treatment technique, ponding is attractive because of its low cost and simplicity of operation. Its disadvantages include: the land area required; the fact that not all the phosphorus and nitrogen will consistently be removed by algal growth; the algae, if not removed from the effluent represent a BOD source which may exceed the influent BOD; and the adverse effect that seasonal weather changes may have on the pond's performance.

3-3.10 The Potential of Ozonation

Ozone has long been recognized as a powerful oxidizing agent, with an electronegative oxidation potential second only to fluorine. Ozone has been utilized for many years in Europe for drinking water treatment associated with color, taste, odor removal, and microorganism disinfection.

In the last two decades in the U.S., a great deal of interest has been focused on the utilization of ozone for wastewater treatment. A number of laboratory and pilot studies have been conducted which indicate that ozonation is most applicable to the tertiary treatment of wastewaters. Also, several large scale studies have been conducted which have demonstrated the efficiency of ozonation of wastewaters. Although few full-scale systems have been installed, ozonation is regarded as an attractive tertiary treatment technique which may be used extensively in the future (Refs. 3-43, 3-44, 3-45).

Ozone Chemistry in Water. In spite of the numerous laboratory studies of ozonation, little is known about the mechanisms and rates of reaction of ozone with other compounds. Such studies are complicated by the nature of ozone: it decomposes spontaneously in water and air; it has a low solubility in water; and it is extremely reactive with oxidizable compounds (Ref. 3-14).

The autodecomposition of ozone occurs as a function of pH and temperature. The halflife of ozone in water is approximately twenty minutes, at normal pH ranges (6.8 to 7.2) and ambient wastewater temperatures.

Although ozone reacts rapidly with compounds in water, it is believed that the reaction occurs primarily at the gas-liquid interface between bubbles containing ozone and the bulk solution. It has been postulated that ozone reactions occur through a free radical mechanism, but evidence of nucleophilic as well as electrophilic reactions indicates that some other mechanisms are likely to be involved. A few of the organic molecule groupings ozone is known to react with include: carbon-carbon double and triple bonds, carbon-nitrogen bonds, aromatic and heterocyclic molecules, carbon-hydrogen bonds in amines, alcohols, ethers, hydrocarbons, etc. (Ref. 3-44). Ozone will also oxidize inorganic compounds.

Ozone Generation. Ozone is generated by electrical discharge in an atmosphere containing oxygen (hence the smell of ozone after lightning strikes). This principle has been adapted commercially to several different ozone generation configurations—Otto plate, tube, and Lowther plate. Either air, oxygen enriched air, or pure oxygen is passed into the generator and through a corona discharge where ozone is generated by the discharge. The percent ozone generated is partially a function of the percent oxygen in the feed gas. Up to 6 percent ozone can be generated with pure oxygen feed.

Since ozone autodecomposes so rapidly, it is not feasible to store and transport it to the point of application. For wastewater treatment, the ozone generator must be located at the treatment plant and the ozone used as soon as it is generated. The ozone generator requires clean and dry feed gas and electric power. Due to the economics of ozone generation, it is usually cheaper to provide a pure oxygen or oxygen enriched air to the generator rather than normal air. Current ozone generation technology is very inefficient in terms of pounds of ozone produced per kilowatt consumed, but it is expected that greater efficiency in production will be achieved in the future as ozone generation technology develops.

Ozonation of Wastewater. Ozone's great oxidation potential makes it attractive for wastewater treatment. Compared to chlorine for the oxidation and disinfection of conventional secondary effluent, ozone has several advantages: (Ref. 3-46)

a. Ozone is a powerful oxidizer so a high degree of oxidation can be expected.
b. Most of ozone's reactions are rapid so less contact time is required.
c. Ozone is a powerful viricide with a faster rate of disinfection of bacteria and viruses than chlorine.
d. Ozone application results in a beneficial oxygen residual in the water.
e. Ozonated effluent products are generally less toxic than chlorinated products.

In addition to these advantages, ozonation possesses other attractive features as a tertiary treatment technique including the reduction of BOD, COD, color, odor, and possibly ammonia nitrification (Ref. 3-47). Ozone will not remove suspended solids by reaction, but it has been observed to produce a foam which concentrates suspended solids and can be separated from the waste stream.

It appears that the critical feature of an ozonation process is the contact between the ozone bearing gas and the wastewater. Since the ozone reaction occurs at the gas-liquid interface, it is necessary that a large interface area be generated. Special diffusers, injectors, and stirrers have been utilized to create a large number of small gas bubbles in the wastewater. The degree of ozone reaction is directly proportional to the degree of the gas-liquid contact.

Disadvantages of Ozonation. The ozonation of wastewater is a new and developing technology and, consequently, there is not a vast history of field experience to draw on. Several manufacturers have entered this field and produced ozonation equipment for wastewater treatment, but there is no great demand at the present time for this treatment technology. Interest is high, however, and many field studies are being planned or performed. Ozonation is highly regarded as possessing future tertiary wastewater treatment potential.

As a consequence of undeveloped technology, the cost of wastewater ozonation is high compared to chlorination. However, it is commonly stated that ozonation is less expensive than activated carbon adsorption. The literature presents cost figures for ozonation varying from 7.7 to 83 cents per 1,000 gallons, which reflects the variable nature of the applications and current technology (Refs. 3-46, 3-48).

Therefore, the two main disadvantages of ozonation are the newness of the technology and attendant lack of field experience on the part of the engineering profession, and the relatively high costs of application.

3-3.11 Land Application of Wastewater

Land application of wastewater has been utilized on a small scale in the U.S. for many years, although the practice has not been widespread. The Federal Water Pollution Control Act Amendments of 1972 encouraged land application as an alternative to conventional and other advanced wastewater treatment techniques in planning wastewater treatment schemes (Refs. 3-49, 3-50). The impetus towards land application thus is burgeoning, although the technology and understanding of the mechanisms of treatment are yet limited.

Land application has been successful with secondary effluent, primary effluent and, in some cases, even raw wastewaters after only gross solids removal. However, it is expected that future regulations governing land application will require at least secondary treatment, either with or without chlorination, prior to application on land.

Land Application Techniques. Roughly defined, land application is any technique which utilizes the interactions between natural soil and vegetation and wastewaters to upgrade the quality of the wastewater. The various application procedures can be classified into three main groups: irrigation, overland flow, and infiltration-percolation (Ref. 3-51). Figure 3-11 depicts each type application, while Table 3-17 compares the characteristics of each procedure.

Irrigation may be defined as the application of wastewater to land to sustain plant growth. If the soil is suitable, salable crops may be grown and harvested; the income from which will partially offset the cost of wastewater application. During irrigation, wastewater is recycled completely to the land, being either evaporated, transpired, incorporated into plants, or percolated into the subsoil. There are three methods of applying irrigation water: by spraying from fixed or mobile sprinklers; by flowing into prepared channels (ridge and furrow); and by alternate flood and drying of land with water.

In the overland flow technique, wastewater is sprayed onto gently sloping ground, which is planted with vegetation. Biological treatment occurs as the wastewater contacts the biota in the ground cover vegetation and the soil. The wastewater is collected in ditches, and then can be discharged to a nearby stream or, possibly, recycled in another way. Some of the wastewater is lost to evaporation and percolation, but the majority runs into the collection ditch. The vegetation produced during this overland flow technique is not usually harvested.

The infiltration-percolation technique is based on wastewater application to very permeable soils, with the purpose of allowing the water to percolate

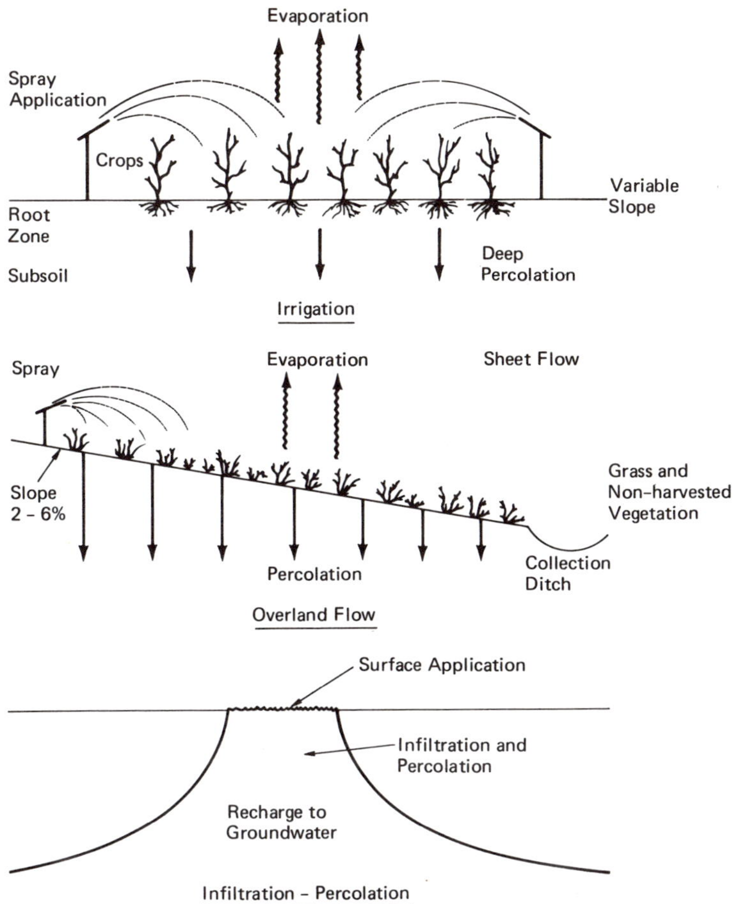

Fig. 3-11. Land application procedures. (Ref. 3-51)

through the soil and enter the groundwater. Vegetative growth is not encouraged. Evaporation is a very small factor in this technique, so nearly all the wastewater is directly recycled to the land. In some instances, deep wells may be installed to intercept the wastewater as it percolates through the soil, and extract it for reuse.

The use of these land application techniques is subject to limitations. The main limitation is the availability and suitability of land. Additionally, climatic conditions affect the possible use of land application by limiting the loading rate and the types of vegetation which can be grown.

Fate of Wastewater Constituents. Since the purposes of land application of wastewaters are to further treat the wastewater and to return it to the envi-

TABLE 3-17 Characteristics of Land Application Procedures (Ref. 3-51).

	Irrigation	Overland Flow	Infiltration-Percolation
Liquid loading rate[a]	0.5 to 4 in./wk	2 to 5.5 in./wk	4 to 120 in./wk
Annual application	2 to 8 ft/yr	8 to 24 ft/yr	18 to 500 ft/yr
Land required for 1 MGD flow	140 to 560 acres plus buffer zones	46 to 140 acres plus buffer zones	2 to 62 acres plus buffer zones
Application techniques	Spray or surface	Usually spray	Usually surface
Soils	Moderately permeable soils with good productivity when irrigated	Slowly permeable soils such as clay loams and clay	Rapidly permeable soils such as sands, loamy sands, and sandy loams
Probability of influencing groundwater quality	Moderate	Slight	Certain
Needed depth to groundwater	About 5 ft	Undetermined	About 15 ft
Wastewater lost to:	Predominantly evaporation or deep percolation	Surface discharge dominates over evaporation and percolation	Percolation to groundwater

[a]Irrigation rates of 4 in./wk are usually seasonal; yearly maximum loads of 8 ft/yr would average about 2 in./wk.

ronment, it is pertinent to know the ultimate fate of the constituents in the wastewater. However, this subject is extremely complex, due to the variety of physical, chemical, and biological reactions which may occur (Ref. 3-49). Sullivan, *et al.* (Ref. 3-52) excellently detail this situation:

> "The fate of these substances during the process of land application will vary with the type of distribution system, the nature of the soil, the rate of application, the climate, the resting periods, and the location and proximity of the groundwater aquifer and the surface water source which receives runoff from the site. The phenomena involved include: the physical condition of entrapment or mechanical filtration; the biological, biochemical, electrochemical and other manifestations in, and in contact with the soil; evaporative factors; atmospheric oxidation; bacteriological, germicidal, and bacteriophage or anti-contamination reactions; and others which are not totally understood even by highly trained and experienced scientists."

In general, it can be said that suspended matter will become entrapped on soil particles or vegetation. It may then be decomposed or become part of the humus in the soil. Colloidal matter likewise may be adsorbed onto solid particles to become part of the soil matrix. Organic dissolved material may be utilized by the biota in the soil or chemically fixed in the soil. Inorganic plant nutrients (principally nitrogen and phosphorus) will be utilized to some degree by plant life, and the remainder may become fixed in the soil through ion exchange and chemical reaction. Dissolved salts, heavy metals, and other trace constituents will probably remain in the soil through adsorption and ion exchange. Bacteria and viruses will be destroyed by environmental changes and by competing or antagonistic microorganisms indigenous to the soil.

These general descriptions are, at best, only estimates of the fate of the wastewater constituents. Certainly they are not exhaustive.

A point to note is that the constituents retained in the soil may accumulate to intolerable levels, so that the land is unsuitable for further wastewater application. This may occur as the assimilative capacity of the soil is exhausted and vegetative growth is restricted by toxic levels of some compounds.

Furthermore, it is possible to leach out the adsorbed and retained constituents under certain conditions (e.g., lowering the pH of the wastewater), so that the potential for groundwater contamination does exist.

Clearly, land application is not the best answer to all the problems of wastewater treatment and disposal. However, land application is a potentially viable technique which is acceptable under the proper conditions of application.

TABLE 3-18 Treatment Efficiencies of Land Application (Ref. 3-51).

Objective	Irrigation	Overland Flow	Infiltration-Percolation
Use as a treatment process with a recovery of renovated water	Impractical	50–60% Recovery	Up to 90% Recovery
Use for treatment beyond secondary:			
1. For BOD and suspended solids removal	90–95%	90–99%	90–99%
2. For nitrogen removal	Up to 90%[a]	70–90%	0–80%
3. For phosphorus removal	80–99%	50–60%	70–95%
Use to grow crops for sale	Excellent	Fair	Poor
Use as direct recycle to the land	Complete	Partial	Complete
Use to recharge groundwater	0–30%	0–10%	Up to 90%
Use in cold climates	Fair[b]	Insuff. Data	Excellent

[a]Dependent upon crop intake.
[b]Conflicting data—woods irrigation acceptable, cropland irrigation marginal.

Treatment Efficiencies. The potential efficiences of wastewater treatment by land application are impressive. Table 3-18 presents the efficiencies of the three types of land application, and compares some of their pertinent operating characteristics. These efficiencies are based on data collected by Pound and Crites (Ref. 3-51).

All three techniques, irrigation, overland flow, and infiltration-percolation, are extremely efficient in BOD and suspended solid removals. They are moderately efficient in removal of nitrogen and phosphorus, with irrigation achieving the highest efficiency. Thus, with a single wastewater treatment step (land application) the major pollutants remaining in conventionally treated wastewater can be significantly reduced.

3-3.12 Disinfection

A thorough description of the basic principles of disinfection is beyond the scope of this chapter; therefore, the purposes of this section are to briefly outline the principles and agents of disinfection and to describe in detail the current disinfectant practices utilized in wastewater treatment. More complete descriptions of disinfection can be found in Frobisher (Ref. 3-53) and in standard environmental engineering references (Refs. 3-1, 3-9, 3-24).

The Purpose of Disinfection. Sterilization is the destruction or removal of all life, while disinfection refers to the destruction or removal of all organisms capable of causing infection (Ref. 3-53). The latter is the goal of disinfectant practices in water and wastewater treatment. In particular, the purpose of wastewater disinfection is to prevent the discharge of pathogenic organisms to the environment, thereby breaking the fecal-oral transmission route of infection.

Since the laboratory procedures for the detection of pathogens are very complicated, the efficiency and degree of disinfection of effluents is determined by measuring the concentration of indicator organisms–coliform bacteria. In general, local public health organizations and governments determine what coliform levels are acceptable in effluents and receiving waters. Ideally, effluents should be sterile, but achieving this would be extremely costly. Experience has shown that public health can be protected adequately by more reasonable standards, although a risk, albeit small, will still exist of disease transmission. Typical standards for receiving waters (and hence indirectly for effluents) range from 1,000 to 5,000 colonies per 100 ml (total coliform).

Destruction of Microorganisms. In order to insure the destruction of microorganisms, an irreversible change must be accomplished in their metabolism or physical structure which renders them nonviable. Any agent capable of causing this change may be considered a disinfectant or sterilant.

Most disinfection/sterilization agents act by one of three mechanisms to kill microorganisms: denaturation, surface tension reduction, or specific chemical toxicity. Denaturation is an irreversible change in the secondary and tertiary structures of proteins so that their functioning is impaired. For example, proteins in a cell wall may lose their strength and allow osmotic pressure to rupture the cell; or proteins essential to enzyme systems may lose their enzymatic properties, thereby halting an entire chemical process within the cell. In general, any agent which induces or causes denaturation is lethal to microorganisms.

Surface tension reduction is a microbicidal property of some chemical agents (surfactants). In solution, these agents adsorb onto the surface of the microorganisms, coating the cell wall and enzyme surfaces. Enzyme action may, therefore, be inhibited and cell wall strength impaired. In some instances, surfactants can even cause cell lysis by emulsifying portions of the cell wall and membrane.

Some chemical agents can act by the mechanism of specific toxic effects on cellular metabolism. These agents may react with substances within the cell, e.g., enzymes, ribosomes, RNA etc., and disrupt metabolism.

These three mechanisms are the prime means through which microorganisms are killed. The details of the exact reactions of the mechanisms or a cellular level are unknown (Ref. 3-53).

Disinfectants. There are four agents by which disinfection and sterilization can be accomplished: heat, chemical, radiation, and mechanical abrasion. The first three have been applied effectively for disinfection and sterilization. The fourth, mechanical abrasion, is of only academic interest at this time.

High temperatures are lethal to all life forms because the organic molecules which support life are extremely heat sensitive. Heat, by denaturing protein, destroys the ability of microorganisms to carry on metabolic processes. Heating is the most common sterilization technique used in laboratories, but it is impractical and too costly for wastewater treatment.

A variety of chemicals are effective disinfectants; phenols, alcohols, acids, bases, halogens, heavy metals and detergents. Table 3-19 lists the mechanisms responsible for the effectiveness of each chemical disinfectant. In many cases the disinfectant mechanisms of the chemicals are only partially understood.

Various types of radiation are effective disinfectants. Ultraviolet radiation (present in sunlight) may kill a cell, retard its growth, or cause a mutation which could lead to death. Ultraviolet energy is adsorbed by proteins and nucleic acids, which leads to the denaturation and disruption of the molecules and metabolic processes. Death or malfunctioning is the result. Gamma and X radiation are also microbicidal. These ionizing radiations cause free radical formation of molecules within the cell, which disrupt cell functioning and lead to death.

Of these three different agents which could be employed for disinfection of

TABLE 3-19 Agents and Mechanisms for Chemical Disinfection.

Agent	Mechanisms (in order of importance)
Phenol and Derivatives	1. Denaturation and coagulation 2. Specific toxic effects 3. Surfactant action
Alcohols	1. Surfactant action 2. Denaturation and coagulation
Acids and bases	1. Denaturation and coagulation
Halogens and halogen compounds	1. Denaturation and coagulation 2. Specific toxic effects
Heavy Metals	1. Denaturation and coagulation 2. Specific toxic effects
Detergents	1. Surfactant action

wastewater, only chemicals are routinely used in wastewater treatment. Heating is too expensive and impractical for large volumes, and radiation has not been proven effective in field applications. Chlorine is by far the most common chemical disinfectant and the traditional choice for wastewater treatment.

Removal of Microorganisms. Viable microorganisms may be removed from wastewater by other means than cellular destruction. Physical treatment processes may physically remove the microorganism by screening, enmeshment in precipitates, and ordinary sedimentation. The microorganisms are then destroyed in the treatment of the separated wastes. Biological treatment processes destroy microorganisms by predation from higher life forms, and by competition in the environment of the biological treatment process. Table 3-20 lists the percent removal or destruction of bacteria from wastewater by physical and biological treatment. The percent removals in this table are not cumulative,

TABLE 3-20 Removal or Destruction of Bacteria in Conventional Wastewater Treatment (Ref. 3-1).

Process	Percent Removal (noncumulative)
Coarse screens	0–5
Fine screens	10–20
Grit chambers	10–25
Plain sedimentation	25–75
Chemical precipitation	40–80
Trickling filters	90–95
Activated sludge	90–98

that is they represent the bacterial removal of each process singly, regardless of the entire wastewater treatment scheme.

The Practice of Chlorination. Chlorine gas in water hydrolyzes through the disproportionation process shown in Equation 3-15, and the resulting hypochlorous acid then partially dissociates to hypochlorite ion as shown in Equation 3-16.

$$Cl_2 + 2H_2O \rightleftharpoons H_3O^+ + HOCl + Cl^- \quad K = 3 \times 10^4 \text{ @ } 15°C \qquad \text{Eq. 3-15}$$

$$HOCl + H_2O \rightleftharpoons H_3O^+ + OCl^- \quad K = 2.2 \times 10^{-8} \text{ @ } 15°C \qquad \text{Eq. 3-16}$$

At normal pH ranges (6.8 to 7.2) the majority of chlorine in water is present as hypochlorous acid. Chlorine also may be added to solution as hypochlorite salts (e.g., $Ca(OCl)_2$, NaOCl). Hypochlorous acid is a strong disinfectant, hypochlorite ion is less effective by one or two orders of magnitude, and chloride ion has no disinfectant power.

Hypochlorous acid and hypochlorite ion are oxidizing agents and will react indiscriminantly with organic matter in solution. Microorganisms are not destroyed preferentially, but merely as a portion of the organic matter. Thus, the disinfection of wastewater containing a substantial amount of organic content requires more chlorine than wastewaters with low organic levels.

Ammonia is normally present in wastewater effluent at concentrations up to 15 mg/l. Hypochlorous acid reacts with ammonia to form chloramines, as indicated in Equations 3-17, 3-18, and 3-19. These equations are representative of the overall process, but the actual reactions are not known.

$$HOCl + NH_3 \rightleftharpoons NH_2Cl + H_2O \qquad \text{(Eq. 3-17)}$$

$$HOCl + NH_2Cl \rightleftharpoons NHCl_2 + H_2O \qquad \text{(Eq. 3-18)}$$

$$HOCl + NHCl \rightleftharpoons NCl_3 + H_2O \qquad \text{(Eq. 3-19)}$$

In the measurement of chlorine forms in water, the chloroamines are referred to as the combined chlorine, while the hypochlorous and hypochlorite forms are the free chlorine.

Since the equilibria of Equations 3-17, 3-18, and 3-19 lie far to the right, chloroamines are stable in water. However, as hypochlorous acid and hypochlorite ion are consumed in oxidizing organic matter, combined chlorine hydrolyzes to free chlorine in order to maintain equilibrium. It is postulated that the microbicidal effects of combined chlorine (which is a less effective disinfectant than free chlorine) are due to the liberation of free chlorine rather than being an inherent property of combined chlorine (Ref. 3-24). Therefore, although free chlorine is consumed by ammonia in water, it is later released as the concentration of free chlorine is reduced. The disadvantage of producing combined chlorine by reaction with ammonia is that with less free chlorine in the wastewater, disinfection is less efficient and pathogenic microorganisms may

TABLE 3-21 Typical Chlorine Dosages for Disinfection (Ref. 3-1).

Effluent from	Dosage in mg/l of Chlorine in the Effluent
Primary sedimentation	5–20
Chemical precipitation plant	2–6
Trickling filter plant	3–15
Activated sludge plant	2–8
Multimedia filter following activated sludge	1–5

survive the disinfection process. Addition of sufficient chlorine to water in the presence of ammonia will oxidize the ammonia to nitrogen gas, a process known as breakpoint chlorination.

The effectiveness of chlorine as a disinfectant is dependent on the chlorine concentration in the water, the time of contact with the water, and the temperature and pH levels. Data has been presented in the literature showing the kill of *E. coli* at specified test conditions (Ref. 3-9), but conventional treatment generally follows rules of thumb based on past experiences. Table 3-21 lists typical chlorine dosages for various wastewater effluents. Chlorine is injected into wastewaters in a chlorine contact tank as a gas or as a hypochlorite solution. Detention times in the chlorine contact chambers range from 15 to 20 minutes. Chlorine dosages are usually predicated on experience or may be determined by trial and error through measurements of coliform bacteria in the effluent. State and local standards regulating the concentration of the chlorine or coliform bacteria effluents or receiving water are utilized as guidelines for disinfection. Conventional biological treatment with disinfection may be expected to achieve in excess of 99.99 percent destruction of bacteria.

Disinfection with Ozone. The properties and potential of ozonation as a wastewater treatment process have been discussed in Section 3-3.10. It was noted that ozone is a much more effective disinfectant than chlorine, and that several applications of ozonation have been made in the U.S.

As a disinfectant it is believed that ozone may function similarly to chlorine-protein denaturation and enzyme inactivation through oxidative reactions. It has also been suggested that ozone decomposition products (free radicals) may play an important part in disinfection (Ref. 3-24).

3-3.13 Lime Coagulation Processes

Lime addition to wastewater may be practiced in conjunction with three wastewater treatment processes: suspended solids removal, phosphate removal, and pH adjustment for ammonia stripping. There are two basic procedures which are employed—single stage and double stage.

Single Stage. In the single stage process, sufficient lime is added to elevate the wastewater pH to about 9.5. At this pH level, all the bicarbonate alkalinity is converted to the carbonate form and reacts with calcium to produce insoluble calcium carbonate, which is separated from the wastewater in a clarifier. Single stage lime clarification can remove about 85 percent of the suspended solids present in secondary effluent. The alkalinity of the wastewater prior to lime addition determines the efficiency of the phosphate removal process. High alkalinities (greater than 150 mg/l) will insure the formation of sufficient calcium carbonate to enmesh and effectively separate the hydroxyapatite from the liquid phase. Waters of low alkalinity will produce too little calcium carbonate to effectively flocculate the hydroxyapatite, which then remains in colloidal dispersion. Ammonia removal by air stripping at pH 9.5 is inefficient, since approximately 40 percent of the ammonia nitrogen exists as ammonium ion at this pH level.

Double Stage. Increased removals of phosphorous and ammonia can be accomplished with a two-stage lime clarification process. In the first stage of the two-stage process, sufficient lime is added to raise the pH to about 11.0 or 11.5. At this pH level, insoluble magnesium hydroxide will form (if magnesium is present in the water). Consequently, a combined floc consisting of calcium carbonate, magnesium hydroxide, hydroxapatite, and suspended solids will settle out of solution. Since most of the ammonia nitrogen exists in the ammonia form at this elevated pH level, stripping can be successfully utilized to remove the ammonia nitrogen. Following lime addition in the first stage of the two-stage process, the water possesses a high pH (over 11) and calcium ions in excess of those necessary to react with the carbonate alkalinity. Carbon dioxide dissolving into the water at this elevated pH level will form hydrogen ions and carbonate, which react with the excess calcium to form insoluble calcium carbonate. This precipitate may be deposited as scale in stripping towers or on valves and pipes in the wastewater treatment system. To avoid the latter problem, and to reduce the pH of the wastewater. The second stage of the two-stage process involves the addition of carbon dioxide to the water (recarbonization) and the precipitation of calcium carbonate. When ammonia stripping is combined with two stage lime clarification, the stripping process is located between the first and second stages. A filtration process is frequently added, after both one and two stage lime clarification systems, to insure complete removal of calcium carbonate and suspended solids.

3-3.14 Solids Treatment and Disposal

In Section 3-2, it was noted that water is never pure in nature and always carries some constituents. If the concentration of any constituent is excessive, then the water may be considered polluted; i.e., it is degraded and unsuitable for some purposes. One purpose of wastewater treatment is to remove the pollutants so that the wastewater effluent will not pollute the receiving water.

The previous sections of this chapter have detailed the techniques available for wastewater treatment. Most of these techniques ultimately produce a solid waste which must be removed from the wastewater. The purified wastewater is treated further or discharged, and the solid waste is processed. This section will describe the techniques for the treatment and disposal of the solids produced during wastewater treatment.

It is commonly assumed that solids treatment encompasses one fourth to one half the total cost of wastewater treatment. The importance of solids treatment cannot be overrated since it is counterproductive to remove pollutant wastes from water only to release them on the land or in the air.

Sources of Solids in Wastewater Treatment. Solids are produced at several locations in a typical domestic wastewater treatment plant (Ref. 3-33). Solids are initially removed by screening at the plant influent. These solids include gross objects such as branches, rags, bottles, toys, cans, and other objects which can be retained on a bar screen with openings of two to four inches. This matter is usually collected in bins and removed by a solid waste collection service. Normally about two to four cubic feet of this material is collected per million gallons of wastewater screened.

The next source of solid matter in a typical domestic wastewater treatment plant is the grit chamber, where inorganic particles such as sand, gravel, and rock chips settle out. In essence, the grit chamber is a small sedimentation basin with a very short detention time (one to two minutes). This is sufficient time for the grit to settle out from the waste flow and collect on the bottom of the chamber. The grit has little or no pollution load per se, but if not collected, it will settle in other treatment units, partially filling them and causing nuisance problems in operation and maintenance. Grit removed from the waste stream is usually washed clean of organic material and then buried. Approximately two cubic feet of grit will be produced per million gallons of typical domestic wastewater. This volume may vary widely as a function of the type and condition of the wastewater collection system, the soil characteristics, and the prevalence of household garbage grinders in the service area (Ref. 3-33).

The third source of solids is the primary settling tank, where skimmings and sludge are removed from the tank surface and bottom, respectively. The skimmings (matter with a specific gravity less than water) consist of wood, plastics, grease, garbage, and rubber products. These are normally collected by a mechanical skimmer and added to the sludge, which is withdrawn from the bottom of the primary settling tank. The volume of skimmings varies widely, based on the living characteristics of the community served, and may range from less than one to as much as seven cubic feet per million gallons of flow.

The solids matter from the bottom of the primary settling tank consists of the settleable solids fraction of the total solids carried by the wastewater. This sludge is primarily organic matter. It is usually grey and slimy in appearance, with a strongly disagreeable odor. It putrefies readily. Approximately 390 cubic feet of primary sludge (2,950 gallons at 5 percent solids) per million gallons of domestic wastewater is collected in a typical primary sedimentation process. The solids content of this sludge may vary from two to six percent (Ref. 3-1).

The last solids' source in a typical activated sludge treatment plant is the waste activated sludge. This represents the dissolved organic matter which has been assimilated by bacteria and metabolized to cell mass. This sludge is normally brown and fluffy with a characteristic but not offensive odor. This sludge is hydrophilic and difficult to dewater. Typically, 2,500 cubic feet of sludge (19,400 gallons at 1.5 percent solids) are produced per million gallons of domestic wastewater, with a typical solids content of less than two percent (Ref. 3-1).

Advanced wastewater treatment techniques may also produce solids which must be treated. Suspended solids removed by filtration and screening are usually returned to the influent stream, where they are more amenable to removal. Chemical sludges may require special disposal techniques, for example, lime sludges may be buried or recalcined to recover the lime.

Primary and secondary sludges must be treated and disposed of by special techniques. A common procedure is to mix primary and secondary sludges to produce a combined sludge.

Ultimate Disposal. After treatment, all solids and sludges must ultimately be returned to either the air, sea, or land. The air receives gases and often ashes from incineration (air pollution codes require strict attention to potential pollutant discharges). The oceans are sometimes utilized for sludge disposal by coastal communities. However, it is likely that this practice will be halted in the near future (Ref. 3-49). In general, land burial is the most common ultimate disposal method utilized for sludges.

As with wastewater treatment, the purposes of sludge treatment and disposal are to protect public health by destroying pathogens, and to preserve environment quality by properly treating the wastes to prevent pollution. Therefore, a prime purpose in sludge treatment is to "stabilize" the sludge. As the term implies, a stabilized waste sludge is one in which the organic material has been oxidized or decomposed so that it will not undergo further autodecomposition after disposal, which might create noxious and polluting conditions.

Sludge Processing Techniques. A variety of physical, chemical, and biological techniques are applicable to sludge treatment. Figure 3-12 presents a general flowsheet of sludge treatment techniques in approximately the order in which they could be applied in a treatment scheme. The paragraphs which follow de-

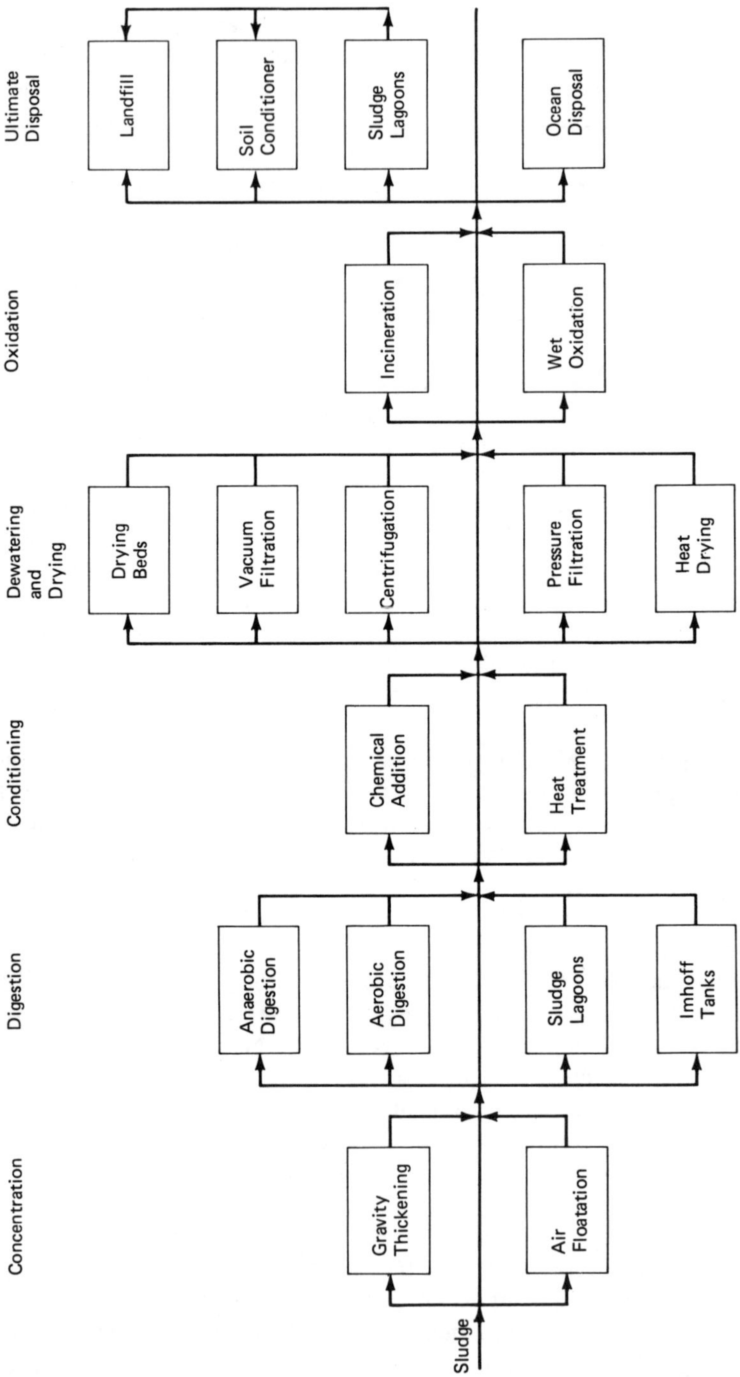

Fig. 3-12. Generalized flowsheet of sludge treatment techniques.

scribe each step in the flowsheet and comment on the techniques within each step. This description will not attempt to detail design parameters, but will concentrate on the qualitative aspects of the techniques.

Concentration. The purpose of concentration techniques is to remove water from the sludge so that the sludge can be handled more efficiently. The extracted water is usually returned to the plant influent. Gravity thickeners are settling tanks in which the sludge from primary and/or secondary clarifiers is allowed to settle. Mechanical arms slowly stir the sludge to gently break up the sludge matrix and allow water to separate from the floc, thereby concentrating the sludge. Primary sludge can be thickened to 8 percent to 10 percent solids, whereas waste-activated sludge can be thickened to 2.5 percent to 3.5 percent and solids and combined sludges to 5 percent to 9 percent solids (Ref. 3-1).

Flotation thickeners are used primarily with secondary sludge, whose specific gravity is very close to that of water. Air is dissolved into the sludge mass under high pressure. When the pressure is released, the air comes out of solution as small bubbles, which collect on the sludge particles and cause them to float. The floating sludge is then collected by skimmers. A sludge with approximately 4 percent solids content can be obtained by this technique.

Digestion. Digestion is a stabilization technique in which the organic matter is digested by bacteria in anaerobic or aerobic aconditions. Properly digested sludge is brown to black in color, with an inoffensive color. It will not further decompose. Most pathogens are destroyed by predation and the environmental conditions in the digestor; however, digested sludge should not be considered totally disinfected.

Anaerobic digestion is usually performed in deep, circular, covered tanks, which may or may not be heated. Under anaerobic conditions, two types of bacteria grow and digest the sludge organic content—acid producers and methane producers. The acid bacteria metabolize organics and produce organic acids as an end product, whereas the methane bacteria metabolize the organic acids and produce methane gas. Thus, the organic material in the sludge is converted primarily to bacterial cell mass and gas. The digestor gas contains about 70 percent methane and the remainder is primarily carbon dioxide and water vapor with occasional traces of hydrogen sulfide, nitrogen, and hydrogen. Gas produced in the digestion process is collected and may be burned freely or compressed and utilized for fuel either at the treatment plant or commercially. Digested sludge is suitable for drying and disposal, whereas supernatant, produced in the digestion process, is usually returned to the plant influent.

Conventional anaerobic digestion may be carried out in one or two stages (high rate digestion). Single stage digestion utilizes one unmixed tank in which digestion and settling take place. The bottom third of the tank is filled with digested solids. The middle third contains the actively digesting solids, and the top is filled with the gas produced by digestion. Between the middle and top thirds is the supernatant layer. Two-stage digestion entails a completely mixed digestion tank prior to the separation tank. Digestion takes place primarily in the first tank, whose contents are mixed to increase the digestion rate. The contents of this first tank are pumped to the second tank where the digested sludge separates into layers of digested solids, supernatant, and gas. Digestor design may be based on the mean cell residence time, loading factors, or population.

In the past, aerobic digestion was rarely used for sludge stabilization, although it has found more application recently, particularly in small plants where there are fewer personnel to properly maintain an anaerobic digestor. The aerobic digestor is basically a tank, large enough to provide up to twenty days detention time for the sludge (into which oxygen is introduced with blowers or mixers). The sludge is digested aerobically by bacteria contained in the sludge. Aerobic digestion is reported to have the following advantages over anaerobic digestion: lower BOD in supernatant; sludge that dewaters better; sludge with more fertilizer value; fewer operational problems; and a lower capital cost. Disadvantages include a higher power cost for supplying oxygen, and the loss of a combustible gas by-product. Experiences with aerobic digestion are limited at the present time, although more applications of the aerobic digestion technique are likely in the future (Refs. 3-1, 3-9).

Sludge lagoons are earthen holding basins which receive sludge flow and act as large shallow digestors. No provisions are made for gas recovery or continuous removal of digested sludge. Eventually, the lagoon is filled with sludge and must be abandoned or excavated for reuse. Large land areas are required for sludge lagoons and odor problems may occur frequently.

An Imhoff tank is an anaerobic digestor suitable for small installations. It consists of a two-story tank in which primary sedimentation occurs in the upper portion and sludge digestion and storage in the bottom portion. The interior design includes baffles to separate the sludge from the supernatant and vents to release the gas produced during the digestion. Although widely used in the past, it is less popular now because of the superiority of other digestion techniques. Small treatment installations may still include Imhoff tanks, since these tanks require no skilled operators and are easily maintained.

Due to their biological nature, all digestion processes are affected by temperature. Usually it is recommended that provision be made in digestor design for heating, in order to provide optimal temperatures for biological activity (90 to 100°F).

Conditioning. Sludge is conditioned to improve its dewatering characteristics. The most popular technique involves chemical addition, in which coagulant chemicals are added to the sludge just prior to dewatering. These chemicals, including ferric chloride, lime, alum, and polymers, coagulate the solids and encourage the release of absorbed water. Their use can be economically justified by increased yields and greater flexibility in the dewatering process.

Heat treatment also conditions sludge by subjecting it to heat under pressure for short time periods. As a result solids coagulate, and the gel structure of sludge breaks down so that water is released. At the same time, the sludge is partially oxidized and completely sterilized. Following heat treatment, sludge can be dewatered to 50 percent solids in some cases; this is a much higher solids level than can be achieved by chemical conditioning. Disadvantages include high capital costs and cost of heating the sludge.

Other conditioning techniques such as freezing and irradiation have been demonstrated to be effective on a laboratory basis, but are uneconomical for full scale application at the present time.

Dewatering and Drying. The purpose of these sludge dewatering and drying techniques is to remove water from the sludge so that it can be efficiently treated or disposed. Drying beds consist of prepared earthen beds, usually filled with a sand bottom and serviced by drains, on which the sludge is poured in an 8 to 12 inch layer. After the sludge drains and dries on the beds, it is mechanically collected and disposed. Drying beds are used primarily for digested sludge.

Vacuum filters may be used to continuously dewater raw or digested sludge, with best results achieved when sludges are conditioned prior to filtration. The vacuum filter consists of a drum partially submerged in a cylindrical vat of sludge. The filter media, which may consist of cloth, coil springs, or wire mesh, is wrapped around the drum. The entire circumference is subjected to an internal vacuum, except for a small portion where the cake is removed from the drum. When the drum rotates, the submerged circumference collects a sludge cake on the filter medium as the filtrate flows through the medium in response to the vacuum. The sludge cake is further dewatered as it rotates out of the sludge vat. At the point in the circumference where there is no vacuum, a doctor blade or similar device removes the sludge cake. The sludge cake solids content may approach 30 percent.

Although centrifugation is widely utilized in the chemical industry to thicken slurries and remove solids, it has only recently been applied to dewater domestic waste sludges. The most common centrifuge for sludge dewatering is the solid-bowl design. Sludge is fed at a constant rate into the centrifuge where it is separated into a dense cake, which contains from 15 to 40 percent solids, and a liquid centrate. The cake is extracted by a screw feeder and the centrate is recycled to the plant influent. Centrifugation is simple in concept and operation and does not require prior sludge conditioning. One disadvantage of centrifugation is the high power costs for operation.

Pressure filtration utilizes filter presses to subject sludges to pressures up to 180 psi for periods as long as three hours. In the pressure filtration process, liquid is forced through the filter medium by pressure, and then the sludge cake is removed from the pressure plate following pressure release. Sludge cake solids content may range from 30 percent to 45 percent under optimal conditions, but prior conditioning is required. Maintenance and filter medium replacement costs are the most significant operating costs associated with pressure filtration. Very little experience has ben acquired in filter pressing wastewater sludges in the U.S.

Heat drying may be utilized in applications where the sludge is to be incinerated or bagged for sale as fertilizer. Multiple hearth driers, rotary kiln driers, and flash driers have been used successfully to dry sludges. The final solids content following heat drying is usually greater than 90 percent. A major problem associated with sludge drying involves the control of gases and ash particles which may be generated in the drying process.

Oxidation. Sludges may be oxidized to reduce organic content, with the consequent destruction of bacteria and a significant volume reduction. Incineration and wet oxidation are the two processes through which sludges may be oxidized. Incineration is most commonly performed in a multiple hearth furnace, although fluidized beds and flash dryers may also be utilized. In any case, the exit gases must be scrubbed to remove ashes and meet air pollution standards. With adequate dewatering (usually to about 30 percent solids) undigested sludges will sustain combustion so that supplementary fuel is required only for start up. Non-dewatered sludges are not commonly incinerated because of the energy demand required to drive off the extra water.

Wet oxidation is a process in which sludge is ground, mixed with a specified quantity of air, and then subjected to high temperature and pressure in a reactor vessel. The sludge oxidizes under these conditions to produce a mixture of gas, liquid, and ash exiting the reactor. Gases are vented from the process, while the ashes produced are settled and disposed, and the liquid is returned to the plant influent. Wet oxidation also may be thermally self-sufficient if undigested sludge is used.

Ultimate Disposal. Following stabilization, disinfection, and dewatering, sludge must be ultimately disposed. The purpose of the sludge treatment process is to prepare sludge for disposal. Some of the organic material may be dispersed to the atmosphere as gas and products of oxidation in the digestion and oxidation steps. The remainder, mainly inorganics and inert ashes, must be deposited on land or at sea.

Ocean disposal is practiced by either pipelining the sludge from shore and dispersing the sludge along the ocean bottom, or by barging the sludge far offshore and discharging it to the ocean surface. Many questions have been raised concerning the potentially detrimental effects of ocean disposal; however, currently there are no clear answers. It is likely that ocean disposal will not be allowed in the future (Ref. 3-54).

There are several suitable land disposal techniques currently available. Dried sludge or ashes may be placed in a landfill for burial. Wet or dry digested sludge may be applied to land and plowed in to act as a soil conditioner and provide nutrients. Wet digested sludge also may be sprayed onto cropland where it functions as irrigation water and fertilizer (Ref. 3-55). Finally, wet sludge may be placed in lagoons; however, when the lagoons fill they must be abandoned. Although sludges are sometimes recommended for their fertilizer value, they actually are very poor fertilizers–their chief value being as soil conditioners. However, sludges can be utilized as a base or filler for fertilizer preparations (Ref. 3-9).

Typical Sludge Treatment Schemes. The sludge treatment techniques previously described may be combined to produce complete sludge treatment schemes. Figure 3-13 and 3-14 present two common sludge treatment schemes.

Figure 3-13 outlines the most common sludge treatment scheme in the U.S.–single stage anaerobic digestion followed by sand bed application and landfilling. In this scheme, the combined sludge from the primary and secondary clarifier is pumped directly, without thickening, to an anaerobic digestor. Following diges-

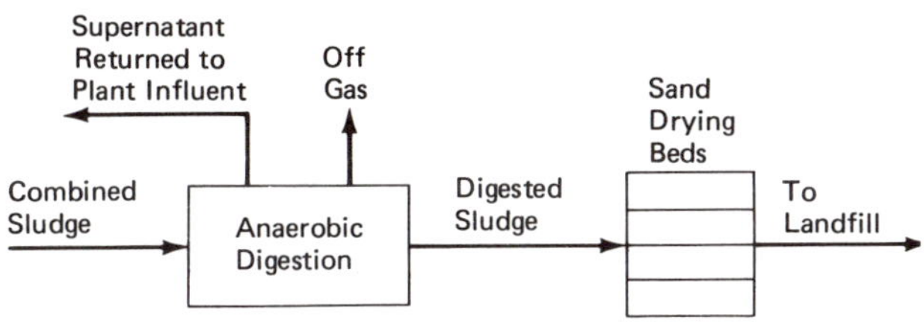

Fig. 3-13. Typical sludge digestion process.

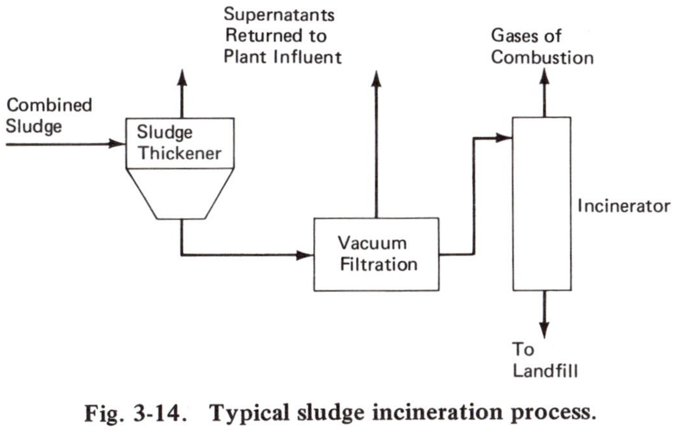

Fig. 3-14. Typical sludge incineration process.

tion, the sludge is dried on sand beds and then transported to a landfill for disposal.

Figure 3-14 depicts the most common incineration scheme currently utilized in the U.S. for sludge disposal. Waste sludge, with or without thickening, is conditioned chemically, vacuum filtered, and incinerated in a multiple hearth furnace. The ashes produced during the incineration are deposited in a landfill.

At present there is a great deal of interest in and research on sludge treatment and disposal. It is likely that the future will see many new techniques and approaches to sludge treatment and disposal.

3-4 WASTEWATER TREATMENT SCHEMES

This section will describe how currently available wastewater treatment techniques can be merged to provide complete treatment schemes. The schemes developed by the authors are not exhaustive, but are an attempt to outline the most feasible and practical treatment alternatives available to meet desired effluent quality levels.

The generation of wastewater treatment schemes is purposely general since it is not possible to consider all the factors that must be considered to select a specific treatment plan at a particular location. Wherever appropriate, additional factors which impact treatment scheme plan selection will be mentioned, however, no attempt will be made to include all these factors in treatment scheme development, since their inclusion would impose limits and reduce the generality of this approach.

The prime factor in the selection of treatment schemes is the overall effluent quality which is to be produced by the facility. Based on the desired effluent quality, certain treatment alternatives can be eliminated immediately as either being insufficient to meet these goals, or too efficient and achieving an effluent quality higher than required. These latter alternatives are not viable since higher effluent quality usually implies higher cost and a treatment scheme that is not cost effective.

Section 3-4.1 considers various factors which determine the required degree of wastewater treatment and effluent quality. Section 3-4.2 presents classifications of feasible effluent quality levels. Finally, Section 3-4.3 describes various treatment schemes which achieve the different effluent quality levels identified in Section 3-4.2.

3-4.1 Effluent Quality Planning Factors

Since the purposes of wastewater treatment are to protect human health and preserve environmental quality, it is mandatory that wastewater treatment plant effluents possess a high quality. However, this is not to say that all effluent discharges must be of the same quality. There are many factors specific to particular locations which may dictate that a higher or lower effluent quality is acceptable.

Governmental Factors. In the 1972 Water Pollution Control Act Amendments, the federal government promulgated standards pertaining to the quality of all effluents discharged into interstate waters (Ref. 3-50). The amendments affect both industrial and municipal point effluent discharges. With respect to municipalities, all municipal wastewater treatment plants must produce secondary quality effluent by July 1, 1977. In addition, the municipal plants are required to utilize the "best practicable control technology" (BPCT) by July 1, 1983. The Act is enforced by the National Pollutant Discharge Elimination Permit System (NPDES) which only allows discharges that meet the requirements of the Act. The U.S. Environmental Protection Agency was authorized by Congress to supervise and regulate the water pollution control activities of the Act (Ref. 3-65, 3-57).

Secondary treatment standards were defined by the EPA in the Federal Register on August 17, 1973, and include maximum acceptable thirty day mean values of BOD_5 (30 mg/l), suspended solids (30 mg/l), fecal coliform (200 colonies/100 ml), and pH (6.0 to 9.0) (Ref. 3-58). These effluent criteria are based on a consideration of the minimum dissolved oxygen level which is required to maintain aquatic biota, and the maximum fecal coliform concentration permissible in primary contact recreation waters. The NPDES requires that all effluents meet these standards and, where necessary, even higher quality in order to preserve the receiving water quality. In particular, effluents must possess such a quality that the level of dissolved oxygen in the receiving water does not fall below the minimum allowable concentration (5 mg/l), usually during a ten year–seven day low flow period (Q10-7). Thus, the allowable effluent quality may vary at different treatment facilities based on the assimilative capacity and flow of the receiving water, but in all cases, must meet at least secondary effluent criteria.

A formal definition of BPCT has not yet been published by the EPA, but BPCT is generally accepted to define the most cost effective treatment alternative to achieve the required effluent quality. Thus, the BPCT sets no minimum standards of quality, but rather requires an analysis of the various treatment alternatives to determine which is the most efficient with respect to economy as well as social and other intangible factors (Ref. 3-60).

Aquatic Factors. The capacity of a receiving water to assimilate wastes is a major factor in the determination of effluent quality guidelines. The size, flow velocity, quality, and location of receiving waters are some of the factors which must be considered in assimilative capacity determinations. Waters which are reaerated rapidly can absorb far more pollutants without violating stream water quality standards for water bodies having minimal reaeration capacity. Similarly, water bodies having a high volume of flow are capable of accepting a higher pollutant load than low flow receiving waters.

The determination of receiving water assimilative capacity is a difficult task. Normally, computers are employed in conjunction with sophisticated mathematical models to simulate the effect of effluent discharges on stream water quality. In this manner, the maximum allowable pollutant content of a given effluent that will result in acceptable receiving water quality can be predicted (see Chapter 4).

Discharge Flow and Facility Location. Other factors to be considered in determining effluent quality levels include discharge flow and treatment facility location. The relation of discharge flow to the receiving water flow determines to some extent the impact the wastewater effluent will exert on the receiving body and consequently, future plans to increase treatment facility size to meet population increases should consider how this increased flow will affect stream quality. It is entirely possible that a large discharge in lieu of a smaller discharge at the same location may have to meet more stringent effluent criteria so that the assimilative capacity of the receiving water will not be surpassed.

Facility location is also a pertinent factor in water quality management planning. A facility located near to other discharges will usually be required to meet more stringent effluent criteria than if there were no nearby discharges, since only a portion of the stream assimilative capacity is allocated to each discharger in a stream segment having multiple discharges.

Areawide wastewater treatment planning requires a more sophisticated development of effluent quality and treatment schemes than plans for only one specific plant. The effect of the effluent quality from each facility on the area's water quality must be considered in conjunction with all other plants in the area. Consequently, the effluent quality levels required of each plant may be regulated in part by other facilities in the planning area.

Public Input Factors. Although the public is represented by federal and state agencies in pollution control activities, the EPA requires consultants and planners of wastewater treatment facilities and regional plans to include the public-at-large and individual citizens in the water quality management planning process. This allows the public to have a direct influence on pollution control decisions.

3-4.2 Potential Effluent Quality Levels

Consideration of various water quality management planning factors results in a specific effluent quality, which must be met by a wastewater treatment facility. The required effluent quality may range from conventional secondary effluent to drinking water quality (although the EPA does not encourage direct water reuse).

The authors have classified effluent qualities into eight levels, ranging from secondary effluent to potentially potable water. The quality levels are based on effluent of three constituents: organic matter (BOD_5), ammonia nitrogen, and phosphorus. It is assumed that the wastewater is typical domestic wastewater. Table 3-22 displays how these classifications are distinguished and, for purposes of nomenclature, each quality effluent is referenced by a letter. It should be noted that phosphorus removal is noted only as yes or no and not as a percent, since degrees of phosphorus removal are not usually controlled precisely. However, there is some variation in the amount of phosphorus remaining in the effluent due to process variations and the operation of the phosphorus removal techniques. Similarly, nitrogen treatment is noted as either nitrification or nitrogen removal through denitrification or ammonia stripping.

The A quality effluent is basically conventional activated sludge process effluent. It is acknowledged that proper operation of the activated sludge process may result in BOD_5 removals to above 90 percent, but typical operation and influent variations usually yield removals ranging from 85 to 90 percent. Some phosphorus and nitrogen are removed in the activated sludge process; however, these removals are typically less than 40 percent. Therefore, the nitrogen and phosphorus concentrations contained in conventional activated sludge effluent may be considered a baseline or minimum against which advanced treatment schemes may be evaluated.

Effluents of quality B through E basically represent activated sludge effluent quality with minor modifications. B quality effluent represents the extended aeration modification of the activated sludge process in which nitrification is accomplished. This effluent quality might be appropriate where a reduction in the ammonia oxygen demand is required, but eutrophication by nitrate-nitrogen is not a consideration. C quality effluent represents increased BOD_5 removal, but

TABLE 3-22 Classification of Effluent Quality.[a]

	Degree of Pollutant Removal[b]		
Class	BOD_5 (%)	Phosphorus	Nitrogen
A	85–90	—	—
B	85–90	No	Nitrification
C	95	No	No
D	95	Yes	No
E	95	No	Nitrification
F	95	Yes	Nitrification
G	95	Yes	Yes
H	99	Yes	Yes

[a]As defined by the authors.
[b]Based on typical domestic wastewater as described in Table 3-4.

there is no additional nitrogen or phosphorus treatment other than that accomplished by the activated sludge biomass. D effluent quality defines both increased BOD_5 and phosphorus removals but no nitrogen removal. E quality effluent represents both nitrification and increased BOD_5 removal, but no additional phosphorus removal.

The remaining three effluent qualities, F through H, represent extremely high quality levels. Effluents of F quality might be necessary where BOD_5 removal is required, in addition to phosphorus removal and nitrification. G quality is similar to E quality except that it also includes denitrification. G quality effluent might be required where eutrophication from both phosphorus and nitrogen was a consideration.

Finally, H quality effluent represents extremely efficient treatment schemes in which virtually all nutrients and all organic matter are removed. This quality effluent might be appropriate as input to a water reuse treatment process.

It is assumed in all these effluent quality levels that bacteria and viruses are reduced to the lowest levels possible with conventional disinfection techniques. Water destined for reuse (H quality) might be expected to undergo a more thorough disinfection process after discharge from the wastewater treatment facility.

3-4.3 Development of Alternative Wastewater Treatment Schemes

This section presents alternative wastewater treatment schemes to meet each of the effluent quality levels, A through H. The treatment techniques included in these schemes are proven techniques, fully consonant with current water quality control technology. In the development of treatment schemes, specific effort has been focused on the elimination of extremely advanced and relatively unproven techniques, although in special situations it is recognized that it may be appropriate to include more advanced techniques. Of course, the treatment alternatives available at any one location are a function of local variables such as land availability, climate, funding, and technical expertise.

Presentation of Schemes. The wastewater treatment schemes devised by the authors to achieve specific effluent quality levels are presented in Table 3-23. These alternatives may be considered the most feasible applications of current wastewater treatment technology. Table 3-23 also includes effluent characterization estimates for the various treatment schemes listed. These estimates have been developed from data presented in the literature which depict treatment scheme operation. Variations in influent flow and constituent concentrations in conjunction with facility operation will significantly affect effluent quality levels. The estimates of operational efficiency, presented in Table 3-23, are intended to indicate the relationships among effluents from different treatment schemes and the pollutant concentrations which might be expected from each treatment scheme, operating under normal loads and at normal efficiencies.

TABLE 3-23 Alternative Wastewater Treatment Schemes.

Effluent Quality	BOD_5	P	N	Alternative Treatment Schemes	Effluent Constituents (mg/l)					
					SS	BOD_5	COD	NH_3	P	UOD
—	—	—	—	Typical Domestic Wastewater	230	200	350	18	13	381
A	85–90%	—	—	1. AS + CHL	20	20	50	15	10	99
B	85–90%	No	NTFN	1. AS + BN + CHL	15	15	40	1	7	27
				2. EA + CHL	20	20	50	1	7	35
C	95%	No	No	1. AS + MMF + CHL	3	8	25	15	10	81
D	95%	Yes	No	1. AS + AL + MMF + CHL	3	8	25	15	1	81
				2. AS + LC1 + MMF + CHL	3	3	23	15	0.2	74
				3. LC1 + MMF + AC + CHL	3	2	5	15	0.2	71
E	95%	No	NTFN	1. AS + BN + MMF + CHL	3	5	20	1	7	12
				2. EA + MMF + CHL	3	8	25	1	7	17
F	95%	Yes	NTFN	1. AS + BN + AL + MMF + CHL	3	5	20	1	1	12
				2. EA + AL + MMF + CHL	3	8	25	1	1	17
G	95%	Yes	Yes	1. AS + BND + AL + CHL	8	8	25	1	1	17
				2. AS + LC2 + AST + MMF + CHL	3	3	22	2	0.2	14
				3. LC2 + AST + MMF + AC + CHL	2	5	8	2	0.2	17
H	99%	Yes	Yes	1. AS + BND + AL + MMF + AC + CHL	1	1	5	1	1	6
				2. AS + LC2 + AST + MMF + AC + CHL	1	1	5	2	0.2	11
				3. LC2 + AST + MMF + AC + CHL	1	2	5	2	0.2	12

Abbreviations:
AC–Activated Carbon.
AL–Alum in Act. Sludge.
AS–Activated Sludge.
AST–Ammonia Stripping.
BN–Biological Nitrification.
BND–Biological Nitrification and Denitrification.
CHL–Chlorination.
EA–Extended Aeration.
LC1–Lime Coagulation–1 stage.
LC2–Lime Coagulation–2 stage.
MMF–Mixed Media Filtration.
NTFN–Nitrification.

Alternatives–Effluent Quality A. A quality effluent is typically produced by the conventional activated sludge process. This quality is the minimum acceptable according to current EPA regulations, and is the baseline against which other effluents can be measured. The UOD of conventional activated sludge effluent is approximately 99 mg/l, of which 69 mg/l is nitrogenous oxygen demand.

Alternatives–Effluent Quality B. Two biological treatment schemes will produce B quality effluent, which essentially is A quality effluent with nitrification of the effluent ammonia nitrogen. The first scheme utilized the conventional activated sludge process with the addition of biological nitrification in a separate aeration tank. Separate nitrification has the advantage of also reducing the effluent's carbonaceous oxygen demand. The second biological scheme produces a nitrified effluent through the extended aeration modification of the activated sludge process.

In both processes the phosphorus level in the effluent is reduced to some extent as a result of the growth of nitrifying bacteria, but not enough to be significant. The effluent UOD levels of the two-stage nitrification and extended aeration systems are estimated to be 27 and 35 mg/l, respectively. It should be noted that nitrification of effluent ammonia nitrogen concentrations results in a large decrease in the effluent UOD.

Alternatives–Effluent Quality C. The only difference in this effluent quality and conventional activated sludge effluent quality is an increase in organic carbon (BOD_5) removal, which is most efficiently accomplished by adding a suspended solids removal step. Mixed media filtration was selected for suspended solids removal rather than microstraining, because of its inherent reliability and efficiency. However, microstraining may be selected as more suitable where space is limited or flow variations are small.

Since only the effluent carbonaceous oxygen demand is reduced during the mixed media filtration process, the effluent UOD remains high (81 mg/l) due to the remaining ammonia nitrogen concentration.

Alternatives–Effluent Quality D. There are two biological and one physical-chemical treatment schemes that will provide effluent quality D, which necessitates increased organic carbon (BOD_5) and phosphorus removals. The first biological scheme involves the addition of alum, either to the aeration tank or to the effluent from the aeration tank of the activated sludge process, in order to remove phosphorus. This is followed by mixed media filtration of the secondary clarifier effluent to remove residual suspended solids. This scheme is essentially the same as that for effluent quality C, except for the addition of alum. Effluent UOD levels may be expected to average approximately 81 mg/l from this treatment scheme.

The second biological process which produces an effluent quality D utilizes one-step lime coagulation following the activated sludge process to remove both phosphorus and colloidal and suspended solids. Mixed media filtration is performed after lime coagulation to insure that all the precipitated phosphorus and coagulated solids are removed from the effluent stream. Following lime coagulation and filtration, the wastewater stream will have a pH of approximately 9.0, and consequently will not be acceptable for discharge without pH adjustment. It is therefore necessary to include recarbonation to reduce the effluent pH level to an acceptable value. Phosphorous removal by lime coagulation is highly efficient, with typical effluent phosphorus concentrations averaging approximately 0.2 mg/l. The effluent UOD of this treatment scheme averages about 74 mg/l.

The physical-chemical process which provides an effluent of D quality utilizes lime coagulation and mixed media filtration to remove solid matter, and activated carbon to absorb dissolved organic compounds. As with most p-chem schemes, this scheme is reliable and able to respond quickly to influent variations. The effluent from this treatment scheme will possess an elevated pH level, necessitating recarbonation following lime coagulation and filtration. Effluent UOD levels from this physical-chemical scheme will average approximately 71 mg/l.

Alternatives—Effluent Quality E. This effluent quality necessitates increased organic carbon (BOD_5) removal and nitrification. This can be accomplished by the addition of mixed-media filtration to the two biological treatment schemes outlined to produce effluent quality B; that is, two-stage nitrification and extended aeration. The effluent UOD from the two-stage nitrification-mixed-media filtration scheme averages 12 mg/l, whereas the extended aeration-multi-media scheme effluent UOD level is expected to average 17 mg/l.

Alternatives—Effluent Quality F. This effluent quality classification requires increased organic carbon (BOD_5) and phosphorus removal and nitrification. The two biological schemes which can meet these requirements are the same as those for effluent quality B (two-stage nitrification and extended aeration), with the addition of mixed media filtration to remove suspended solids, and alum to remove phosphorus. As an alternative to alum for phosphorus removal, lime coagulation can be utilized. In both schemes, the lime coagulation would be placed after the nitrification step to prevent a high pH level from inhibiting biological activity. However, phosphorus removal with lime coagulation as a separate process would be more complicated than alum addition, and would require recarbonation to avoid high effluent pH levels.

The effluent UOD from the two sludge nitrification-phosphorus removal filtration scheme and the extended aeration-phosphorus removal-filtration scheme are estimated to be 12 and 17 mg/l, respectively.

Alternatives–Effluent Quality G. Effluent quality G is similar to D, except that nitrogen removal is also included. The first biological scheme capable of producing an effluent of G quality incorporates a denitrification step, following nitrification, to convert nitrates to nitrogen gas. Since the wastewater settles in three clarifiers during this treatment scheme (following the conventional aeration, nitrification, and denitrification tanks) the effluent suspended solids and BOD_5 levels are very low. Phosphorus may be precipitated in this system in the denitrification clarifier by means of alum addition, and effluent UOD levels are expected to average 17 mg/l. It is assumed that the alum will have no detrimental effect on the denitrification biomass. Phosphorus removal can also be accomplished in a separate treatment step following denitrification. Another alternative is to remove phosphorus by means of alum addition in the first aeration basin, while leaving sufficient phosphorus levels in the wastewater stream to support biological growth in the subsequent nitrification and/or denitrification tanks.

A second biological scheme includes activated sludge followed by two-stage lime coagulation and ammonia stripping to eliminate nitrogen and phosphorus. Mixed media filtration is used to complete suspended matter and organic carbon (BOD_5) removal. The effluent UOD of this treatment scheme is predicted to average 14 mg/l.

The physical-chemical scheme capable of producing effluent quality G is similar to the second biological scheme except that the activated sludge process at the beginning of the treatment train is replaced with activated carbon adsorption following mixed media filtration. The effluent UOD level for this treatment scheme is approximately 12 mg/l.

Alternatives–Effluent Quality H. This classification requires extremely high BOD_5 removal (greater than 99%) and phosphorus and nitrogen removals. Two very complicated activated sludge schemes can be devised to produce this effluent quality. They differ only in the means of nitrogen and phosphorus removal and both utilize activated sludge, mixed media filtration, and activated carbon adsorption.

The first process includes biological nitrification and denitrification followed by alum addition for phosphorus removal. An effluent UOD of 6 mg/l may be expected from this scheme. The second system removes nitrogen and phosphorus in a two-stage lime coagulation and ammonia stripping process. Since ammonia stripping is not as efficient as biological nitrogen removal, the estimated effluent UOD is approximately 11 mg/l.

Other Alternatives. As mentioned previously, treatment techniques can be combined in various ways to produce different treatment schemes. Since it is not possible to elaborate all possible treatment schemes only the most practical schemes were selected to generate each effluent quality classification. However, several alternate treatment schemes should be mentioned.

Two common treatment techniques are not normally considered viable treatment alternatives in water quality management planning–trickling filters and package plants. Both techniques have been extensively utilized for wastewater treatment in the past, and, consequently, water quality management planners are often faced with the problem of how to deal with already existing wastewater treatment plants utilizing these techniques. Decisions to either replace or upgrade these treatment facilities, based on a logical and cost-effective analysis, must be made.

It is not likely that water quality management planners would include trickling filters as a feasible biological treatment technique, since the activated sludge process is more efficient and dependable, though slightly more expensive. It is conceivable, however, that trickling filters might be utilized in a roughing application to reduce the organic content of an unusually strong wastewater prior to activated sludge or p-chem treatment.

Existing trickling filter plants may be upgraded by addition of appropriate treatment techniques. In particular, suspended solids, dissolved organics, and phosphorus removal techniques may be appended to trickling filter processes to upgrade effluent quality. Additionally, biological nitrification and denitrification or p-chem nitrogen removal may be added to eliminate nitrogen constituents from the effluent stream.

In summary, trickling filters are not the biological treatment of choice for future wastewater treatment facilities and were not included in treatment schemes presented in this chapter. However, conventional upgrading techniques can be utilized to allow existing trickling filter plants to meet secondary treatment standards (Refs. 3-80, 3-81).

Since package plants are small installations serving small populations, they are disadvantaged by economy of scale. Therefore, many upgrading techniques suitable for larger installations are too expensive for addition to package plants. Manufacturers of package plants have developed two upgrading techniques which may be added to existing package plants, or with new orders to achieve higher effluent quality (Ref. 3-82).

The first is an automatic sand filter unit which filters the effluent prior to discharge, thus reducing the suspended solids concentration. The second is chemical addition (often alum) for phosphorus removal in the clarifier. Some suspended solids removal is also accomplished by the chemical addition.

In addition, the manufacturers often advise that careful attention to the operation and maintenance procedures will increase package plant efficiencies. This is so, although increased attention actually represents an upgrading technique compared to a typical package plant installation, since a higher O + M cost is incurred to provide additional manpower.

A common upgrading technique is the construction of a lagoon to receive the package plant effluent and provide additional detention time before discharge. Solids in the effluent may settle, and some degree of biological treatment may occur in the lagoon. A disadvantage to lagooning is the presence, during the summer, of algal growth that may actually increase the BOD_5 and solids in the effluent.

Finally, ozonation has been suggested as an upgrading technique. Ozone would react with soluble organics in the effluent, thereby reducing the effluent BOD_5. Ozonation would also eliminate the need for chlorination for disinfection. However, there is very limited data extant which suggests the economic feasibility of ozonation as a package plant upgrading technique.

The extended aeration process is often utilized in small package plants, because it accomplishes satisfactory BOD_5 removal and a partial degree of sludge digestion without extensive operation and maintenance costs. However, the extended aeration process is more expensive than activated sludge in all but the smallest plants; it is very temperature dependent for nitrification, and requires more land area than a separate nitrification tank would. Therefore, extended aeration is not likely to be utilized for nitrification in plants with flows larger than one million gallons per day.

Small plants would probably not be built to include denitrification, since economies of scale would dictate the construction of larger, regional plants in this instance. Regional plants which include denitrification would probably not utilize extended aeration for nitrification (for the reasons cited above) and consequently, the extended aeration process was not included in effluent classifications G and H.

Another treatment alternative involves different methods of phosphorus removal. In all schemes presented, lime or alum is utilized to precipitate phosphorus from the wastewater; however, ferric chloride or other salts can also be utilized to accomplish phosphorus precipitation in the primary sedimentation tank or in a separate tertiary treatment process. The phosphorus removal techniques incorporated into the treatment schemes presented in this chapter are for the most part based on popular practice. Obviously, local considerations may indicate that a different approach be investigated for phosphorus removal.

Solids Treatment and Disposal. In evaluating treatment scheme feasibility, it is necessary to consider the treatment and disposal of solids produced by the scheme. There are three distinct types of solids produced by the treatment schemes presented in Table 2-23—primary, secondary, and chemical sludges. The primary and secondary sludges may be treated by any of the techniques described in Section 3-3.14. Chemical sludges may be oxidized to recover the chemical, or may be landfilled.

Normally, solids treatment and disposal is a secondary consideration in devising alternative wastewater treatment schemes. The primary consideration is to provide a scheme which will meet effluent quality standards. Then, after a scheme is generated, solids treatment processes can be evaluated to determine the most efficient approach for the particular wastewater treatment scheme. If more than one treatment scheme is applicable at one location, then the most effective solids treatment and disposal process for each scheme can be developed and the schemes can be compared on the basis of overall wastewater and solids treatment effectiveness. Clearly, the ultimate impact of a treatment scheme includes the appropriate solids handling process.

The determination of the best solids treatment process and ultimate disposal is a function of local factors, land availability, fuel costs, nearby disposal locations, etc. Since many variables in the selection of an appropriate solids treatment process and ultimate disposal are peculiar to the local conditions, it is not possible in this general approach to include optimum solids handling and ultimate disposal in the development of wastewater treatment schemes. Rather, it must be noted that for specific applications, the solids treatment disposal scheme suitable to the locale must be included in the overall evaluation of the cost and effectiveness of the total water quality management system.

3-5. WASTEWATER TREATMENT ALTERNATIVE COSTS

This section presents and describes cost data for various wastewater treatment techniques and schemes as a function of treatment plant size. Since there is no extant single source which details costs for alternative treatment schemes, it was necessary to consult various literature sources to develop treatment alternative cost estimates. In some cases, significant variances in treatment costs were found in the literature, and consequently, it was necessary to discriminate between cost variations according to some reasonable methodology. To this end recent sources were considered most reliable, and sources which described treatment schemes in detail and explicity delineated specific process costs were considered more reliable than those which did not.

Whereas an overwhelming number of cost estimate sources exist for some treatment techniques (e.g., activated sludge), there is a paucity of reliable information describing the costs of many other treatment methods (e.g., water reuse). Consequently, the accuracy of cost estimates of these latter treatment alternatives suffers accordingly.

The cost curves presented by the authors were synthesized from a wide variety of sources and subjected to detailed analysis and comparison. The final composite treatment alternative cost curves are believed to be the best estimates of wastewater treatment costs available to date.

The discussion of wastewater treatment alternative costs is divided into four sections. Section 3-5.1 delineates the cost types associated with wastewater treatment, while 3-5.2 describes various aspects of cost estimation and explains the development of the cost curves in this chapter. The third section, 3-5.3, presents the cost estimates for wastewater treatment techniques, and section 3-5.4 combines these technique costs to produce cost estimates of entire wastewater treatment schemes.

3-5.1 Wastewater Treatment Cost Estimate Development

Wastewater treatment costs can basically be divided into two categories: capital costs and operation and maintenance (O + M) costs. Capital costs are incurred during the construction of a wastewater treatment facility, whereas O + M costs are those necessary to operate and provide upkeep for the plant following construction. Capital costs are typically expressed in millions of dollars as a one time expense. Costs for operation and maintenance may be expressed as total annual expenditures or as a cost per unit of wastewater treated (e.g., cents per thousand gallons). The latter is most common in current cost estimates, and this convention is followed in this chapter.

A major portion of the capital cost outlay associated with a treatment facility entails the actual labor and material construction costs which may account for up to 80 or 90 percent of the total capital costs. The remaining portion of the capital cost is associated with land acquisition, administration, initial financing, and legal and engineering services. The O + M costs associated with a wastewater treatment facility includes operating and maintenance labor, materials, supplies, and power costs. Labor accounts for approximately 40 to 60 percent of the total O + M cost.

3-5.2 Elements of Cost Estimation

This section describes aspects of cost estimation such as sources, reliability, estimating procedures, types of estimates, indexing, scale economies, and methods of comparing cost estimates for cost effectiveness.

Sources of Treatment Alternative Cost Estimates. Wastewater treatment alternative cost estimates may be found in the literature, in government research reports, and occasionally may be obtained from equipment manufacturers. The accuracy and reliability of these estimates depend both on the procedure utilized to develop the estimate and the purpose for which the estimate is being made. In addition, the time period during which the estimate was developed and the locale for which it was specified may influence the quality of the estimate.

There are many difficulties associated with compiling cost estimates from various sources. A common problem in the lack of descriptive detail concerning the exact treatment technique for which the estimate is made. For example, an estimate may be presented for the "activated sludge process" without specifying whether the estimate includes sludge hauling and disposal, treatment technique modifications, or influent wastewater characteristics.

Another difficulty associated with treatment technique cost estimates involves technological changes in a treatment technique which may occur after an estimate is compiled. Costs estimates may be significantly affected by technological progress and resulting process revisions.

In recognition of these and other problems, some authors have produced summaries for these wastewater treatment processes. In particular, Smith (Ref. 3-61) and Smith and McMichael (Ref. 3-62) have produced comprehensive summaries. In addition, Bechtel (Ref. 3-63) developed cost estimating guidelines under a government contract, and Patterson and Banker (Ref. 3-64) developed a manual for activated sludge plant cost estimation. Useful summaries of cost estimates may also be found in planning reports for regional water quality management, such as that by Bishop (Ref. 3-33).

The summaries listed above, plus a wide variety of original sources and personal contacts, were utilized to develop the cost curves presented in this chapter (Refs. 3-57, 3-65 through 3-79).

Types of Wastewater Treatment Alternative Cost Estimates. Cost estimates fall into two general categories: those constructed from data reported for existing plants, and those projected from basic estimates of different portions of the overall treatment technique. The former may be termed deductive estimates and the latter inductive estimates.

Deductive estimates are usually of a general nature. They often fail to adequately identify all the parameters of a treatment process, and may combine several variations of a treatment technique together in one estimate. For example, a cost estimate for activated sludge treatment may be developed from data sources through the entire United States which depict a wide variety of treatment modifications and sludge handling techniques. Such an estimate is intended to be a general guideline which is not really applicable to any one locale or treatment modification.

Deductive estimates usually encompass a wide spectrum of plant sizes (normally ranging from 1 to 100 mgd). In some instances, insufficient data may exist to describe accurately the entire 1 to 100 mgd range and cost estimates may be extrapolated by scaling costs up or down from the reliable data. Naturally, the portion of these estimates which is predicated on scaling is less reliable than those portions which reflect genuine data.

Inductive estimates involve costs for specific treatment modifications which are developed by estimating unit treatment process costs and then summing the partial estimates to arrive at an estimate of the entire treatment technique. Thus, the inductive estimate attempts to develop cost data for a specific treatment technique from known cost data estimates. Patterson and Banker (Ref. 3-64) have utilized this approach in developing treatment scheme costs.

Both deductive and inductive treatment cost estimates may be presented as either a cost versus flow curve or as an equation. Since deductive estimates are often developed by curve fitting procedures applied to existing data, they are often presented in equation form. A popular cost estimate equation, which is commonly utilized because it reflects the economy of scale which might be expected, is as follows:

$$Y = aX^{b}, \qquad \text{(Eq. 3-20)}$$

where Y is the treatment cost, X is a flow rate, and a and b are empirical constants.

Economy of Scale and Treatment Costs. Economy of scale is the phenomena that a unit of capacity normally costs less in large plants than in small plants. This economy is due to the decrease in unit costs of equipment and manpower at higher throughput. Wastewater treatment plants show substantial economies of scale, and the difference in treatment cost per unit volume between large and small plants may be dramatic, as much as an order of magnitude.

Graphically, economy of scale may be shown by plotting cost versus volume (daily flow in wastewater treatment). Logarithmic plots are commonly utilized to display the data, and, as seen in most of the graphs in this section, the plotted lines are nearly straight, having a slope less than one. The magnitude of the slope is an indication of the degree of economy of scale. Thus, a plot of slope one has no economy of scale and, therefore, four times the capacity requires four times the cost, and the per unit capacity cost remains unchanged. A line of slope 0.5 shows substantial scale economy and an increase to four times the capacity can be achieved for only twice the cost and, therefore, the unit capacity cost is reduced by half. The slope of the plotted line is often referred to as the coefficient of economy of scale or cost exponent, and the mathematical expression is shown in Equation 3-21 (Ref. 3-70).

$$C = KQ^{m}, \qquad \text{(Eq. 3-21)}$$

where C is the cost of the item with capacity Q, K is the base cost factor, and m is the measure of economy of scale. This is precisely the same relationship as that shown in Equation 3-21. Wastewater treatment plants are very similar to chemical process plants in economy of scale with cost exponents of approximately 0.6 for individual pieces of equipment or techniques (Ref. 3-70).

Indices and Wastewater Treatment Cost Estimate Updating. Treatment estimates are necessarily associated with costs prevalent at the time they were devised. As costs change it is necessary to update these estimates. Rather than completely revising cost estimates at specific time intervals, indices of construction, labor, and products prices are utilized to accomplish cost estimate updating. These indices are prepared monthly by various sources, notably the U.S. Government and *Engineering News Record* (ENR). The most popular indices for wastewater treatment costs include: the Labor Cost Index issued by the U.S. Department of Labor; the Sewage Treatment Plant Construction Cost Index by EPA; the ENR Construction Cost Index published by ENR; the ENR Building Cost Index issued by ENR; the Wholesale Price Index for Industrial Commodities developed by the U.S. Department of Labor; and the Consumer Price Index published by the U.S. Department of Labor. These indices attempt to describe price variations in a particular segment of the economy. Cost estimates can be updated by the ratio of the index at the time desired to the time of the original cost estimate (See APPENDIX 3-1, "COST INDICES").

Ideally, each element in the capital and O + M cost estimate should be updated with the appropriate index; however, prices usually tend to rise uniformally in most economic segments and this degree of specificity is not necessary in reality. Since construction costs comprise the major expense associated with capital outlay, capital costs may be updated by reference to a construction cost index. Either the Sewage Treatment Plant Construction Cost Index or the ENR Construction Cost Index may be utilized for wastewater treatment plant capital cost adjustment.

O + M costs associated with wastewater treatment facilities can be adjusted by utilizing the Labor Cost Index (Water, Steam, and Sanitary Nonsupervisory Works), although for more precise estimates the materials portion of O + M costs may be adjusted separately with the Wholesale Price Index for Industrial Commodities. It is also possible to utilize the Consumer Price Index for O + M updating.

In this chapter, cost estimates were updated, primarily, by utilizing the Sewage Treatment Plant Construction Cost Index, and the Labor Cost Index (Water, Steam, and Sanitary Nonsupervisory Works). Where possible, the Wholesale Price Index for Industrial Commodities was also utilized to update the materials portion of O + M costs.

Cost Effectiveness Analysis. Pursuant to the Water Pollution Control Act Amendments of 1972 (Ref. 3-50) and subsequent EPA guidelines, the most cost effective treatment techniques and schemes must be utilized to abate water pollution by publicly owned treatment works. Therefore, a cost effectiveness analysis must be conducted to determine which treatment alternative of the feasible pollution abatement schemes is most cost effective. In the cost effectiveness analysis all costs must be identified in each treatment scheme and all nonmonetary factors evaluated. The costs of each scheme must then be compared on a common basis, and the scheme which will meet the pollution standards with the lowest cost (while considering nonmonetary factors) is considered the most cost effective (Refs. 3-56, 3-57).

Two methods of comparing costs are currently approved by EPA—annual worth and present worth analysis. The annual worth is defined as the equivalent annual cost of a treatment scheme in which both the capital and O + M costs are paid over the planning period. The annual cost is calculated by summing the annual O + M cost with the result from multiplying the initial total capital cost by an amortization factor.

The present worth of a treatment scheme is that amount of money required at the beginning of the planning period to pay the capital costs immediately and the O + M costs annually for the entire planning period. The present worth cost is calculated by adding the capital cost to the result from multiplying the annual O + M cost by a present worth factor (See APPENDIX 3-2, "PRESENT WORTH CALCULATIONS").

In order to conduct either the present or annual worth analysis, a planning period and interest rate must be determined. The EPA has specified that a 20 year planning period and a 6 1/8 percent interest rate must be utilized for alternative wastewater treatment planning for the second quarter of 1976 (Ref. 3-57).

The present worth analysis was utilized in this chapter for cost comparisons of treatment schemes. The calculated present worth values are not absolutely accurate, because they are based on generalized cost estimates. They serve to delineate the relative costs among different treatment alternatives and, consequently, identify those which are most cost effective.

3-5.3 Cost Estimates of Wastewater Treatment Techniques

This section presents cost estimates for the treatment techniques discussed in section 3-3. Equations for the cost curves are tabulated in APPENDIX 3-3, "COST CURVE ANALYSIS".

Index Values. The cost estimates presented in this chapter are keyed to the second quarter, 1976 cost level. Since that time, the rate of inflation may have affected the quantitative accuracy of these estimates; however, the relative economies of the various treatment alternatives are believed to remain unchanged.

The index values utilized to update wastewater treatment technique cost estimates in the literature are listed below:

Sewage Treatment Plant Construction Cost Index (1957–1958 = 100)	273.4
Labor Cost Index (W, S, S non-supervisory workers)	\$5.20/hr
Wholesale Price Index for Industrial Commodities (1967 = 100)	170

The present worth analysis factor utilized for $6\frac{1}{8}$ percent and 20 years was 11.36.

Activated Sludge. Since the activated sludge treatment technique has been utilized widely during the last 50 years, there currently exists a large number of cost estimates describing this technique and its various modifications (Refs. 3-33, 3-61, 3-64, 3-65, 3-66, 3-67, 3-68, 3-71, 3-77, 3-79).

The capital costs for an activated sludge facility can be divided into money allocated for construction of the wastewater treatment equipment, the solids treatment equipment, and incidentals. Incidentals include the cost of land, yard work, engineering, legal and administrative fees, fiscal cost, etc. all of which are not basic elements of either the wastewater or solids treatment processes, but must nevertheless be accounted for in the overall capital cost. Incidentals commonly represent 25 to 30 percent of the total capital cost of an activated sludge facility.

Most cost projections for activated sludge systems include estimates for a complete treatment facility, including solids handling, chlorination, and incidental fees. These data are adequate for estimates of the activated sludge technique itself, but too inclusive for tertiary treatment schemes where different solids treatment techniques may be utilized and incidental costs are a function of the

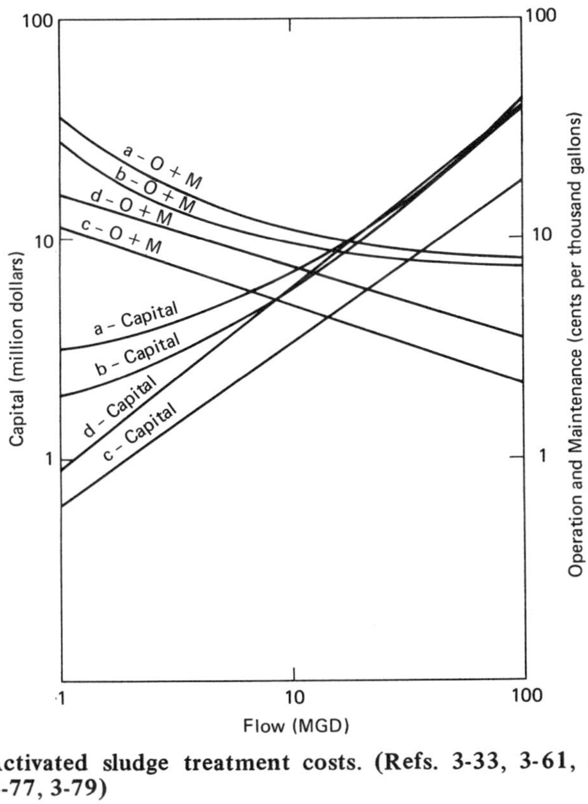

Fig. 3-15. Activated sludge treatment costs. (Refs. 3-33, 3-61, 3-64 through 3-68, 3-71, 3-77, 3-79)

overall tertiary treatment scheme. Therefore, additional cost estimates for the wastewater treatment portion of the activated sludge technique have been prepared and are presented in addition to cost projections for complete activated sludge treatment systems.

Cost estimates for the activated sludge process are presented in Figure 3-15, and are labeled in this figure with letters as defined below:

a. Cost of complete plant, including preliminary treatment, primary treatment, activated sludge, chlorination, vacuum filtration, incineration, administration, lab, garage, and shop facilities, and incidental costs, excluding the costs of land acquisition and the ultimate disposal of ash.
b. Cost of complete plant as in "a" above, with anaerobic sludge digestion and sludge drying beds instead of vacuum filtration and incineration, excluding the costs of land acquisition and the ultimate disposal of dried sludge.
c. Cost of only wastewater treatment; i.e., preliminary treatment, primary treatment, and activated sludge, excluding solids treatment, chlorination, incidental land, and ultimate disposal costs.

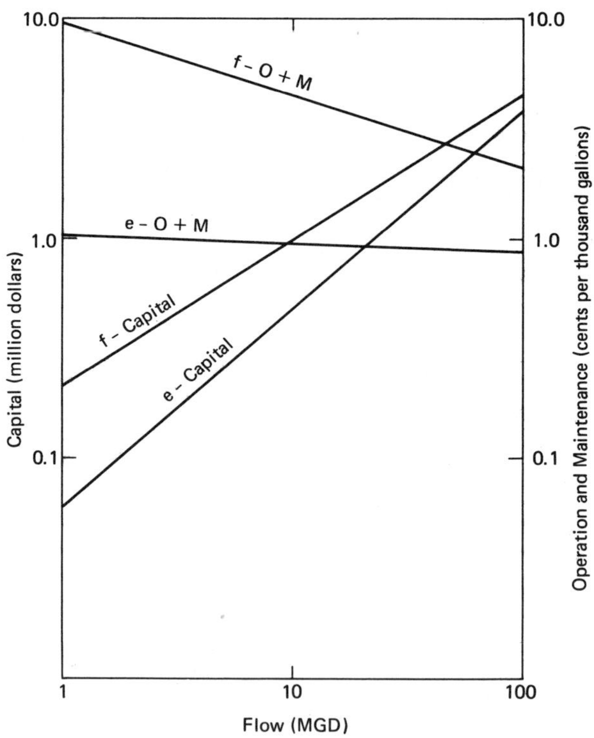

Fig. 3-16. Microstraining and mixed media filtration treatment costs. (Refs. 3-33, 3-61, 3-65, 3-67)

d. Cost of only extended aeration activated sludge treatment; i.e., preliminary treatment, primary treatment, and extended aeration activated sludge, excluding solids treatment, chlorination, incidental land, and ultimate disposal costs.

Suspended Solids. The costs of suspended solids removal by microstraining and mixed media filtration are depicted in Figure 3-16 and are labeled with letters as defined below: (Refs. 3-33, 3-61, 3-65, 3-67)

e. Cost of microstraining, including equipment and piping and excluding the cost of structures to house the microstrainers.
f. Cost of mixed media filtration of secondary effluent at 4 gpm/ft^2, excluding chemical costs.

Dissolved Solids. Figures 3-17 and 3-18 present dissolved solids removal costs associated with activated carbon adsorption, ion exchange, electrodialysis, and reverse osmosis.

Activated Carbon. Activated carbon process costs are depicted in Figure 3-17 and include cost projections for activated carbon adsorption, utilized as

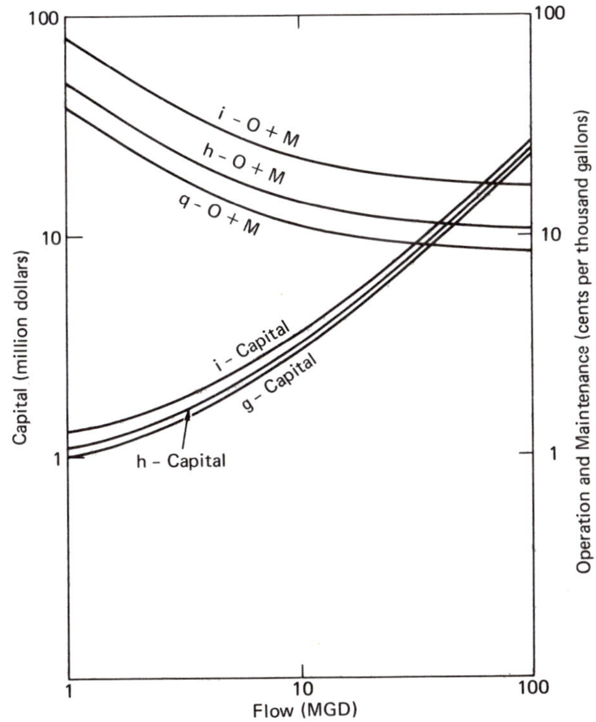

Fig. 3-17. Activated carbon adsorption treatment costs. (Refs. 3-33, 3-61, 3-65, 3-66, 3-67, 3-71, 3-75)

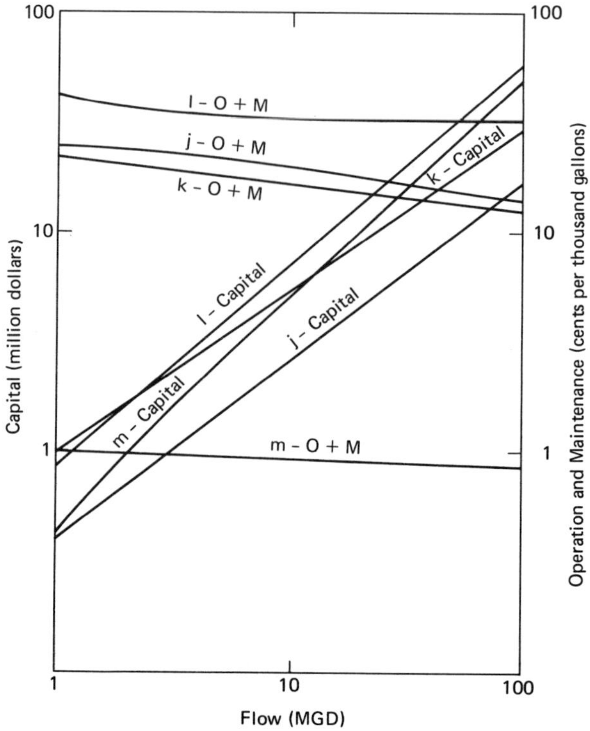

Fig. 3-18. Demineralization process costs for ion exchange. (Refs. 3-66, 3-76, 3-78), electrodialysis (Refs. 3-33, 3-61, 3-66, 3-76, 3-78), reverse osmosis (Refs. 3-66, 3-76, 3-78), and brine disposal (Ref. 3-65).

both a tertiary and secondary treatment technique. The various carbon adsorption cost curves are labeled with letters as defined below: (Refs. 3-33, 3-61, 3-65, 3-66, 3-67, 3-71, 3-75)

g. Cost of activated carbon utilized as a tertiary treatment process for secondary effluent, including downflow pressure contactors, 40 minutes contact time, once a day backwash, makeup carbon, and regeneration.
h. Cost of activated carbon adsorption utilized as a secondary treatment process to remove up to 95 percent BOD_5, including all of "g" plus additional carbon and regeneration requirements.
i. Cost of activated carbon adsorption utilized as a secondary treatment process to remove 99 percent BOD_5, including all of "h" plus additional carbon and regeneration requirements.

Demineralization. Figure 3-18 describes demineralization treatment costs associated with ion exchange (Refs. 3-66, 3-76, 3-78), electrodialysis (Refs. 3-33, 3-61, 3-66, 3-76, 3-78), and reverse osmosis (Refs. 3-66, 3-76, 3-78). Cost curves are labeled with letters as defined below:

j. Cost of ion exchange of secondary effluent, including regeneration and replacement of resin and regenerant and excluding brine disposal.
k. Cost of electrodialysis of secondary effluent, including membrane replacement at 5 year intervals and excluding brine disposal.
l. Cost of reverse osmosis of secondary effluent, including membrane replacement at 2 year intervals and excluding brine disposal.
m. Cost of brine disposal from above processes in evaporation ponds, including pond construction cost but excluding land acquisition cost.

Nitrogen Removal. Nitrogen removal costs are presented in Figures 3-19 and 3-20, with the ammonia stripping cost estimate (Refs. 3-33, 3-62, 3-65, 3-71) depicted in Figure 3-19 and biological nitrification and denitrification process costs (Refs. 3-33, 3-57, 3-64, 3-66, 3-77) outlined in Figure 3-20. Biological nitrification and denitrification treatment costs depicted include costs incurred in addition to the costs of conventional activated sludge treatment. Nitrogen removal process cost curves are labeled with letters as defined below:

n. Cost of ammonia stripping including pumps, tower, basin, and scale removal, and excluding chemical costs and equipment for pH adjustment.

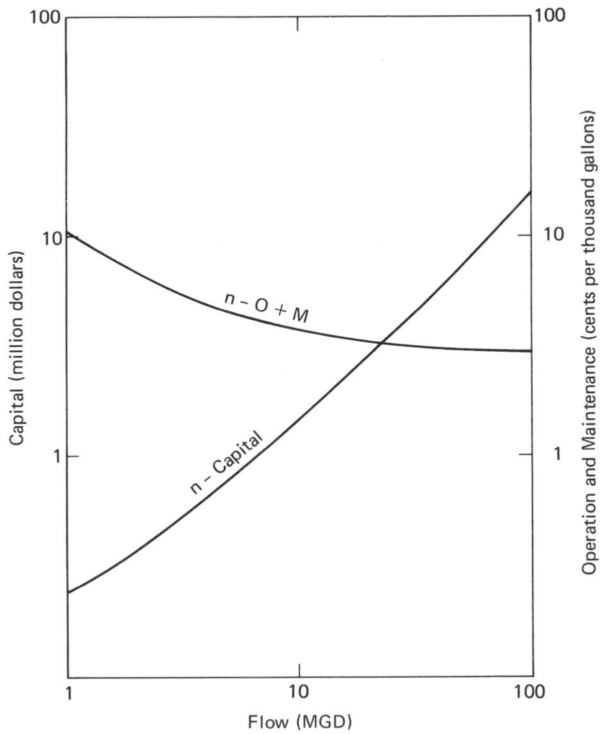

Fig. 3-19. Ammonia stripping treatment costs (Refs. 3-33, 3-62, 3-67, 3-71)

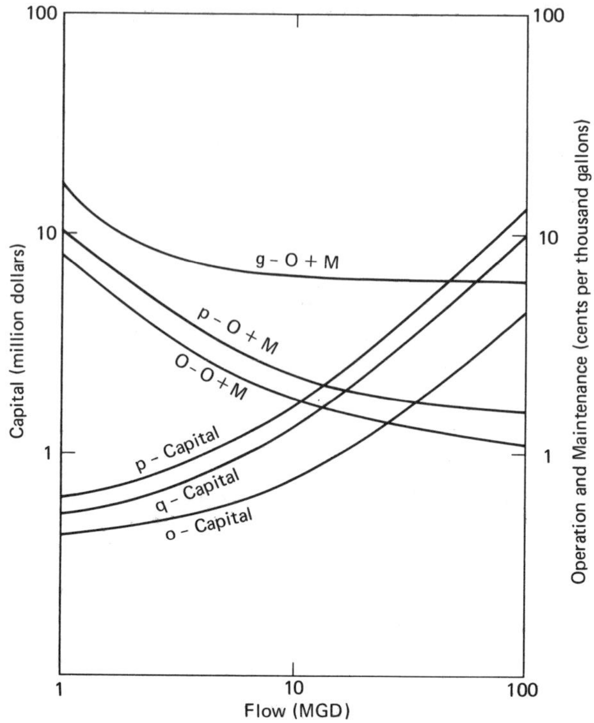

Fig. 3-20. Biological nutrification and denitrification treatment costs. (Refs. 3-33, 3-57, 3-64, 3-66, 3-77)

o. Cost of nitrification in the conventional activated sludge process, including additional air requirements and excluding all costs normally associated with the activated sludge process.

p. Cost of separate nitrification, including aeration tank, air supply, clarifier, and sludge recycle.

q. Cost of suspended growth denitrification including tank, mixing, methanol feed, clarifier, and sludge recycle.

Phosphorus Removal. The cost of phosphorus removal techniques vary significantly as a function of the chemicals utilized and the location of chemical application (Refs. 3-33, 3-72) Figure 3-21 depicts the cost of alum addition to the activated sludge aeration tank, under normal operating conditions, to accomplish phosphorus removal. This cost projection includes 0.5¢/1000 gal for the disposal of excess sludge procuded through chemical addition. The cost curve for phosphorus removal by alum addition is labeled r and defined as below:

r. Cost of phosphorus removal by alum addition in the aeration tank, including structure, process equipment chemicals, all recovery controls, and excess sludge disposal.

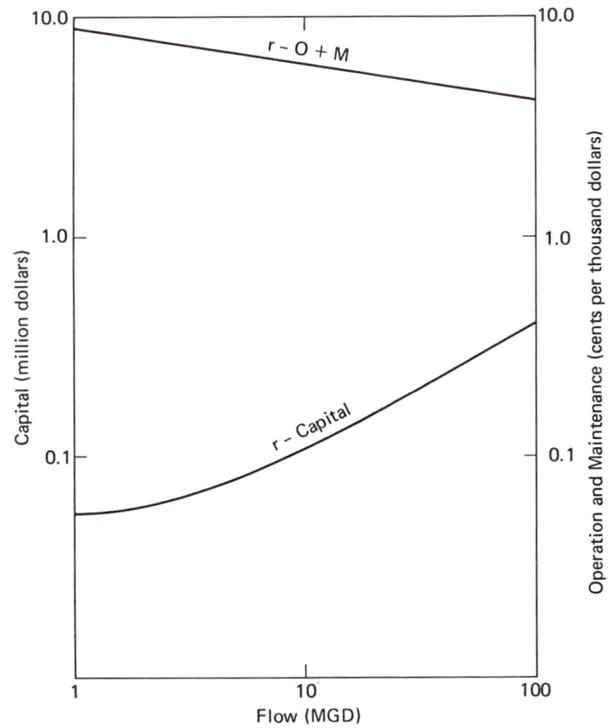

Fig. 3-21. Costs associates with phosphorus removal by alum addition. (Refs. 3-33, 3-72)

Stabilization Ponds. Although cost estimates for wastewater stabilization ponds are available in the literature, they vary significantly, as pond design is based on local climatic conditions and construction needs, as well as flow rate (Refs. 3-64, 3-79). For this reason, and the fact that it is unlikely that wastewater stabilization ponds would be designed for consideration as a best practicable waste treatment technique, a cost estimate for stabilization ponds is not included in this chapter.

Ozonation. Since ozonation is not a commonly utilized treatment technique in the U.S., there is little valid date for developing cost projections for ozonation application. Manufacturers of ozone equipment suggest process costs that often vary by an order of magnitude. Consequently, no cost projections for ozonation are included in this chapter.

Land Application of Wastewater. Experience in wastewater treatment and/or disposal by land application is limited in the U.S.; however, costs can be estimated from design data, and are as presented in Figure 3-22 (Refs. 3-51, 3-79).

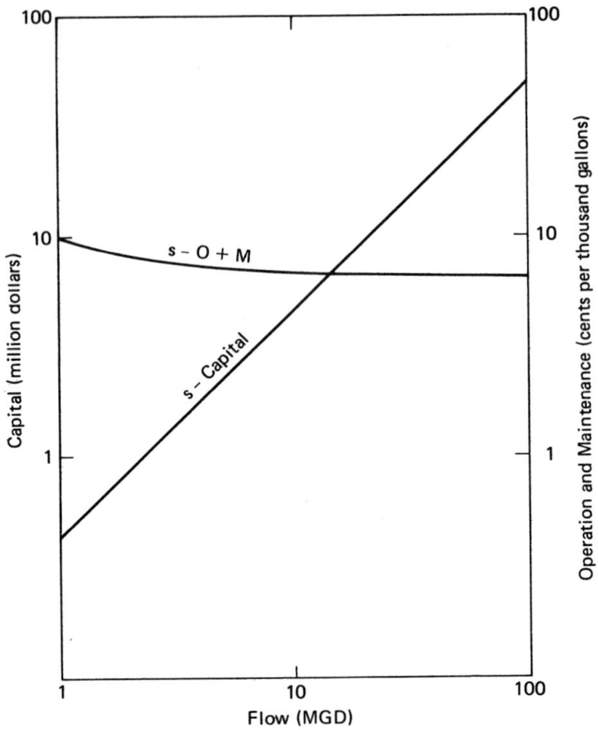

Fig. 3-22. Cost of land application of wastewater. (Refs. 3-51, 3-79).

These cost projections represent the cost of treatment by overland flow, with no specific degree of wastewater quality required for application, although it is expected that conventional secondary treatment would be required of effluents applied to the land. Land application costs are labeled s and defined as below:

s. Cost of land application of wastewater by the overland flow technique, including all piping and pumping but excluding land costs.

Disinfection. The cost of wastewater treatment plant effluent disinfection by chlorination is presented in Figure 3-23. The curve is labeled t and defined as follows: (Refs. 3-61, 3-64, 3-65, 3-66)

t. Cost of chlorination of secondary effluent, including contact chambers (15 minute detention time), chemical feed system, control facilities, and chemical costs.

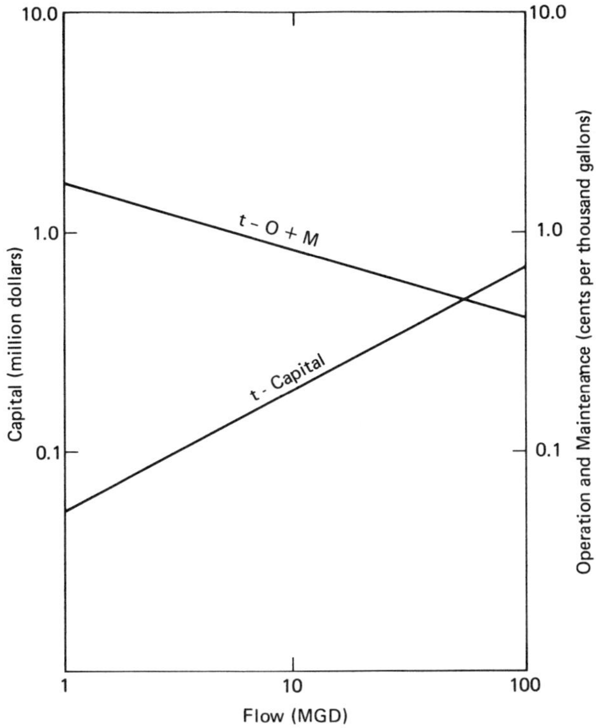

Fig. 3-23. Effluent disinfection costs using chlorine. (Refs. 3-61, 3-64, 3-65, 3-66)

Lime Coagulation Processes. Figure 3-24 includes cost estimates for both single and double stage lime coagulation processes (Refs. 3-33, 3-61, 3-65, 3-67, 3-71). Cost projections have been developed under the assumption that the lime sludge is recalcined to recover and reuse the lime. Although it is also possible to landfill lime sludge, landfilling is economically feasible only in treatment facilities having capacities less than 5 mgd. Operation of recalcining operations becomes economically sound at treatment facilities larger than 5mgd. Lime coagulation cost curves are labeled u, v, and w and are defined as below:

u. Cost of one stage lime coagulation of secondary effluent, including chemicals, feeding equipment, thickeners, centrifuge, furnace or kiln for recalcination, clarifiers, and recarbonation.

v. Cost of two stage lime coagulation of secondary effluent, including all equipment as in "u" above.

w. Cost of recalcining for one and two stage operation, including furnace or kiln, and sludge handling equipment.

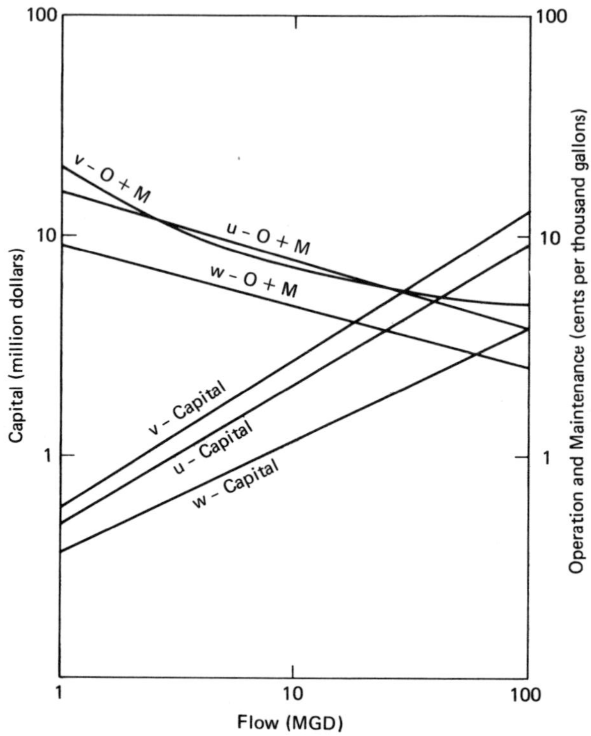

Fig. 3-24. Lime coagulation treatment costs. (Refs. 3-33, 3-61, 3-65, 3-67, 3-71).

Solids Treatment and Disposal. Figures 3-25 and 3-26 depict cost estimates for solids treatment and disposal (Refs. 3-49, 3-64). Land costs are not included in any estimates since these vary widely. Solids treatment and disposal cost curves are labeled with letters as defined below:

x. Cost of vacuum filtration and incineration, including sludge pumping, holding tanks, filter, and incineration and excluding land acquisition and ultimate disposal costs.
y. Cost of anaerobic digestion, including sludge pumping, holding tanks, and digestors and excluding land acquisition and disposal of digested sludge.
aa. Cost of sludge drying beds, including construction of beds and excluding land acquisition and ultimate disposal of dried sludge.
bb. Cost of vacuum filtration and incineration of anaerobically digested sludge, excluding land acquisition, ultimate disposal, and all costs associated with anaerobic digestion.
cc. Cost of land application of digested sludge, assuming 6% solids, truck

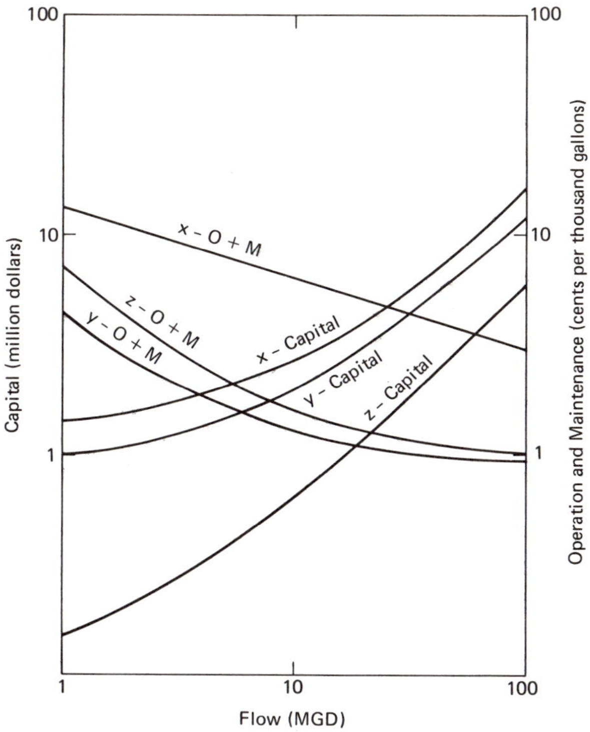

Fig. 3-25. Sludge stabilization costs. (Ref. 3-49, 3-64)

transportation for 20 miles at 10¢/mile dry ton and excluding capital costs and all costs of digestion.

dd. Cost of land application of digested sludge, assuming vacuum filtration to 25% solids and transportation and application costs as above.

3-5.4 Cost Estimates of Wastewater Treatment Schemes

The cost estimates for wastewater treatment techniques can be combined to provide costs for alternative treatment schemes. Table 3-24 presents a comparison of treatment techniques for 1, 10, and 100 mgd flows, which have been synthesized from Figures 3-15 through 3-26. Incidental capital costs are not included in the wastewater treatment technique present worth values presented in Table 3-24.

Cost data from Table 3-24, in conjunction with estimated incidental capital costs, have been utilized to predict total capital, O + M, and present worth costs of the alternative wastewater treatment schemes derived in section 3-4 to meet differing effluent quality criteria. These wastewater treatment scheme costs data

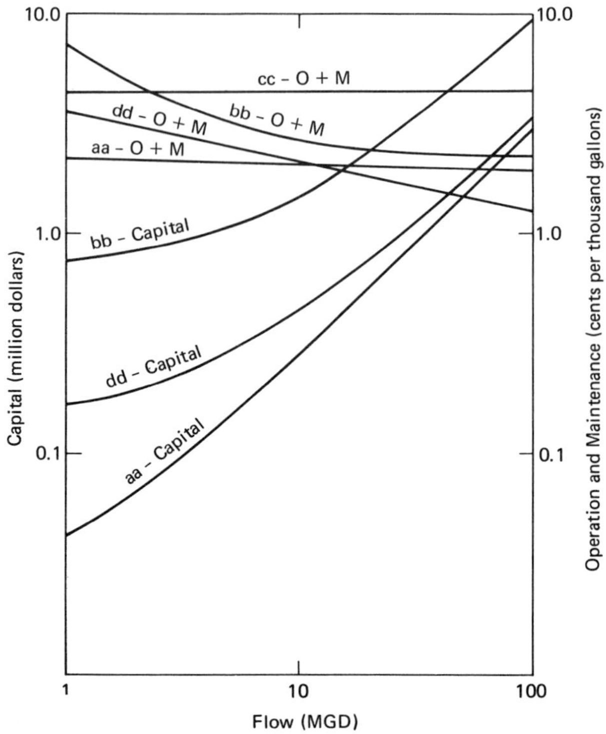

Fig. 3-26. Solids treatment and disposal costs. (Refs. 3-49, 3-64)

are presented in Table 3-25. Since the choice of an appropriate solids treatment and disposal scheme is highly dependent on local conditions, the costs for wastewater and solids treatment have been separated in Table 3-25. Thus, the costs of the wastewater treatment schemes in Table 3-25 do not include the costs of treatment and disposal of the solids produced by the treatment scheme. Consequently, the cost of an appropriate solids handling scheme may be combined with the costs of a desired wastewater treatment alternative to provide an estimated total treatment alternative cost tailored to a specific location.

Finally, Table 3-26 summarizes the present worth costs associated with various wastewater treatment scheme alternatives, and presents estimates of treatment alternative effluent quality. For the purpose of treatment alternative cost analysis comparison, it was assumed that solids are anaerobically digested and dried on sludge drying beds. Lime sludges produced in various treatment alternatives are assumed to be recalcined. Utilization of other solids treatment schemes in lieu of the assumed anaerobic digestion process would alter the treatment costs presented.

Tables 3-24, 3-25, 3-26 and the individual technique cost curves presented earlier in this chapter are important inputs to water quality management planning processes. Together with other data (sewer costs, land planning, and costs) the wastewater treatment costs provide a means of economic comparison of alternative wastewater treatment plans and ultimately contributes to water quality management planning.

3-5.5 Treatment Alternatives Evaluation Case Study

The following example case study briefly explains how the wastewater treatment cost relationships and planning methodology presented in this chapter may be utilized to evaluate wastewater treatment systems.

A southern community now possesses an old primary treatment system. New permit limits recently issued by EPA require the community to treat its wastewater so that it meets secondary standards plus nitrification under Q10-7 conditions. The existing primary treatment facility is over 25 years old and is assumed to have no salvage value. Climatic characteristics of the study area will allow year round land application and/or ammonia stripping. Soil characteristics and local land use of the study area make land application a feasible treatment alternative.

Table 3-27 summarizes the following eight treatment alternatives which will achieve the required effluent quality together with the associated costs and estimated effluent qualities of each treatment alternative:

a. Trickling filter, separate biological nitrification, and chlorination.
b. Trickling filter, two-stage lime coagulation, ammonia stripping, and chlorination.
c. Extended aeration and chlorination.
d. Activated sludge, nitrification in the activated slugde process, and chlorination.
e. Activated sludge, two-stage lime coagulation, ammonia stripping, and chlorination.
f. Trickling filter, land application, and chlorination.
g. Activated sludge, land application, and chlorination.
h. Two-stage lime coagulation, ammonia stripping, mixed-media filtration, activated carbon adsorption (secondary), and chlorination.

The costs associated with solids treatment have been purposely omitted from this example since the selection of an appropriate solids treatment and disposal process for any community is usually affected to a larger degree by local constraints than by the selection of wastewater treatment processes.

A review of Table 3-27 indicates that for a 1 MGD plant, treatment alternatives a, c, and f appear to be most cost effective. However, the effluent quality

TABLE 3-24 Comparison of Wastewater Treatment Technique Costs.[a]

Treatment Technique	Capital (Million $)			O + M (¢/1000 gal)			Present Worth (Million $) ($6\frac{1}{8}$% and 20 years)		
	MGD 1	MGD 10	MGD 100	MGD 1	MGD 10	MGD 100	MGD 1	MGD 10	MGD 100
a. Activated sludge plant w/incineration	3.28	6.88	42.90	37.50	11.10	8.49	4.83	11.50	78.10
b. Activated sludge plant w/anaerobic digestion	2.08	5.94	44.60	29.50	9.86	7.90	3.30	10.00	77.40
c. Activated sludge only w/o solids treatment	0.63	3.37	18.00	11.60	5.13	2.27	1.11	5.50	27.40
d. Extended aeration w/o solids treatment	0.98	6.18	38.90	16.50	7.70	3.59	1.67	9.37	53.80
e. Microstraining	0.06	0.47	3.67	1.02	0.94	0.86	0.10	0.86	7.22
f. Mixed media filtration	0.22	0.97	4.33	9.54	4.53	2.16	0.62	2.85	13.30
g. Activated carbon as tertiary treatment	1.07	3.22	24.70	39.60	11.20	8.40	2.71	7.86	59.50
h. Activated carbon as secondary treatment (95% BOD_5)	1.14	3.37	25.70	50.20	14.30	10.70	3.22	9.30	70.10
i. Activated carbon as secondary treatment (99% BOD_5)	1.32	3.73	27.90	80.90	22.90	17.10	4.67	13.20	98.80
j. Ion exchange	0.42	2.62	16.30	28.40	20.10	19.20	1.60	11.00	75.20
k. Electrodialysis	1.01	5.42	29.10	22.60	16.80	12.50	1.95	12.40	80.90
l. Reverse osmosis	0.91	7.12	55.90	42.50	34.00	33.10	2.67	21.20	193.00
m. Brine disposal	0.60	5.34	47.90	1.06	0.95	0.85	0.64	5.73	51.40

n.	Ammonia stripping	0.25	1.53	14.40	11.20	3.81	3.07	0.71	3.11	27.10
o.	Nitrification in activated sludge	0.44	0.80	4.39	8.02	1.80	1.18	0.77	1.54	9.28
p.	Separate nitrification	0.65	1.75	12.80	10.50	2.34	1.56	1.08	2.72	19.10
q.	Denitrification	0.54	1.37	9.71	17.10	6.64	6.26	1.25	4.12	35.70
r.	Phosphorus removal w/alum	0.06	0.12	0.39	9.01	6.19	4.25	0.43	2.69	18.53
s.	Land application-overland	0.48	4.77	47.70	9.91	7.09	6.81	0.89	7.71	75.90
t.	Chlorination	0.05	0.19	0.67	1.71	0.84	0.42	0.13	0.54	2.39
u.	Lime coagulation, One stage	0.52	2.14	8.84	15.90	7.84	3.87	1.18	5.39	24.90
v.	Lime coagulation, two-stage	0.61	2.75	12.50	20.90	7.41	6.06	1.47	5.48	37.60
w.	Recalcining	0.38	1.20	3.81	9.08	4.81	2.55	0.76	3.19	14.40
x.	Incineration	1.46	2.84	16.60	13.40	6.41	3.07	2.02	5.50	29.30
y.	Anaerobic digestion	1.06	2.08	12.30	4.55	1.30	0.98	1.25	2.62	16.40
z.	Aerobic digestion	0.16	0.67	5.84	7.29	1.59	1.02	0.46	1.33	10.10
aa.	Sludge drying beds	0.05	0.32	3.03	2.24	2.11	1.99	0.14	1.19	11.30
bb.	Incineration of digested sludge	0.80	1.60	9.54	7.64	2.84	2.36	0.40	2.78	19.30
cc.	Land application, 6% sludge	–	–	–	4.63	4.63	4.63	0.19	1.90	19.00
dd.	Land application, 25% sludge	0.18	0.48	3.45	3.68	2.18	1.29	0.33	1.38	8.80

[a]Equations for these cost curves may be found in APPENDIX 3-3, "Cost Curve Analysis."

TABLE 3-25 Comparison of Wastewater Treatment Scheme Costs.

Treatment Schemes	Capital Million $ mgd			O + M ¢/1000 gals mgd			Incidental Million $ mgd			Present Worth Million $ mgd		
Wastewater Treatment excluding solids treatment	1	10	100	1	10	100	1	10	100	1	10	100
AS + CHL	0.68	3.56	18.67	13.31	5.97	2.69	0.17	0.89	4.67	1.41	6.93	30.68
AS + BN + CHL	1.33	5.31	31.47	23.81	8.31	4.25	0.33	1.33	7.87	2.65	10.09	56.76
EA + CHL	1.03	6.37	39.57	18.21	8.54	4.01	0.26	1.59	9.89	2.06	11.50	66.08
AS + MMF + CHL	0.90	4.53	23.00	22.85	10.50	4.85	0.23	1.13	5.75	2.09	10.02	43.09
AS + AL + MMF + CHL	0.96	4.65	23.39	31.86	16.69	9.10	0.24	1.16	5.85	2.53	12.74	61.72
AS + LC1 + MMF + CHL	1.42	6.67	31.84	38.75	18.34	8.72	0.36	1.67	7.96	3.40	15.95	70.20
LC1 + MMF + AC + CHL	1.93	6.67	39.54	77.35	27.51	17.15	0.48	1.67	9.89	5.63	19.75	120.58
AS + BN + MMF + CHL	1.55	6.28	35.80	33.35	12.84	6.41	0.39	1.57	8.95	3.33	13.18	65.39
EA + MF + CHL	1.25	7.34	43.90	27.75	13.07	6.17	0.31	1.84	10.98	2.73	14.60	80.47
AS + BN + AL + MMF + CHL	1.61	6.40	36.19	42.36	19.03	10.66	0.40	1.60	9.05	3.77	15.90	84.02
EA + AL + MMF + CHL	1.31	7.46	44.29	36.76	19.26	10.42	0.32	1.87	11.08	3.17	17.32	99.10
AS + BND + AL + CHL	1.93	6.80	41.57	49.92	21.14	14.76	0.48	1.70	10.39	4.48	17.27	113.51
AS + LC2 + AST + MMF + CHL	1.76	8.81	49.90	54.95	21.72	13.98	0.44	2.20	12.48	4.71	20.81	120.27
LC2 + AST + MMF + AC + CHL	2.27	8.81	57.60	93.55	30.89	22.41	0.57	2.20	14.40	6.95	24.61	164.89
AS + BND + AL + MMF + AC + CHL	3.29	11.14	71.60	109.66	39.97	27.62	0.82	2.79	17.90	8.66	30.51	204.42
AS + LC2 + AST + MMF + AC + CHL	2.83	12.03	74.60	94.55	32.92	22.38	0.71	3.00	18.65	7.69	29.47	185.94
LC2 + AST + MMF + AC + CHL	2.20	8.66	56.60	82.95	27.79	20.11	0.55	2.17	14.15	6.42	23.14	154.04

Solids Treatment												
Vacuum filtration incineration	1.46	2.84	16.60	13.40	6.41	3.07	0.37	0.71	4.15	2.39	6.21	33.45
Anaerobic digestion sludge beds	1.11	2.40	15.33	6.79	3.41	2.97	0.60	3.83	3.83	1.67	4.41	31.53
Anaerobic digestion, vacuum filtration, incineration	1.86	3.68	21.84	12.19	4.14	3.34	0.47	0.92	5.46	2.12	6.32	41.16
Aerobic digestion sludge beds	0.21	0.99	8.87	9.53	3.70	3.01	0.05	0.25	2.22	0.65	2.77	23.62
Anaerobic digestion, vacuum filtration, land application 25% solids	1.24	2.56	15.75	8.23	3.48	2.27	0.31	0.64	3.94	1.89	4.64	29.14

Abbreviations:

AC–Activated Carbon.
AL–Alum addition.
AS–Activated Sludge.
AST–Ammonia Stripping.
BN–Biological Nitrification.
BND–Biological Nitrification and Denitrification.
CHL–Chlorination.
EA–Extended Aeration.
LC1–Lime Coagulation, 1 stage.
LC2–Lime Coagulation, 2 stage.
MMF–Mixed Media Filtration.
NTFN–Nitrification.

TABLE 3-26 Summary of Wastewater Treatment Scheme Costs and Associated Effluent Qualities.

Present Worth $6\frac{1}{8}$% 20 years Million $ MGD	1	10	100																		
				SS	20	15	20	3	3	3	3	3	3	3	3	8	3	2	1	1	1
				BOD	20	15	20	8	8	3	2	5	8	5	8	8	3	5	1	1	2
				COD	50	40	50	25	25	23	5	20	25	20	25	25	22	8	5	5	5
				NH_3-N	15	1	1	15	15	15	15	1	1	1	1	1	2	2	1	2	2
				PO_4-P	10	7	7	10	1	0.2	0.2	7	7	1	1	1	0.2	0.2	1	0.2	0.2
				UOD	99	27	35	81	81	74	71	12	17	12	17	17	14	17	6	11	12
Activated sludge	1.11	5.50	27.40		X	X		X	X	X		X		X		X	X		X	X	
Extended aeration	1.67	9.37	53.80				X						X		X						
Bio. nitrification	1.08	2.72	19.10			X						X		X		X			X		
Bio. denitrification	1.25	4.12	35.70													X			X		
Lime coagulation, 1 stage	1.18	5.39	24.90							X	X										
Lime coagulation, 2 stage	1.47	5.48	37.60														X	X		X	X
Ammonia stripping	0.71	3.11	27.10														X	X		X	X
Alum addition	0.43	2.69	18.53						X					X	X	X			X		
Mixed media filtration	0.62	2.85	13.30					X	X	X	X	X	X	X	X		X	X	X	X	X
Activated carbon tert.	2.71	7.86	59.50																X	X	
Activated carbon sec.	3.22	9.30	70.10								X							X			X
Chlorination	0.13	0.54	2.39		X	X	X	X	X	X	X	X	X	X	X	X	X	X	X	X	X
Anaerobic dig.-beds	1.39	3.81	27.70		X	X	X	X	X	X		X	X	X	X	X	X		X	X	
[a]Present worth M				1	3.08	4.32	3.73	3.76	4.20	5.07	7.30	5.00	4.40	5.44	4.84	6.15	6.38	6.95	10.33	9.36	6.42
G				10	11.34	14.50	15.91	14.43	17.15	20.36	24.16	17.59	19.01	20.31	21.73	21.68	25.22	24.61	34.92	33.88	23.14
D				100	62.21	88.29	97.61	74.62	93.25	101.73	152.11	96.92	112.00	115.55	130.63	145.04	151.80	164.89	235.95	217.47	154.04

[a]Including incidental capital costs.

generated by these treatment alternatives must also be evaluated with respect to treatment cost and possible upgrading requirements. It is interesting to note that the most cost effective treatment alternative changes significantly as the flow treated increases from 10 MGD to 100 MGD. At a flow of 100 MGD, activated sludge with nitrification (alternative d) appears to become a viable cost option.

Consequently, a table of treatment alternative costs (similar to Table 3-27) together with information concerning local solids treatment and disposal options, can enable an engineering planner to determine the most cost effective overall treatment strategy.

3-5.6 EPA Cost Estimates

Comprehensive cost data for wastewater treatment processes are also available from the U.S. Environmental Protection Agency. Over the past 5 years EPA has sponsored two studies which resulted in generalized wastewater treatment cost data. The first is Technical Report EPA430/9-75-002 "A Guide to the Selection of Cost-Effective Wastewater Treatment Systems," published in July 1975. This report developed costs for unit processes; i.e., raw water pumping, preliminary treatment, primary treatment, alternative activated sludge modifications, chlorination, etc. By use of cost curves and equations which define unit costs as a function of fixed cost and flow related variable costs, the cost of an entire wastewater treatment system can be estimated. The costs in the above referenced study were determined utilizing February 1973 as a basis, and must be updated utilizing indices for estimates of current costs.

The second EPA sponsored cost study resulted in a document "An Analysis of Construction Cost Experience for Wastewater Treatment Plants." This report provides estimates of construction costs for treatment plants in six categories of effluent quality. Cost curves in this study were derived from analyzing the construction costs for over 150 treatment plants, and the costs are keyed to December 1975 cost indices.

A comparison of predicted total costs for activated sludge treatment was developed for costs from the EPA cost curves and the cost curves developed by the authors which are presented in this chapter (Ref. 3-85). This comparison is shown in Table 3-28. For the purpose of the comparison, solids treatment was not considered. Therefore, the costs reflect the alternate predictions for activated sludge treatment without solids treatment and without chlorination. The costs are keyed to the second quarter 1976. It is interesting to note that these two independent sources of information agree closely for medium to large flows.

Generalized cost curves cannot adequately reflect the effect of local conditions on construction and operating costs and, consequently, they cannot be utilized

TABLE 3-27 Summary of Case Study Wastewater Treatment Alternatives and Associated Costs and Effluent Qualities.

	Present Worth $6\frac{1}{8}$% 20 Years Million $ MGD											
				SS	30	5	20	20	5	3	3	2
				BOD	30	5	20	20	5	3	2	5
				NH_3-N	1	2	1	2	2	3	3	2
				PO_4-P	10	0.2	7	10	0.2	5	5	0.2
	1	10	100	UOD	50	15	35	40	15	9	8	17
Trickling filter	0.70	2.29	15.80		X	X				X		
Activated sludge	1.11	5.50	27.40					X	X		X	
Extended aeration	1.67	9.37	53.80				X					
Biological nitrification	1.08	2.72	19.10		X							
Nitrification in activated sludge	0.77	1.54	9.28					X				
Lime coagulation, 2 stage	1.47	5.48	37.60			X			X			X
Ammonia stripping	0.71	3.11	27.10			X			X			X
Mixed-media filtration	0.62	2.85	13.30									X
Activated carbon sec.	3.22	9.30	70.10									X
Land application	0.89	7.71	75.90							X	X	
Chlorination	0.13	0.54	2.39		X	X	X	X	X	X	X	X
		[a]Present worth M		1	2.19	3.34	2.06	2.28	3.80	1.96	2.41	6.75
		G		10	6.43	12.90	11.50	8.67	16.60	12.20	15.80	23.60
		D		100	43.80	93.00	66.10	44.80	106.00	109.00	122.00	165.00

[a]Including incidental capital costs.

TABLE 3-28 Comparison of EPA Cost Information with Cost Curves Developed by Pavoni and Perrich (Ref. 3-85) for Activated Sludge Treatment.

	1 MGD		10 MGD		100 MGD	
	EPA	Ref. 3-85	EPA	Ref. 3-85	EPA	Ref. 3-85
Capital \$ million	1.17	0.632	3.41	3.37	19.90	18.0
O + M ¢/1000 gallons	17.80	11.6	5.95	5.13	2.47	2.27
PW \$ million	1.93	1.11	5.88	5.50	30.20	27.4

to accurately predict the costs of individual treatment processes/systems for a given locale. However, generalized cost curves may be effectively utilized to evaluate the relative costs among alternative wastewater treatment systems. With such an approach, the relative costs and efficiencies of alternative treatment systems may be determined in a structured manner resulting in an opportunity for rational wastewater treatment strategy planning.

REFERENCES

3-1. Metcalf and Eddy Inc., *Wastewater Engineering*, 1972, McGraw-Hill.

3-2. Clark, Viessman, and Hammer, *Water Supply and Pollution Control*, 1971, International Textbook Co.

3-3. *Contaminants Concentration Expected Down Stream of Municipal Wastewater Treatment Processes*, EPA, NERC, AWTRL, Cincinnati, Ohio.

3-4. Sawyer and McCarty, *Chemistry for Sanitary Engineers*, 2nd ed., 1967, McGraw-Hill.

3-5. American Public Health Association, *Standard Methods for the Examination of Water and Wastewater*, 13th ed., 1971.

3-6. *Chemical Analyses for Water Quality*, EPA, February, 1973.

3-7. *Methods for Chemical Analyses of Water and Wastes 1971*, EPA, 1971.

3-8. Hem, *Study and Interpretation of the Chemical Characteristics of Natural Water*, Geological Survey Water Supply Paper 1473, U.S. GPO, Washington, D.C., 1970.

3-9. Fair, Geyer, and Okun, *Water and Wastewater Engineering*, 1968, John Wiley & Sons, Inc.

3-10. Benson and Abram, *Control of Communicable Diseases in Man*, 11th ed., American Public Health Association, New York, 1970.

3-11. Goldreich, "Fecal Coliform Concepts in Stream Pollution," *Water and Sewage Works*, Reference Number 1967.

3-12. Long and Bell, "Health Factors and Reused Water," *J. Am. Water Works Assoc.*, April, 1972.

3-13. Taylor, "Viruses–What is their Significance in Water Supplies," *J. Am. Water Works Assoc.*, May, 1974.

3-14. Kugelman, I. J., *Status of Advanced Waste Treatment*, EPA, NERC, AWTRL, Cincinnati, Ohio, May, 1972.

3-15. Helfgott and Hunter, "The Removal of Phosphates from Wastewaters by Electrodialysis", Water–1969, Chemical Engineering Progress Symposium Series, **65**, No. 97.

3-16. Stanier, Doudoroff, and Adelberg, *The Microbial World*, 3rd ed., 1970, Prentice-Hall, Inc.

3-17. Tenney, Echelberger, Guter, and Carberry, "The Removal of Carbonaceous, Phosphatic, and Nitrogenous Pollutants by Biological Treatment," Presented at Symposium on Nutrients in Natural Waters, 161st ACS National meeting, Los Angeles, California, April, 1971.

3-18. Pavoni, *Fractional Composition of Microbially Produced Exocellular Polymers and their Relationship to Biological Flocculation,* Ph.D. Dissertation, 1970, University of Notre Dame.

3-19. Albertsson, McWhirter, Robinson, and Vahldieck, *Investigation into the Use of High Purity Oxygen Aeration in the Conventional Activated Sludge Process*, FWPA, AWTRL, Cincinnati, Ohio, May, 1970.

3-20. Burns and Roe Inc., *Suspended Solid Removal*, EPA, Technology Transfer, October, 1971.

3-21. Stumm and O'Melia, "Stoichiometry of Coagulation," *J. Am. Water Works Assoc.*, May, 1968.

3-22. Black, Birkner, and Morgan, "Destabilization of Dilute Clay Suspensions with Labeled Polymers," *J. Am. Water Works Assoc.*, December, 1965.

3-23. Birkner and Morgan, "Polymer Flocculation Kinetics of Dilute Colloidal Suspensions," Journal AWWA, February, 1968.

3-24. Weber, *Physicochemical Processes For Water Quality Control*, 1972, John Wiley & Sons, Inc.

3-25. Swindell–Dressler Co., *Carbon Adsorption*, EPA, Technology Transfer, October, 1971.

3-26. Channabasappa, "Reverse Osmosis Process for Water Reuse Application," Water-1969, Chemical Engineering Progress Symposium Series, **65**, No. 97, 1969.

3-27. Evans and Wilson, "Capital and Operating Costs—AWT," *J. Water Pollution Control Federation*, January, 1972.

3-28. Ehreth and Barth, *Control of Nitrogen in Wastewater Effluents*, EPA, NERC, Cincinnati, Ohio, March, 1974.

3-29. *Ammonia Removal in a Physical-Chemical Wastewater Treatment Process*, EPA, Office of Research and Monitoring, Washington, D.C., November, 1972.

3-30. *Handbook of Chemistry and Physics*, 45th ed., 1964, Chemical Rubber Company, Cleveland, Ohio.

3-31. O'Farrell, Bishop, and Cassel, *Nitrogen Removal by Ammonia Stripping*, EPA, Office of Research and Monitoring, Washington, D.C. September, 1973.

3-32. Smith, *Design of Ammonia Stripping Towers for Wastewater Treatment*, EPA, NERC, AWTRL, Cincinnati, Ohio August, 1970.

3-33. William Bishop Consulting Engineers, *Palm Beach County Water Quality Management Plan*, September, 1973.

3-34. Sawyer, Wild, and McMahon, *Nitrification and Denitrification Facilities*, EPA, Technology Transfer, August, 1970.

3-35. Barth, Brenner and Lewis, "Chemical-Biological Control of Nitrogen and Phosphorus in Wastewater Effluent," *J. Water Pollution Control Federation*, December, 1968.

3-36. Guter, Echelberger and Tenncy, "Removal of Organic and Inorganic Pollutants from Wastewater by Combined Chemical and Biological Treatment," Proceedings of the Third Annual Environmental Engineering and Science Conference, University of Louisville, Louisville, Kentucky, March, 1973.

3-37. Gavis, *Wastewater Reuse*, NTIS, National Water Commission, Arlington, Virginia, July, 1971.

3-38. Mulbarger, "Nitrification and Denitrification in Activated Sludge Systems," *J. Water Pollution Control Federation*, October, 1971.

3-39. Echelberger and Tenney, "Wastewater Treatment for Complete Nutrient Removal," *Water and Sewage Works*, October, 1969.

3-40. Ames, "Alumina Columns for Selective Removal of Phosphorus from Wastewater," Cincinnati, Ohio, March, 1969.

3-41. Black and Veatch Consulting Engineers, *Phosphorus Removal*, EPA, Technology Transfer, October, 1971.

3-42. Hais, Stamberg, and Bishop, *Alum Addition to Activated Sludge with Tertiary Solids Removal*, EPA, NERC, Cincinnati, Ohio, August, 1973.

3-43. Wynn, Kirkand, and McNabney, *Pilot Plant for Tertiary Treatment of Wastewater with Ozone*, EPA, Washington, D.C., January, 1973.

3-44. Evans, *Ozone in Water and Wastewater Treatment*, 1972, Ann Arbor, Science Publishers.

3-45. Tittlebaum, Pavoni, Spencer, and Fleischman, "Ozone Disinfection of Viruses," Presented at the Institute on Ozonation in Sewage Treatment, University of Wisconsin, Milwaukee, Wisconsin, November, 1971.

3-46. Rosen, "*Use of Ozone and Oxygen in Advanced Wastewater Treatment*," *J. Water Pollution Control Federation*, December, 1973.

3-47. Hewes and Davison, "Renovation of Wastewater by Ozonation," Water–1972, Chemical Engineering Progress Symposium Series, **69,** No. 129, 1972.

3-48. "Ozone Bids for Tertiary Treatment," *Environmental Science and Technology*, November, 1970.

3-49. Proceedings of the Joint Conference on Recycling Municipal Sludge and Effluents on Land, EPA, Washington, D.C., July, 1973.

3-50. *Federal Water Pollution Control Act Amendments of 1972*, Public Law 92-500, October 18, 1972.

3-51. Pound and Crites, *Wastewater Treatment and Reuse by Land Application*, **I & II**, EPA, Washington, D.C., August, 1973.

3-52. Sullivan, Cohn and Baxter, *Survey of Facilities Using Land Application of Wastewater*, EPA, Washington, D.C., July, 1973.

3-53. Frobisher, *Fundamentals of Microbiology*, 8th ed., 1968, W. B. Saunders Co.

3-54. "Chicago Reclaiming Strip Mines with Sludge," *Civil Eng.*, June, 1974.

3-55. Bascom, "The Disposal of Waste in the Ocean," *Sci. Am.*, August, 1974.

3-56. *Alternative Waste Management Techniques for Best Practicable Waste Treatment*, EPA, Washington, D.C., March, 1974.

3-57. *Federal Register*, July 30, 1975, Water Resources Council, "Change in Discount Rate."

3-58. *Federal Register*, August 17, 1973, EPA, "Water Programs, Secondary Treatment Information."

3-59. Paul Wagner, personal communication, EPA, Atlanta Regional Office, 1974.

3-60. *Federal Register*, September 10, 1973, EPA, "Cost Effectiveness Analysis Guidelines."

3-61. Smith, *Cost of Conventional and Advanced Treatment of Wastewaters*, FWPCA, AWTRL, Cincinnati, Ohio, July, 1968.

3-62. Smith and McMichael, *Cost and Performances Estimates for Tertiary Wastewater Treating Processes*, FWPCA, AWTRL, Cincinnati, Ohio, June, 1969.

3-63. Bechtel Corp., *Cost Estimating Guidelines for Wastewater Treatment Systems*, FWQA, Cincinnati, Ohio, July, 1970.

3-64. Patterson and Banker, *Estimating Costs and Manpower Requirements for Conventional Wastewater Treatment Facilities*, EPA, October, 1971.

3-65. Eilers, *Condensed One Page Cost Estimates for Wastewater Treatment*, EPA, AWTRL, Cincinnati, Ohio, November, 1970.

3-66. Smith, *Costs of Wastewater Renovation,* EPA, AWTRL, Cincinnati, Ohio, November, 1971.

3-67. Eckenfelder and Adams, "Design and Economics of Joint Wastewater Treatment," *J. ASCE*, Sanitary Engineering Division, February, 1972.

3-68. Michel, "Costs and Manpower for Municipal Wastewater Treatment Plant Operation and Maintenance 1965-1968," *J. Water Pollution Control Federation*, November, 1970.

3-69. Eckenfelder and Ford, "Economics of Wastewater Treatment," Chemical Engineering, August 25, 1969.

3-70. Berthovex, "Evaluating Economy of Scale," *J. Water Pollution Control Federation*, November, 1972.

3-71. Smith and Eilers, *Cost to the Consumer for Collection and Treatment of Wastewater*, Office of Research and Monitoring, EPA, Washington, D.C., July, 1970.

3-72. Kuman and Clesceri, "Phosphorus Removal from Wastewaters: A Cost Analysis," *Water and Sewage Works*, March, 1973.

3-73. Walter McMichael, Personal Communication, EPA, NERC, AWTRL, Cincinnati, Ohio, 1974.

3-74. Robert Smith, Personal Communication, EPA, NERC, AWTRL, Cincinnati, Ohio, 1974.

3-75. John English, Personal Communication, EPA, NERC, AWTRL, Cincinnati, Ohio, 1974.

3-76. John Westrick, Personal Communication, EPA, NERC, AWTRL, Cincinnati, Ohio, 1974.

3-77. Edwin Barth, Personal Communication, EPA, NERC, AWTRL, Cincinnati, Ohio, 1974.

3-78. Dryden, "Mineral Removal by Ion Exchange, Reverse Osmosis, and Electrodialysis," presented at the Workshop on Wastewater Reuse sponsored by the University of California, Berkeley, at South Lake Tahoe, California, June 25-26, 1970.

3-79. Tchobanoglous, "Wastewater Treatment for Small Communities" Parts I and II, *Public Works*, July and August, 1974.

3-80. Roy F. Weston, Inc. *Upgrading Existing Wastewater Treatment Plants*, EPA, *Technology Transfer*, October, 1971.

3-81. Brenner, *Upgrading Trickling Filters*, EPA, NERC, AWTRL, Cincinnati, Ohio, March, 1974.

3-82. Richard Smith, Personal Communication, Clow Corporation, Florence, Ky., 1974.

3-83. Schimpeler-Corradino Associates, *Water Quality Management Planning Study–Louisville SMSA*, 1974.

3-84. J. L. Pavoni, *Package Treatment Plant Survey*, University of Louisville, Louisville, Ky., June 4, 1974.

3-85. Perrich, J. R., "Economic Analysis of Wastewater Treatment Alternatives," Master of Engineering Thesis, University of Louisville, Louisville, Kentucky, December 1974.

APPENDIX 3-1
COST INDICES

Cost estimates can be updated by use of indices as shown in the equation below:

$$C_2 = C_1 \frac{I_2}{I_1},$$

where

C = cost at times 1 and 2;
I = value of index at times 1 and 2.

Use of this equation assumes that the cost will vary in value at the same time-rate as the value of the index. With the wide variety of indices available, it is usually possible to find a suitable index.

If a cost has two or more components which vary in value at different time-rates, then two or more indices should be utilized for updating.* The equation below is appropriate for updating a cost with two components.

$$C_2 = C_{a1}\frac{I_2}{I_1} + C_{b1}\frac{K_2}{K_1},$$

where

C_a, C_b = components of cost C;
I, K = indices appropriate to each cost component.

APPENDIX 3-2
PRESENT WORTH CALCULATIONS

Present worth analysis is a method of comparing the total costs of alternative projects or processes in which their costs are unequal in time. In essence, the technique brings all anticipated future costs back to the present time and expresses them as a lump sum of money. It is assumed that the lump will be deposited at interest and future expenditures will be drawn from it as they fall due. At the end of the planning time period, all the money is spent.

Given an interest rate, a time period, and the projected future expenditures, the present worth is calculated as below:

$$PW = K_1FW_1 + K_2FW_2 + K_3FW_3 \text{ - - - } K_nFW_n,$$

where

PW = present worth;
FW = expenditures (i.e., future worth) at time n = 1 to n;
K = present worth factor for time n = 1 to n, at the given interest rate.

Present worth factors may be found in most engineering and business economic references.

*An example is the operation and maintenance costs associated with a particular wastewater treatment system. A labor cost index could be utilized for the labor portion and a price index for the materials portion.

If the future expenditures are equal for all time $n = 1$ to n, the equation simplifies to:

$$PW = K'_n \, FW,$$

where

FW = the equal future expenditures,
K'_n = the present worth factor for n equal future expenditures for times $n = 1$ to n.

APPENDIX 3-3
COST CURVE ANALYSIS

The cost curves presented in Chapter 3 were developed and displayed according to EPA procedure. These costs were keyed to second quarter, 1976 cost levels. As price levels in the economy change, and as technology advances, the cost relationship depicted in the cost curves will change.

The capital cost curves and O + M cost curves are the basic cost relationships from which total cost and other cost curves can be developed. These additional cost relationships can then be utilized for treatment cost comparison and water quality planning.

Chapter 3 and Appendix 3–2 demonstrated how total present worth data can be generated and utilized to project wastewater treatment costs. Chapter 3 also presented a simple example of the present worth cost data to water quality planning decision making. It was noted that present worth costs depend on an interest rate and a specified planning period in addition to the capital and O + M cost relationships.

This Appendix demonstrates the generation of more sophisticated wastewater treatment cost relationships. These additional relationships provide the water quality planner with more detailed and explicit cost information about wastewater treatment costs and relationship between costs of different techniques.

Table A3-1 presents equations describing the treatment technique cost data included in Chapter 3. The equations were generated through computer analysis, utilizing the CURFIT program. The equation with the highest index of determination for each treatment cost relationship was selected as representative of that cost relationship. For example, the equations of the cost curves for the activated sludge treatment technique without solids treatment (see curve C in Figure 3-15) are:

$$\text{Capital: } Y = 0.632(X)^{0.727}$$
$$\text{O + M: } Y = 11.6(X)^{-0.354},$$

where Y is the cost ($ million or ¢ per 1000 gallons) and X is the flow rate (million gallons per day).

TABLE A3-1 Cost Equations for Wastewater Treatment Techniques

In these equations, Y is the cost–either capital ($ million) or O + M (¢ per 1,000 gallons). The factor X is the wastewater flow in million gallons per day from 1 to 100 mgd.

Treatment Technique[a]	Equation
a. Activated sludge plant w/incineration	
Capital	$Y = 2.88 + 0.4X$
O + M	$Y = 8.2 + 29.3/X$
b. Activated sludge plant w/anaerobic digestion	
Capital	$Y = 1.65 + 0.429X$
O + M	$Y = 7.68 + 21.8/X$
c. Activated sludge only w/o solids treatment	
Capital	$Y = 0.632(X)^{0.727}$
O + M	$Y = 11.6(X)^{-0.354}$
d. Extended aeration w/o solids treatment	
Capital	$Y = 0.982(X)^{0.799}$
O + M	$Y = 16.5(X)^{-0.331}$
e. Microstraining	
Capital	$Y = 0.0612(X)^{0.889}$
O + M	$Y = 1.02(X)^{-0.0383}$
f. Mixed media filtration	
Capital	$Y = 0.219(X)^{0.648}$
O + M	$Y = 9.54(X)^{-0.323}$
g. Activated carbon as tertiary treatment	
Capital	$Y = 0.833 + 0.239X$
O + M	$Y = 8.09 + 31.5/X$
h. Activated carbon as secondary treatment (95% BOD_5)	
Capital	$Y = 0.890 + 0.248X$
O + M	$Y = 10.3 + 39.9/X$
i. Activated carbon as secondary treatment (99% BOD_5)	
Capital	$Y = 1.05 + 0.268X$
O + M	$Y = 16.5 + 64.4/X$
j. Ion exchange	
Capital	$Y = 0.422(X)^{0.793}$
O + M	$Y = 28.4(X)^{-0.150}$
k. Electrodialysis	
Capital	$Y = 1.01(X)^{0.730}$
O + M	$Y = 22.6(X)^{-0.129}$
l. Reverse osmosis	
Capital	$Y = 0.907(X)^{0.895}$
O + M	$Y = 33.0 + 9.51/X$
m. Brine disposal	
Capital	$Y = 0.595(X)^{0.953}$
O + M	$Y = 1.06(X)^{-0.0485}$
n. Ammonia stripping	
Capital	$Y = 0.103 + 0.143X$
O + M	$Y = 2.99 + 8.24/X$
o. Nitrification in activated sludge	
Capital	$Y = 0.396 + 0.0399X$
O + M	$Y = 1.11 + 6.91/X$

TABLE A3-1 *(Continued)*

Treatment Technique[a]	Equation
p. Separate nitrification	
Capital	$Y = 0.523 + 0.123X$
O + M	$Y = 1.43 + 9.10/X$
q. Denitrification	
Capital	$Y = 0.445 + 0.0926X$
O + M	$Y = 1.25X/(-0.128 + 0.201X)$
r. Phosphorus removal w/Alum	
Capital	$Y = 0.0507e^{0.0896}X$
O + M	$Y = 9.01(X)^{-0.163}$
s. Land application-overland flow	
Capital	$Y = 0.477(X)^{1.0}$
O + M	$Y = 6.78 + 3.13/X$
t. Chlorination	
Capital	$Y = 0.0543(X)^{0.545}$
O + M	$Y = 1.71(X)^{-0.307}$
u. Lime coagulation, one stage	
Capital	$Y = 0.516(X)^{0.617}$
O + M	$Y = 15.9(X)^{-0.307}$
v. Lime coagulation, two stage	
Capital	$Y = 0.606(X)^{0.657}$
O + M	$Y = 5.91 + 15.0/X$
w. Recalcining	
Capital	$Y = 0.381(X)^{0.5}$
O + M	$Y = 9.08(X)^{-0.276}$
x. Incineration	
Capital	$Y = 1.31 + 0.153X$
O + M	$Y = 13.4(X)^{-0.320}$
y. Anaerobic digestion	
Capital	$Y = 0.946 + 0.113X$
O + M	$Y = 0.941 + 3.61/X$
z. Aerobic digestion	
Capital	$Y = 0.0973 + 0.0574X$
O + M	$Y = 0.959 + 6.33/X$
aa. Sludge drying beds	
Capital	$Y = 0.0154 + 0.0302X$
O + M	$Y = 2.24(X)^{-0.0259}$
bb. Incineration of digested sludge	
Capital	$Y = 0.713 + 0.0883X$
O + M	$Y = 2.31 + 5.33/X$
cc. Land application 6% sludge	
Capital	–
O + M	$Y = 4.63$
dd. Land application 25% sludge	
Capital	$Y = 0.147 + 0.033X$
O + M	$Y = 3.68(X)^{-0.227}$

[a]Treatment techniques coded (a, b, c, etc.) as in Chapter 3. See Chapter 3 for further explanation.

The present worth cost of this treatment technique is generated by converting the O + M costs into annual expenditures by applying a conversion factor:

$$\frac{¢}{1{,}000\ \text{gallons}} \times \text{mgd} \times 365\,\frac{\text{days}}{\text{yr}} \times \frac{\$1}{100¢} = 3650.$$

The present worth factor for 6⅛ percent and 20 years is 11.36. Therefore, the total present worth at any given flow rate is calculated as:

$$PW = \text{cap} + (11.36)\,(3650)\,(X\ \text{mgd})\,(\text{O} + \text{M in } ¢/1000\ \text{gal}).$$

This calculation is easily performed, and the resultant present worth cost curve for the activated sludge treatment technique without solids treatment is:

$$PW = 1.09\,(X)^{0.699}$$

An average cost curve can be generated from the present worth curve simply by dividing by the flow rate, X. The resulting curve is:

$$\text{Avg cost} = 1.09\,(X)^{-0.301}.$$

The average cost is simply the cost per mgd of flow that the treatment technique costs at any flow rate. As might be expected, the larger the flow the more efficient and economical the plant becomes, and so the average cost decreases with size.

Another important cost relationship is the marginal cost of treatment. This is the extra cost incurred at any given level of flow to treat the next incremental addition to the flow. For example, the additional cost necessary to treat 11 mgd versus 10 mgd is a marginal cost. Mathematically, the marginal cost is the differential of the total present worth cost with respect to the flow. In this example the marginal cost is:

$$\text{M.C.} = \frac{d\,PW}{dx} = 0.762\,(X)^{-0.301},$$

where M.C. is the marginal cost in \$ million per mgd.

The present worth, average, and marginal cost relationships can be derived for all the wastewater treatment techniques and schemes described in Chapter 3 by procedures identical to this example.

Water quality management planners can utilize the present worth, average and marginal cost relationships, in conjunction with other data such as sewer installation costs, extent of sewer service, projected wastewater flows, etc., to assist in decision making. Planners may utilize these cost curves to optimize wastewater treatment over a wide area by determining the most economical location, size and degree of treatment at wastewater treatment facilities.

Given the basic capital and O + M cost data in Chapter 3, updated as necessitated by price and technology change, a planner can readily generate the present worth, average, and marginal costs of treatment schemes by the procedures demonstrated in this Appendix.

4

Utilization of modeling in water quality management studies

Edward G. Foree

Department of Civil Engineering
University of Kentucky
Lexington, Kentucky

John S. Tapp

Commonwealth Technology, Inc.
Lexington, Kentucky

4-1 INTRODUCTION

4-1.1 General Concept of Modeling

In the most basic form, modeling can be thought of as using some form of mathematical relationship to describe the relationships among a set of variables or natural processes. The ability of a model to represent a given phenomenon depends on how well the mathematical relationships in the model describe the processes which control the event being predicted. The level of complexity of the processes being described determines the degree of sophistication of mathematical relationships and computing techniques required. For the type of modeling generally encountered in a water quality management study of significant scope, a digital computer will most likely be the optimum tool to use. The high speed with which the modern digital computer can solve complex mathematical relationships allows a large range of alternative solutions to be evaluated at a relatively low cost.

Under actual conditions, the physical, chemical, and biological processes to be simulated by water quality models are generally subject to random variations. Models which incorporate this variability into their structure are termed stochastic models. Because little data are available to describe the probability distributions associated with variations in water quality parameters, most models currently used are based on mean parameter values. These are termed deterministic models, and should be considered as producing results approximating average conditions.

Unfortunately, a water quality model sometimes becomes the end rather than the means through which to arrive at the end. In its proper context a model is only a tool which can be used to help in the solution of a problem. Therefore, the model should be geared to the level of detail of the problem, and this interrelationship should then dictate the proper hardware necessary for a solution to the problem. In using a model of any type, whether it be a simple equation or a complex computerized model, one should have a thorough understanding of the process or processes to be simulated. A lack of this understanding can result in arriving at a solution which has no meaning in a "real world" sense.

Because many of the relationships which the models used in water quality management studies attempt to describe are complex (containing many variables), the interrelationship between variables may be unknown or difficult to describe, and the models must generally be calibrated or verified using actual data. The terms calibration and verification are used somewhat loosely in the modeling field. Calibration usually denotes the testing of the model against real data for a small number of events, while verification generally means testing the model over the entire range of events or processes which the model might be designed to simulate. In its purest sense, true verification would probably never be attained. The calibration-verification process is essentially the process by which the model user determines the ability of the model to simulate the necessary processes, and allows the model user to develop confidence in the model or to reject the model because of its inability to simulate the necessary processes.

Since most water quality management studies are complex in nature, the models to be discussed herein are those which can be utilized in these studies to focus in on the problems normally encountered and will, by nature, be of the type for which a solution by digital computer is generally the most efficient technique.

4-1.2 Objectives of Computer Modeling in Water Quality Management Studies

The term "water quality management study" can take on various meanings, depending on the particular interests of the individual viewing the term. A "water quality management study" may be distinguished from a "water quality study" from the standpoint that a "water quality management study" not only deals with a study of the water quality of an area, but further deals with studying the management framework and interrelationships necessary to solve the water quality problems identified in the water quality study. The main utilities of using models in water quality management studies lie in their ability to project some future condition based on measurements of past conditions and the speed and efficiency with which a range of alternatives may be evaluated. In many cases, exact solutions are not attainable, but a model easily allows the determination of relative impacts based on a wide range of conditions.

The three main types of ongoing water quality management studies are those conducted under sections 303, 208, and 201 of Public Law 92-500. The objectives of the studies conducted under each section of the act vary and, therefore, the tools or models used are somewhat different.

303–River Basin Plans. The River Basin Water Quality Management Plans and the planning process established under section 303 of PL 92-500 were designed to be the primary means utilized by a State Water Pollution Control Agency in managing water quality control efforts in a river basin. In general, the 303 plans contain basin inventories and other information necessary to establish the quality of an effluent that could be discharged to a given stream with the expectation of allowing instream water quality standards to be maintained. The primary tools utilized in establishing effluent limitations are water quality models—either stream or estuarine, depending on the location of the discharger. In general, the main parameters considered in the 303 analysis are dissolved oxygen and the various oxygen demanding constituents. In some instances, other parameters are considered. The complexity of the modeling effort is geared to the level of detail necessary to establish confidence in the results and to reflect the capital expenditure which would be based on the results. A "Type A" analysis is applied to a noncritical situation, where a minimum of expenditure for wastewater treatment would be required and a one dimensional steady state model based on little or no actual field data is used in establishing the effluent limitation. A "Type B" analysis is applied to a more critical situation, and uses a one dimensional model which is calibrated and/or verifed, based on actual field measurements. A "Type C" analysis is applied to a complex situation, and a two or three dimensional model, based on a large amount of field data, is used. In utilizing the above analyses, the 303 plans establish the effluent limitations that are the basis for the more detailed 201 Facility Plans. Both the 303 plans and the 201 plans are primarily concerned with point sources of wastewater.

201–Facility Plans. The Facility Plans prepared under section 201 of Public Law 92-500 are the "hardware plans," and are used to evaluate a wide range of alternatives within the water quality constraints developed by the 303 plans. They are designed to arrive at the most cost effective design of wastewater interception and treatment facilities for a given area. The Facility Plans are normally based on the twenty year projected needs. Therefore, projections of population are important, and a range of alternatives should be evaluated based on different potential growth and land use patterns. Computer modeling can allow the evaluation of a range of growth and land use patterns with a minimum of effort. Computer modeling is also useful in the Facility Plans to optimize or arrive at a least cost solution to the location and sizing of treatment plants, force mains, pump stations, and interceptors. In the classical pump or treat scheme, modeling easily allows a range of alternatives based on the land use and population projection schemes that are developed. Another type of computer model useful in the preliminary design of wastewater treatment plants, which is a part of the Facility Plan, is one which allows a range of treatment processes to be evaluated and sized, based on a range of influent parameters and varying assumptions concerning the process constants used in the design.

208–Water Quality Management Plans. There are two types of water quality management plans to be conducted under section 208 of Public Law 92-500–areawide and statewide. Areawide Plans are to be conducted on complex urban-industrial areas to develop coordinated waste treatment schemes for a given area or region. Statewide water quality management plans are to be conducted for the rest of a given state, with the majority of the resources put into known problem areas.

The 208 plans, unlike the 303 or 201 plans, are to concentrate on both point and nonpoint sources of pollution. Thus, 208 studies look at a wide range of problems. The 201 study can actually be a part of the 208 study and the 201 computer modeling described above is also applicable to 208 studies. In general, the 208 studies are conducted within the framework of the 303 plans. However, where sources of pollution not considered in the 303 plans, such as nonpoint sources, are important, the 208 studies will further the analyses conducted through the 303 plans. Due to the complexity and variability of the nonpoint source problem in general, and the urban runoff problem in particular, computer models are necessary to test a wide range of assumptions and evaluate relative impacts. Urban runoff may be distinguished from general nonpoint source runoff when the former is collected in drainage systems. However, in cases where storm sewer systems typically cover small areas (for example, a few square blocks) and discharge at numerous points in a study area, they are sometimes considered as general nonpoint sources in 208 studies. Since the nonpoint source/urban runoff problem is event oriented, it is dynamic in nature and the computer models tend to become somewhat more complex than those used in the 303 plans. A wide range of computer models is generally used, including nonpoint source/urban runoff generation and, possibly, routine and complex water quality models which may dynamically simulate the effects of these rainfall related events on receiving water quality.

4-2 GENERAL MATHEMATICAL FORMULATIONS FOR WATER QUALITY MODELING

In order to simulate the behavior of a substance of interest in a water body, a differential equation is first formulated which takes into account all factors significantly affecting the behavior of the substance. Under certain conditions, an explicit solution of the differential equation may be possible, and the resulting integrated equation is incorporated directly into a model for subsequent analysis and simulation. In other cases, an explicit solution of the governing differential equation may not be possible and the differential equation itself must be incorporated into the model. The differential equation is then solved for analysis and simulation by utilizing an appropriate mathematical solution technique, such as finite difference.

The following presentation of equations, typically applicable to water quality modeling, begins with a general case and proceeds to more limited cases in which some factors may be insignificant and, therefore, may be deleted from the general equation. The following general equation, as given by Water Resources Engineers (Ref. 4-57), describes the behavior of a substance in water in one dimension (X) in which its concentration is affected by dispersion, advection, growth or decay, and other sources and sinks. The concepts associated with this and the following equations may be extended to two or three physical dimensions to develop the corresponding two or three dimensional equations.

$$(A\, dX)\frac{\partial C}{\partial t} = \frac{\partial\left(AD_L \dfrac{\partial C}{\partial X}\right)}{\partial X}\, dX - \frac{\partial (AUC)}{\partial X}\, dX + (A\, dX)\frac{dC}{dt} \pm S \quad \text{(Eq. 4-1)}$$

where

C = concentration of substance, M/L^3
X = distance, L
t = real time, T
A = cross-sectional area perpendicular to X, L^2
D_L = dispersion coefficient, L^2/T
U = water velocity perpendicular to X, L/T, and
S = source or sink of substance, M/T.

A solution to this equation would typically yield a relationship expressing the dependent variable, C, as a function of the independent variables, X and t. To make such a solution, the other variables (A, D_L, U, S) must be replaced in Eq. 3-1 by appropriate expressions, in which they are expressed as functions of C, X, t, and constant terms and coefficients. For example, A may be a function of X and t; U may be a function of X and t, sources and sinks (S) may be functions of C, X, and t; and D_L may be a constant.

4-2.1 Basic System Characteristics

The particular type of model and the basic equations to be used to simulate a system should be a function of the characteristics of that system. For example, a steady state solution to the basic equation given above may be sufficiently accurate for many systems. In others a real time solution may be necessary, due to continual variation in key elements of the system. In addition to the time characteristics of a system, the advective/dispersive nature of the system is important. In some systems either advection or dispersion alone may be the primary mechanism by which substances are transported while in others both advection and dispersion may play key roles. A third characteristic of importance is the number of dimensions in which a system must be simulated to accurately reflect the real situation. Models can be developed in one, two, or three dimensions, as appropriate. In some systems, no variations will exist in any dimension, and the system can be said to be completely mixed.

Time Varying Systems. Where the system is continually changing and a steady state condition is never reached, a real time simulation may be necessary. The time step to be used in this type of simulation varies, depending on how fast the system changes. In some situations, time steps on the order of minutes may be necessary, while other situations may call for time steps on the order of months. For the time varying situation, the basic equation remains as Eq. 4-1. Real time simulations may be necessary for high flow conditions in streams, for situations dealing with rainstorm events, for situations where pollutant inputs constantly change, for systems influenced by flow regulation, and for many estuaries.

Steady State Systems. Where the system being simulated does not change with time, the system is said to be at steady state and the local derivative $\partial C/\partial t$ = 0. For this situation, the basic equation reduces to:

$$\frac{\partial\left(AD_L \dfrac{\partial C}{\partial X}\right)}{\partial X}\,dX - \frac{\partial(AUC)}{\partial X}\,dX + (A\,dX)\frac{dC}{dt} \pm S = 0 \qquad \text{(Eq. 4-2)}$$

The assumption that a system is at steady state is generally applicable for flowing streams at moderate to low flows, for estuaries where constituents do not change over the tidal cycle, and, in many cases, for lakes and reservoirs.

Systems Characterized by Advection Plus Dispersion. The combined effects of advection and dispersion can probably best be illustrated schematically. For this purpose, Figures 4-1A and 4-1B describe the behavior of a substance after its discharge into a receiving stream under the following set of conditions: 1) the

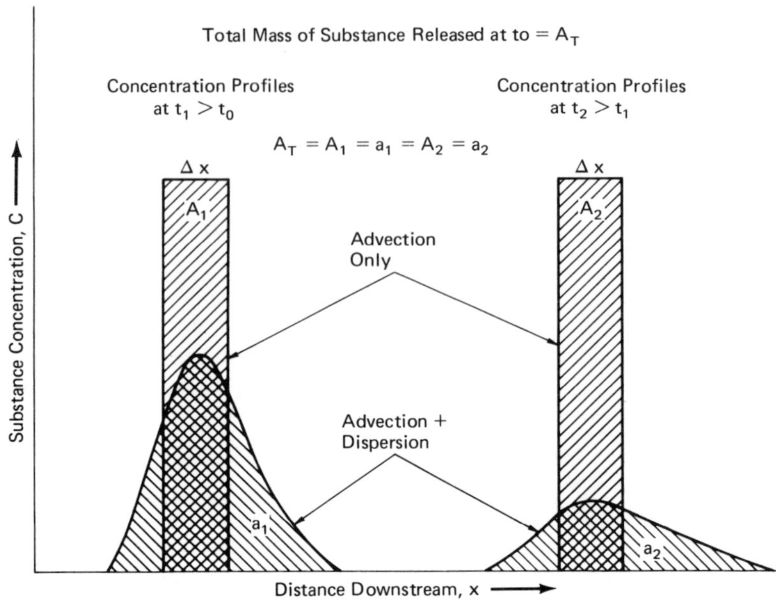

Fig. 4-1A. Effects of advection and dispersion on concentration profiles resulting from the introduction of a slug of conservative substance into a stream.

substance is discharged at a constant rate over a finite time so that it is initially distributed over a longitudinal distance, ΔX, 2) downstream from the discharge, all properties of the stream, such as flowrate and velocity, are constant, 3) the stream is one dimensional (as subsequently defined in this section), 4) net velocity is positive in the downstream (+) X direction, and 5) Figure 4-1A represents the case for a conservative substance and Figure 4-1B represents the case for a nonconservative substance. A nonconservative substance decays as a function of time, while a conservative substance does not.

For both Figures 4-1A and 4-1B, the total mass of substances initially released (designated A_T is the same, and the area under the concentration profile curves (designated A_1, a_1, etc.) represents the total mass present at the corresponding time after release (t_1, t_2, etc). The effect of advection is reflected in both figures by the displacement downstream of the concentration profiles. The effect of dispersion is reflected in both figures by the spreading of the concentration profiles and by the lowering of the peak concentrations. The combined effects of advection and dispersion are reflected in the skewness of the concentration profiles in the downstream direction. The effects of substance decay are reflected by comparison of corresponding areas under the concentration profile curves between Figures 4-1A and 4-1B, and by the equalities and inequalities shown.

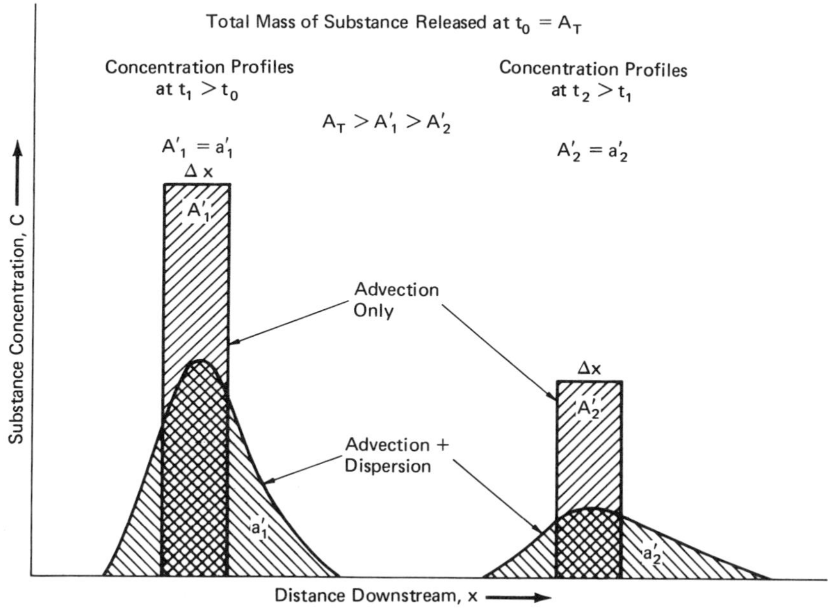

Fig. 4-1B. Effects of advection and dispersion on concentration profiles resulting from the introduction of a slug of nonconservative substance into a stream.

The effects of advection and dispersion may also be illustrated for the case of a continuous point source discharge to a one dimensional stream with the constant properties assumed above. Figures 4-2A and 4-2B schematically show the instream concentration profiles resulting from the continuous discharge at $X = 0$ of a conservative and nonconservative substance, respectively, at a constant rate. Note that for a conservative substance (Figure 4-2A) the effect of dispersion is to produce a concentration profile upstream from the discharge point, while identical concentration profiles are produced downstream from the discharge point for both advection only and for advection plus dispersion. For a nonconservative substance (Figure 4-2B) the effect of dispersion also produces an upstream concentration profile, but is diminished by decay as compared with Figure 4-2A. However, for the nonconservative substance the maximum concentrations (at $X = 0$) and the shapes of the downstream concentration profiles are not the same for the advection only and advection plus dispersion cases. These are both a function of the dispersion coefficient.

For systems characterized by both advection and dispersion, Equation 4-1 would apply for time varying situations and Equation 4-2 would apply for systems which can be assumed to be at steady state. Systems which are normally

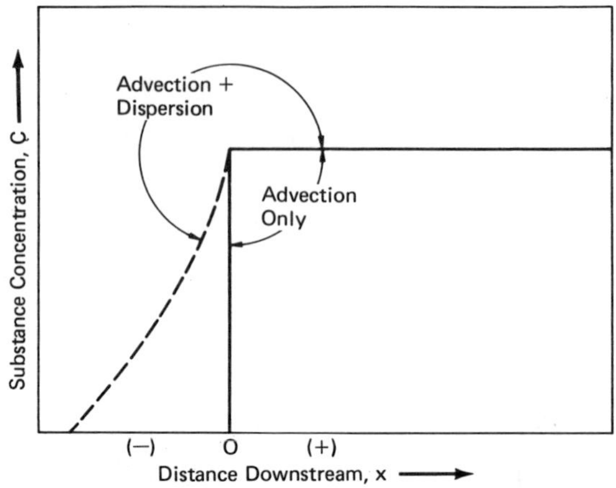

Fig. 4-2A. Effects of advection and dispersion on concentration profiles resulting from a constant continuous discharge of conservative substance to a stream.

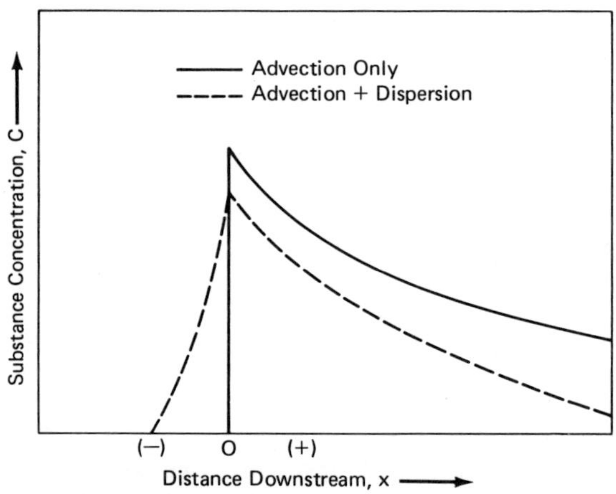

Fig. 4-2B. Effects of advection and dispersion on concentration profiles resulting from a constant continuous discharge of nonconservative substance to a stream.

simulated as advective plus dispersive are major rivers (both tidally and nontidally influenced).

Systems Characterized by Advection Only. In practice, an advection only system is probably nonexistent, but in many systems, the effects of dispersion are so minor that they may be neglected. For these systems D_L = 0 and Equation 4-1 reduces to:

For steady state simulations:

$$-\frac{\partial(AUC)}{\partial X}\,dX + (A\,dX)\frac{dC}{dt} \pm S = 0 \qquad \text{(Eq. 4-3)}$$

For time varying simulations:

$$(A\,dX)\frac{\partial C}{\partial t} = -\frac{\partial(AUC)}{\partial X}\,dX + (A\,dX)\frac{dC}{dt} \pm S \qquad \text{(Eq. 4-4)}$$

Systems which can generally be simulated as advection only are many free-flowing streams and some large tidal rivers.

Systems Characterized by Dispersion Only. Systems where the velocity of flow is not significant and substances are transported only by diffusion from regions of higher concentration to regions of lower concentration can be simulated using dispersion only transport of materials. If the velocity $U = 0$, Equation 4-2 becomes:

For time varying simulations:

$$(A\,dX)\frac{\partial C}{\partial t} = \frac{\partial\left(AD_L\dfrac{\partial C}{\partial X}\right)}{\partial X}\,dX + (A\,dX)\frac{dC}{dt} \pm S \qquad \text{(Eq. 4-5)}$$

For steady state simulations:

$$\frac{\partial\left(AD_L\dfrac{\partial C}{\partial X}\right)dX}{\partial X} + (A\,dX)\frac{dC}{dt} \pm S = 0 \qquad \text{(Eq. 4-6)}$$

Systems for which dispersion only may apply are some large estuaries and lakes.

Completely Mixed Systems. A system is defined as completely mixed if the concentration in the outflow is the same as the concentration in the system itself. If the assumptions are made that the rates of inflow and outflow are constant with time, the substance decay follows first order kinetics, and complete mixing of substance and system water are occurring, the following expression may be developed to express substance concentration as a function of time:

$$C = \frac{W}{Q + Vk}\left\{1 - e^{-t[(1/t_o) + k]}\right\} + C_o e^{-t[(1/t_o) + k]} \qquad \text{(Eq. 4-7)}$$

where

C = concentration of substance in the system at time t;
W = mass rate of substance inflow to the system;

Q = flowrate out of the system;
V = system volume;
k = first order reaction rate coefficient;
C_o = concentration of substance in the system at time ($t = 0$); and
t_o = hydraulic retention time of the system = V/Q.

The system will eventually reach equilibrium, at which time the mass lost by outflow and decay would equal the mass added, or the concentration in the system would not change with time. Equilibrium would theoretically occur at time ($t = \infty$), and equilibrium concentration, C_e, may be calculated as:

$$C_e = \frac{W}{Q + Vk}. \qquad \text{(Eq. 4-8)}$$

Because equilibrium is theoretically never reached, the period of time required for the concentration to reach 99 percent of the equilibrium concentration may be taken as the practical time required for equilibrium. The time to reach practical equilibrium t'_e can be calculated as follows for the case where $C_o = 0$:

$$t'_e = \frac{\ln\left(\frac{1}{1 - C/C_e}\right)}{\frac{1}{t_o} + k}. \qquad \text{(Eq. 4-9)}$$

Examples of systems where complete mixing analysis may be appropriate are some lakes, estuaries, and harbors.

Dimensions. The number of dimensions in which a system must be modeled to accurately reflect its characteristics is most important. When the differential equations presented above must be solved in more than one dimension, a much more sophisticated solution technique, generally requiring a more sophisticated computational machine, is necessary. A good discussion of solution techniques for modeling in more than one dimension is given by Thomann (Ref. 4-43). In general, a one dimensional model will simulate stream and river behavior with sufficient accuracy. An exception might be near field effects in the vicinity of the discharge for a large river. Tidal rivers may in some cases be simulated with a one dimensional model in the longitudinal direction, but in many cases a two dimensional model simulating the longitudinal and the lateral directions may be necessary. Due to the salt wedge in tidal rivers, a two dimensional model in the longitudinal and vertical direction or a three dimensional model may sometimes be required. Estuaries generally require a two dimensional model in the longitudinal and lateral direction or a three dimensional model. Lakes may need to be simulated as completely mixed, one dimensional in the vertical direction, two dimensional in the longitudinal and vertical direction, or three dimensional.

While simulating certain dimensions of a system may be necessary to very accurately represent the system, the availability of data, the ability to solve the differential equations using a practical amount of computer time, and other controls may necessitate that the problem solver settle for something less than optimum. Three dimensional models are rare today as are tidal river and estuary models which simulate in the vertical direction. Lake models which simulate in the longitudinal as well as the vertical direction are also rare.

4-2.2 Types of Constituents

Constituents may be classified into two general categories, nonconservative and conservative. Nonconservative substances are those for which the mass changes as a function of time, while conservative substances are those for which the mass is independent of time. A given constituent may be conservative in certain systems and nonconservative in others. For example, in an aquatic system characterized by aerobic conditions and the absence of significant plant growth, nitrate would behave as a conservative substance; in a system characterized by aerobic conditions and significant growth of algae or other aquatic plants, nitrate might behave as a nonconservative substance, its concentration being subject to change by plant synthesis, and; in a system characterized by anaerobic conditions, nitrate might behave as a nonconservative substance, its concentration being subject to change by denitrifying bacteria.

Nonconservative Substances. For most water quality applications, the time rate of change of concentration or mass of nonconservative substances is described by either first order kinetics or kinetics of the Michaelis-Menten type. In first order kinetics, the time rate of change of concentration of a substance is formulated as proportional to the concentration, C, present at time, t, as follows:

$$\frac{dC}{dt} = kC. \qquad \text{(Eq. 4-10)}$$

Integration of this expression yields:

$$C_t = C_o e^{kt} \qquad \text{(Eq. 4-11)}$$

where

C_t = concentration of substance at time $t, M/L^3$
C_o = initial concentration of substance at $t = 0, M/L^3$
t = time, T, and
k = first order rate coefficient, T^{-1}.

For substance decay (decrease of concentration with time) k is algebraically negative, and for growth (increase of concentration with time) k is algebraically positive. First order kinetics are typically used in water quality modeling applications to describe the loss of biochemical oxygen demand, the die-off of coliform bacteria, decay of radioactive substances, atmospheric reaeration (oxygenation), and similar phenomena.

Kinetics of the Michaelis-Menten type are frequently used to describe microbial growth and substrate utilization in biological systems, and may be formulated as:

$$\mu = \hat{\mu}\left(\frac{C}{k_C + C}\right), \qquad \text{(Eq. 4-12)}$$

where

μ = specific substrate utilization rate = mass of substrate removed per unit time per unit microbial mass present

$\hat{\mu}$ = maximum specific substrate utilization rate, occurs at large substrate concentration (see Figure 4-3)

C = substrate concentration, and

k_C = "half saturation concentration," also sometimes termed the Michaelis-Menten coefficient or constant, equal to the substrate concentration corresponding to a specific substrate utilization rate equal to one-half of the maximum rate, $\hat{\mu}$ (see Figure 4–3).

For small substrate concentrations ($C << k_C$), the specific substrate removal rate is first order with respect to substrate concentration ($\mu \simeq C/k_C$); and for

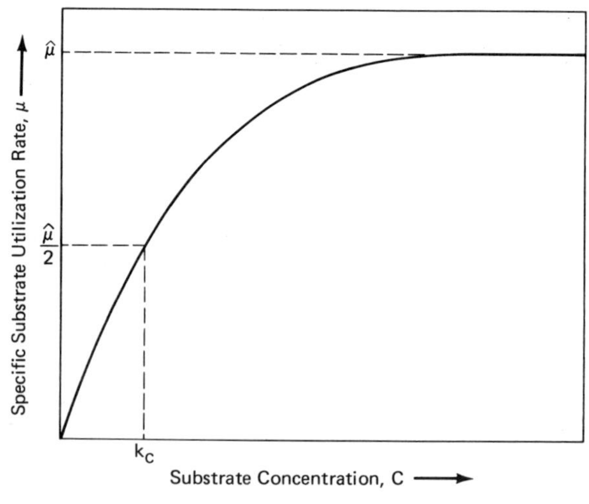

Fig. 4-3. Schematic representation of Michaelis–Menten kinetics for microbial substrate utilization.

large substrate concentrations ($C >> k_C$), the specific substrate removal rate is zero order with respect to substrate concentration ($\mu \simeq \hat{\mu}$).

Equation 4-12 applies for situations in which microbial growth is limited by the availability of a single substrate; *i.e.*, all other substances necessary for growth are assumed to be present in abundance. To describe situations in which more than one substance may be limiting growth, the following expression has been postulated, in which the numerical subscripts denote different substances (Ref. 4-56):

$$\mu = \hat{\mu}\left(\frac{C_1}{k_{C_1} + C_1}\right)\left(\frac{C_2}{k_{C_2} + C_2}\right)\cdots\left(\frac{C_n}{k_{C_n} + C_n}\right) \qquad \text{(Eq. 4-13)}$$

Conservative Substances. Simulation of the behavior of conservative substances generally requires a less sophisticated approach than that required for nonconservative substances. Since, by definition, the mass of a given conservative substance does not change with time, an accounting of the inputs and the outputs of the substance to the system is all that is generally necessary. However, in complex systems, the actual distribution of a conservative constituent may be significant.

In the foregoing general differential equations, the time dependency of constituent concentration is reflected in the (AdX) (dC/dt) term. For nonconservative substances, the appropriate kinetic expressions should be incorporated into this term; and, for conservative substances, it becomes zero.

4-3. DISSOLVED OXYGEN (DO) AS A PRIMARY PARAMETER FOR MODELING IN WATER QUALITY MANAGEMENT STUDIES

Of the numerous parameters for which permissible limits are established in typical surface water quality standards, DO concentration is generally considered the most significant. From a surface water quality standpoint, the presence of sufficient DO is desirable for aesthetic reasons and for its desirable reactions with other chemical constituents. However, its greatest significance lies in the necessity of certain minimum concentrations to support and maintain desirable aquatic biota.

From a water quality management standpoint, its significance lies in the fact that much of the planning for future water quality management and the resultant expenditure of funds for implementation of the plans will center around management of point and nonpoint sources of pollution to minimize their detrimental effects on the DO resources of the receiving waters. Comparisons of the results of current surface water quality monitoring with the applicable water quality standards typically shows that the frequency of violation of standards is much greater for DO than for any other parameter.

Not all surface water quality standards specify the same permissible limits and criteria for DO concentration, due to geographical location, regulatory agency philosophy, and other reasons. However, they are normally established along the lines of the following recommendations, prepared by the Federal Water Pollution Control Administration in 1968 (Ref. 4-8).

> "The following environmental conditions are considered essential for maintaining native populations of fish and other aquatic life.
>
> (1) For a diversified warm-water biota, including game fish, DO concentration should be above 5 mg/l, assuming normal seasonal and daily variations are above this concentration. Under extreme conditions, however, they may range between 5 and 4 mg/l for short periods during any 24-hour period, provided that the water quality is favorable in all other respects. In stratified lakes, the DO requirements may not apply to the hypolimnion. In shallow unstratified lakes, they should apply to the entire circulation water mass.
>
> These requirements should apply to all waters except administratively established mixing zones. In lakes, such zones must be restricted so as to limit the effect on the biota. In streams, there must be adequate and safe passageways for migrating forms. These must be extensive enough so that the majority of plankton and other drifting organisms are protected (see section of zones of passage).
>
> (2) For the cold-water biota, it is desirable that DO concentrations be at or near saturation. This is especially important in spawning areas where DO levels must not be below 7 mg/l at any time. For good growth and the general well-being of trout, salmon, and their associated biota, DO concentrations should not be below 6 mg/l. Under extreme conditions, they may range between 6 and 5 mg/l for short periods provided the water quality is favorable in all other respects and normal daily and seasonal fluctuations occur. In large streams that have some stratification or that serve principally as migratory routes, DO levels may range between 4 and 5 mg/l for periods up to 6 hours, but should never be below 4 mg/l at any time or place.
>
> (3) DO levels in the hypolimnion of oligotrophic small inland lakes and in large lakes should not be lowered below 6 mg/l at any time due to the addition of oxygen-demanding waste or other materials."

The maintenance of a desirable DO concentration in surface waters is hindered by several factors. Oxygen is only a slightly soluble gas in water, saturation concentration being on the order of 8 mg/l under summer conditions. The driving force for solution of oxygen from the atmosphere into water is the concentration deficit or the difference between the actual water concentration and the saturation concentration. To maintain desirable DO conditions, this driving force must by definition be small, typically less than 3 mg/l (8 mg/l saturation minus 5 mg/l minimum desirable). Elevated temperature adversely affects dissolved oxygen concentration in two important ways: 1) the saturation concentration decreases with increasing temperature; and 2) the rates of both biochemical and chemical reactions which result in DO depletion generally increase with increasing temperature.

The following sections describe various sources and sinks of DO. A source is defined as a phenomenon which results in a positive contribution to the water DO concentration, and a sink is a phenomenon which results in a depletion from or a negative contribution to the water DO concentration. In order to mathematically incorporate sources and sinks into overall rate expressions typically utilized in DO modeling applications, it is necessary to describe each with a rate function. Most applications require a volumetric rate function having the dimensions of mass per unit volume per unit time. The following rate functions are presented on this basis and can be thought of as the type that would be applied for DO modeling of a water body such as a segment of a flowing stream.

4-3.1 Sources of Dissolved Oxygen in Surface Waters

There are two general sources which significantly contribute to DO concentrations in surface waters: atmospheric reaeration and photosynthesis by aquatic plants.

Atmospheric Reaeration. Atmospheric reaeration is the term normally used to describe the transfer of oxygen from the gaseous phase in the atmosphere to the solution phase in water in contact with the atmosphere. The time rate of DO deficit depletion may be described as:

$$\frac{\mathrm{d}D}{\mathrm{d}t} = -k_a D \qquad \text{(Eq. 4-14)}$$

$$D = C_s - C, \qquad \text{(Eq. 4-15)}$$

where, D is the DO deficit concentration, t is time, k_a is the atmospheric reaeration rate coefficient, C_s is the DO saturation concentration, and C is the actual water DO concentration. This formulation is universally applied to describe atmospheric reaeration in DO modeling applications. Formulations utilizing the surface transfer rate coefficient, k_L, are sometimes used to describe atmospheric reaeration. For a water body which is essentially mixed throughout its depth, H, k_a may be determined from k_L as:

$$k_a = \frac{k_L}{H}. \qquad \text{(Eq. 4-16)}$$

Several methods have been used for determination of values of k_a. Values have been reported from less than 0.10/day base e for large sluggish rivers, to greater than 50/day for steep turbulent streams. Since C_s is essentially constant for a given modeling application, the volumetric rate of contribution to DO concentration from reaeration may be expressed as:

$$\left(\frac{\partial C}{\partial t}\right)_a = k_a D = k_a\,(C_s - C). \qquad \text{(Eq. 4-17)}$$

Special consideration must be given in DO modeling at points in streams where unique reaeration phenomena occur. An example is the reaeration occurring as a result of turbulence created at a dam or waterfalls. Various formulations have been used to describe these type phenomena.

Photosynthesis. Photosynthesis by aquatic plants represents the other significant source of DO to surface waters. Although many types of aquatic plants contribute DO to water throughout their photosynthetic activity, algae are generally the most important in DO modeling applications. Photosynthesis is an autotrophic metabolic process requiring light as an energy source (solar energy in natural surface waters) and carbon dioxide as a carbon source for synthesis. Thus, in natural surface waters, active photosynthesis occurs only during light periods. In the absence of light, algae carry on respiration (essentially the reverse of photosynthesis), a process in which the cellular protoplasm is consumed for maintenance energy. During respiration, oxygen is consumed and carbon dioxide is produced as a by-product. In DO modeling, respiration may be considered mathematically as a negative source or as a sink of DO.

The behavior of the two sources of DO, reaeration and photosynthesis, may be contrasted with regard to their time dependency. While the reaeration rate, as characterized by k_a, may generally be considered time independent for a given water body over the period for which the modeling is being performed, photosynthesis rates change drastically over a 24-hour period, due to the dependency upon solar energy.

O'Connor (Ref. 4-34) has defined P as the overall rate at which oxygen is released by photosynthesis. He assumed the photosynthetic rate to vary as sunlight intensity during the light and to be zero during the dark, defining this relationship by the following periodic function:

$$P(t) = P_m \sin \frac{t}{p}\,\pi, \quad \text{for } 0 \leqslant t \leqslant p \qquad \text{(Eq. 4-18)}$$

$$p(t) = 0, \text{ for } (1 - p) \leqslant t \leqslant 1 \text{ day} \qquad \text{(Eq. 4-19)}$$

If the photosynthetic period is assumed to be 12 hours (0.5 day), then this function may be described as a Fourier Series which may be approximated by the first three terms. Using this approximation, the volumetric rate of contribution to DO concentration from photosynthesis may be expressed as:

$$\left(\frac{\partial C}{\partial t}\right)_p = P_m \left[\frac{1}{\pi} + \frac{1}{2} \sin \pi \frac{t}{p} - \frac{2}{3\pi} \cos 2\pi \frac{t}{p}\right], \qquad \text{(Eq. 4-20)}$$

where P_m is the amplitude of the wave, the maximum rate over the period, p. In this expression p must be equivalent to 12 hours, and t is the time since the beginning of the photosynthetic period expressed in the same units as p. Values for P_m are highly variable, ranging from zero to 30 mg/l/day or greater.

4-3.2 Sinks of Dissolved Oxygen in Surface Waters

Several DO sinks have been identified in surface waters, as discussed below. However, seldom do all of these exert a significant influence in a single water body at the same time, and thus, some may normally be eliminated for a specific DO modeling application.

Carbonaceous Biochemical Oxygen Demand (CBOD). CBOD results from the oxidation of organic materials by indigenous organisms. The reaction is normally considered to be first order and described as:

$$\frac{dL}{dt} = -k_d L, \qquad \text{(Eq. 4-21)}$$

where L is the ultimate CBOD (that amount of CBOD exerted in an infinite time) and k_d is the CBOD deoxygenation rate coefficient. The volumetric rate of depletion of DO concentration from CBOD may be expressed as:

$$\left(\frac{\partial C}{\partial t}\right)_l = -k_d L. \qquad \text{(Eq. 4-22)}$$

Nitrogenous Biochemical Oxygen Demand (NBOD). NBOD results from the oxidation of reduced forms of nitrogen by indigenous organisms. The reduced forms of nitrogen are inorganic ammonia and nitrite, and organic nitrogen, which is normally liberated as ammonia upon decomposition of the nitrogen-containing organic material. The overall oxidation of ammonia nitrogen (lowest oxidation state) to nitrate nitrogen (highest oxidation state, stable form in aerobic environments) is a two-step process accomplished by two groups of "nitrifying" bacteria. In the first step, ammonia is oxidized to nitrite by *Nitrosomonas* bacteria, and in the second step, nitrite is oxidized to nitrate by *Nitrobacter* bacteria.

The kinetics to accurately describe this overall two-step reaction are quite complicated. However, it has been found that in most surface waters, the first step is the overall rate limiting step (as evidenced by the fact that nitrite is seldom found to be present in other than small concentrations). Based on this and other observations, NBOD is usually approximated for water quality modeling purposes by a first order reaction in which the rate of decay of ammonia is proportional to its concentration. For DO modeling, it is necessary to convert ammonia concentrations to their oxygen demanding equivalency. The theoretical quantity of oxygen required to completely oxidized ammonia nitrogen to nitrate nitrogen is 4.57 gm of oxygen per gm of ammonia nitrogen. A conversion factor somewhat less than this is normally used in practice because some of the ammonia nitrogen is synthesized by the microbes and is not available for oxidation. However, the ammonia nitrogen available for oxidation may be increased by that liberated from the decomposition of nitrogen containing organic materials. This is not accounted for when NBOD is computed as proportional to the ammonia concentration. Alternatively, NBOD may be computed as proportional to total Kjeldahl nitrogen (organic plus ammonia nitrogen). In this case the NBOD may be overestimated, because a portion of the organic nitrogen may not be converted to ammonia nitrogen.

Assuming a first order reaction and proportionality to ammonia concentration, NBOD is typically formulated for DO modeling purposes as:

$$\frac{dN}{dt} = -k_n N \qquad \text{(Eq. 4-23)}$$

$$N = (4.5)\,(NH_3\text{-N}), \qquad \text{(Eq. 4-24)}$$

where N is the ultimate NBOD, NH_3-N is the ammonia nitrogen concentration expressed as N, and k_n is the NBOD deoxygenation rate coefficient. Values of k_n typically range from 0.05 to 0.50/day base e. The volumetric rate of depletion of DO concentration from NBOD may be expressed as:

$$\left(\frac{\partial C}{\partial t}\right)_n = -k_n N. \qquad \text{(Eq. 4-25)}$$

Algal Respiration. The phenomenon of algal respiration has been discussed along with photosynthesis in the foregoing part of this chapter. The DO depletion rate resulting from algal respiration is normally designated as R, and considered as constant for DO modeling of a particular water body; e.g., a stream segment. R is sometimes considered a constant value during the dark period and zero during the light period of the day. O'Connor (Ref. 4-34) has formulated R as a constant over the entire 24-hour period. Thus, in his formulation the net contribution to DO during the light photosynthetic period is the difference between the positive contribution of photosynthesis and the negative contribution of respiration. Using this formulation, the volumetric rate of depletion of DO concentration from respiration may be expressed as:

$$\left(\frac{\partial C}{\partial t}\right)_r = -R. \qquad \text{(Eq. 4-26)}$$

Values of R typically range from zero to 5 mg/l/day or greater.

Benthic Deposits. Oxygen depletion in the overlying water may result from the decomposition of organic materials in bottom sediments. There are two primary mechanisms through which this occurs. When aerobic conditions exist within the upper layer of the sediment, oxygen diffuses into this layer for aerobic microbial oxidation. In anaerobic sediment zones, which in many cases may comprise the entire sediment layer, anaerobic microbial degradation produces reduced substances which diffuse upward and are subsequently oxidized in the overlying aerobic zones, with a corresponding oxygen demand. The production of volatile organic acids by anaerobic bacteria and their subsequent oxidation by aerobic bacteria is a typical example. The rate of benthic deposit demand is typically determined from laboratory or field studies on an areal basis, and typical units of expression are gm/m^2/day. If this areal rate is expressed as B', then the volumetric demand rate is calculated as B'/H for situations of complete mixing through the depth, H. Values of B' typically range from zero up to a few gm/m^2/day. The demand rate for a given modeling application is normally considered constant, and the volumetric rate of depletion of DO concentration from benthic deposits may be expressed as:

$$\left(\frac{\partial C}{\partial t}\right)_b = -B. \qquad \text{(Eq. 4-27)}$$

Oxidation of Reduced Inorganic Substances. The biochemical oxidation of reduced nitrogen has been discussed separately due to its special significance. Numerous other reduced inorganic substances are readily oxidized chemically and/or biochemically under the conditions prevalent in many surface waters and, therefore, exert an oxygen demand. Ferrous iron and sulfide are typical examples. Ferrous iron is highly resistant to oxidation at low pH values, such as those typical of a steel pickling liquor wastewater or acid mine drainage, but is readily oxidized chemically to ferric iron at pH values typical of surface waters. These phenomena must be addressed in DO modeling when known to be of significance in a particular application. The frequency of occurrence of any particular one is not considered sufficiently great to warrant further discussion here.

4-3.3 Advection

Advection has sometimes been described as a source or sink of DO. However, it is not a true source or sink in the sense of those described above, but is a mathematical quantity which represents the net sum or difference of the effects of all the sources and sinks of DO in a particular water body. In simple terms, the net rate of DO advection to a water body is the difference between the rate at which DO is entering in the water flowing into it and the rate at which DO is leaving in the water flowing out of it. At a point X in a stream where the velocity is U (positive U in the positive downstream direction of X), the advective volumetric rate contribution to DO concentration may be expressed as:

$$\left(\frac{\partial C}{\partial t}\right)_v = -U\frac{\partial C}{\partial X}. \qquad \text{(Eq. 4-28)}$$

4-3.4 An Expression for the Behavior of DO in One Dimensional Nondispersive Streams

In many DO modeling applications, free flowing streams are considered to be one dimensional (uniformly mixing across the cross-section) and nondispersive (no mixing in the direction of flow). From these assumptions, from Equation 4-1, and from the foregoing discussions, a DO balance for a stream segment may be expressed in general form as:

$$\frac{\partial C}{\partial t} = -U\frac{\partial C}{\partial X} + k_a(C_s - C) + P - k_nN - R - B. \qquad \text{(Eq. 4-29)}$$

In general, U, P, L, N, R, and B may all be functions of both X and t. In utilizing this equation as the basis for modeling with the digital computer, stream segments are normally selected to be spacially small enough so that several of the parameters may be held constant within a given segment and the expression is applied on a segment-by-segment basis with appropriate constant values for parameters for each segment. U, R, and B are typically assumed constant within a segment, independent of both X and t. The photosynthesis rate, P, may be expressed as independent of X, but as a function of t as given in Equation 4-20. L and N may be expressed as independent of t, but as a function of X by integration of Equation 4-21 and 4-23, respectively, with the following results:

$$L = L_o e^{-k_d(X/U)} \qquad \text{(Eq. 4-30)}$$

$$N = N_o e^{-k_n(X/U)}, \qquad \text{(Eq. 4-31)}$$

where L_o and N_o are the ultimate CBOD and NBOD concentrations, respectively, at the beginning of the segment ($X = 0$), and X/U represents the time of travel downstream through the reach 0 to X. Making these substitutions and changing the dependent variable from DO concentration, C, to DO deficit, $D(D = C_s - C$ and $\partial C/\partial t = -\partial D/\partial t)$, Equation 4-29 becomes:

$$\frac{\partial D}{\partial t} = -U\frac{\partial D}{\partial X} - k_a D - P_m\left[\frac{1}{\pi} + \frac{1}{2}\sin \pi\frac{t}{p} - \frac{2}{3\pi}\cos 2\pi\frac{t}{p}\right] + k_d L_o e^{-k_r(X/U)} + k_n N_o e^{-k_n(X/U)} + R + B. \quad \text{(Eq. 4-32)}$$

With the appropriate boundary conditions, Equation 4-32 integrates to the following:

$$D = D_o e^{-k_a(X/U)} + \frac{k_a L_o}{k_a - k_d}(e^{-k_d(X/U)} - e^{-k_a(X/U)}) + \frac{k_n N_o}{k_a - k_n}(e^{-k_n(X/U)} - e^{-k_a(X/U)}) + \frac{B + R - P_m/\pi}{k_a}(1 - e^{-k_a(X/U)})$$
$$- \frac{P_m}{2(k_a^2 + \pi^2/p^2)}\left\{\left[k_a \sin \pi\frac{t}{p} - \frac{\pi}{p}\cos \pi\frac{t}{p}\right] - \left[k_a \sin\frac{\pi}{p}\left(t - \frac{X}{U}\right) - \frac{\pi}{p}\cos\frac{\pi}{p}\left(t - \frac{X}{U}\right)\right]e^{-k_a(X/U)}\right\} + \frac{2P_m}{3\pi(k_a^2 + 4\pi^2/p^2)}$$
$$\cdot\left\{\left[k_a \cos 2\pi\frac{t}{p} + \frac{2\pi}{p}\sin 2\pi\frac{t}{p}\right] - \left[k_a \cos 2\pi\frac{t}{p} + \frac{2\pi}{p} \cdot \sin\frac{2\pi}{p}\left(t - \frac{X}{U}\right)\right]e^{-k_a(X/U)}\right\}. \quad \text{(Eq. 4-33)}$$

Note that Equation 4-33 expresses D as a function of constant (time independent) initial values (at $X = 0$), D_o, L_o, N_o; other constant parameters and coefficients; and two independent variables X and t. To reiterate, X represents the distance downstream from the beginning of the segment and t represents real or clock time measured from the beginning of the photosynthetic period. By selecting a unique set of X and t, the DO deficit concentration may be calculated for a given time of day at a particular point. By fixing a constant value of X and varying t over a 24-hour period, the behavior of D over the 24-hour period at a single station within the segment may be determined. By fixing a value of t and varying X along the length of the segment, a profile of D vs. distance may be determined for a particular time of day.

A time lag in the exertion of NBOD in the presence of relatively high concentrations of CBOD is a well documented phenomenon. The two stage BOD curve resulting from the conventional BOD test on wastewaters containing ammonia, in which the beginning of NBOD exertion typically lags the beginning of CBOD exertion by 5 to 10 days, is an example. Differences of opinion exist as to the significance of this lag in DO modeling applications, particularly in streams immediately below point source discharges. As more stringent effluent limitations result in lower instream CBOD concentrations, the NBOD lag will generally become of less significance. Also, conditions for the establishment of viable nitrifying bacterial populations on fixed surfaces are generally more favorable in streams than in conventional laboratory BOD tests. Field tests to determine the extent (if any) of the NBOD lag are, of course, desirable, but seldom possible for wide scaled DO modeling applications. The time lag of NBOD exertion may be accounted for mathematically in Equation 4-33 by setting $k_n = 0$ through the segment(s) corresponding to the time of travel equivalent to the desired lag time, and then setting k_n equal to the appropriate positive values for downstream segments.

O'Connor (Ref. 4-34) distinguishes between the rate at which carbonaceous material is removed from water in a stream, k_r, and the rate at which oxygen is consumed in the oxidation of carbonaceous material, k_d. This difference may exist in a stream due to such factors as sedimentation, scour, or volatilization. Sedimentation and volatilization would result in k_r being greater than k_d, while scour would result in k_r being less than k_d. In the formulation of Equation 4-33 it has been assumed that $k_r = k_d$. If this is known not to be approximately the case, k_r should be substituted for k_d in the denominator and in the exponent of the second term on the right hand side. These coefficients, k_d and k_r, should always be determined directly or estimated from field data, in that they might be substantially different from the rate coefficient determined from the conventional laboratory BOD test, commonly termed k_1. O'Conner notes that this difference is due primarily to the various physical and biochemical characteristics of the two environments. As discharges of suspended organic solids are decreased due to increasingly stringent effluent limitations, the factors contributing to the difference between k_d and k_r will generally become less significant and they in many cases may be assumed to be equal, at least within the accuracy of determination of either.

When atmospheric reaeration is considered the only DO source and carbonaceous BOD the only DO sink, Equation 4-33 then reduces to the following familiar Streeter-Phelps equation:

$$D = D_o e^{-k_a(X/U)} + \frac{k_d L_o}{k_a - k_d}\left(e^{-k_d(X/U)} - e^{-k_a(X/U)}\right). \qquad \text{(Eq. 4-34)}$$

4-4. DETERMINATION OF PARAMETERS FOR STREAM DO MODELING

Determination of values for numerous parameters is required for a typical modeling application. The scope of most modeling studies for planning purposes will generally allow for experimental determination of only a few, if any, of these parameters. Data and predictive methodologies from the published results of previous studies are commonly utilized as a basis of parameter determinations.

4-4.1 Hydrologic Considerations

One of the most critical factors in DO modeling of streams, particularly for point source discharges under low flow conditions, is the amount of flow in the stream upstream from the point of discharge available for dilution. Thus, it is essential to determine the magnitude of this flow as accurately as possible for the condition to be modeled.

Stream Flow Records, Frequency-Duration Determinations. The best source of historical stream discharge records in the United States is the United States Geological Survey. These records result from data collected from a comprehensive stream gaging network. Continuous records for many stations are available for a period of 30 years or more. These data are summarized on a water year basis (October–September) and have been published by U.S.G.S. in various forms over the years. Beginning with the 1975 water year, water data have been published on a state-boundary basis in final form and will not be republished in the water supply paper series as in previous years. In addition to stage and discharge records for stream gaging sites, these reports also typically contain records of water quality of streams, stage and volume of lakes, and water levels and water quality of wells and springs (Ref. 4-50).

The U.S.G.S. and other stream gaging records are stored in various computer data banks for easy access. For example, the Kentucky Department for Natural Resources and Environmental Protection maintains a set of computer tapes containing the complete historical records for all U.S.G.S. stream gaging stations in the state. These are updated yearly and are available for access for various water studies.

In order to perform DO modeling for a particular stream, the stream discharge regime must be defined. This may be either a specific discharge selected for a particular modeling application or it may be a regime determined by a statistical analysis of streamflow records. For river basin (303) or facilities planning (201) purposes, DO modeling is usually performed to simulate critical conditions to occur under an extreme low flow regime. This regime is typically defined as the 10-year, 7-day low flow as determined from stream-flow records by an acceptable statistical technique. A 10-year, 7-day low flow is the lowest average 7-day discharge which would statistically not be exceeded with a frequency of once in 10 years. Implicit in establishing a frequency-duration flow regime for modeling for planning purposes is the statistical probability that the stream water quality standards may be violated for the frequency-duration selected. For example, if the permissible effluent quality for a point source discharge is determined so that minimum stream water quality standards are just met under a 10-year, 7-day low flow, then the stream standards will probably be violated when the stream discharge is less than the 10-year, 7-day value. This will statistically occur for a duration of seven days once in ten years, but more frequently for durations of less than seven days.

Stream low flow frequency-duration analysis is normally conducted using the statistical frequency distribution which best fits the low flow data for the stream being analyzed. Many different statistical distributions are available and a discussion of the theories and details of application of these distributions is beyond the scope of this text, but may be found in many publications on statistical hydrology. Typical results are shown in Table 4-1. These statistical techniques are based on the assumption that variations in stream discharge are due to natural phenomena, and the techniques therefore cannot theoretically be applied for gaging stations above which the stream discharge is regulated by artificial means. Examples of artificial regulation are artificial discharges and withdrawals for

TABLE 4-1 Results of Statistical Low-Flow Frequency-Duration Analysis For Cumberland River Near Harlan, Kentucky. Analysis Based on 30 Years of Continuous Record at U.S.G.S. Gaging Station.

Duration (days)	Recurrence Interval (years)	Discharge Magnitude (cu ft/sec)
1	50	4.7
1	20	6.2
1	10	7.9
1	5	10.8
1	2	19.8
7	50	6.1
7	20	7.9
7	10	10.1
7	5	13.6
7	2	24.4
15	50	6.8
15	20	9.1
15	10	11.8
15	5	16.2
15	2	29.9
30	50	8.5
30	20	11.6
30	10	15.1
30	5	21.0
30	2	38.8
120	50	23.1
120	20	30.2
120	10	38.4
120	5	51.7
120	2	93.0

municipal and industrial purposes and reservoir releases. However, due to the high frequency of occurrence of artificial regulation, it is often desirable to perform a frequency-duration analysis for a station below artificial regulation, and to use judgment based on the nature and significance of the regulation in interpretation of the results.

Low flow regimes other than the 10-year, 7-day magnitude are sometimes used for planning purposes. A 20-year, 3-day flow which is sometimes used is typically lower than the corresponding 10 year, 7-day value, as can be inferred from Table 4-1.

Many point source discharges are located on streams at points where the upstream (base) 10-year, 7-day low flow is zero. In some of these cases the upstream flow may be zero for extended periods with regular frequency (in extreme cases, up to several months almost every year). These typically represent critical situations for water quality management and applying more stringent limitations to discharges to such streams than to streams with non-zero low flows may be desirable due to the long duration of critical conditions in the former.

Reservoir Releases, Rule Curves, Routing. For modeling in streams in which the flow is significantly affected by releases from upstream reservoirs, the effects of these releases must be taken into account in establishing the appropriate flow regime. If the reservoir has been in operation for a period of several years, records from downstream gaging stations may be utilized as a guideline. Most reservoirs of significant size with controlled releases operate in accordance with established release rule curves. Rule curves generally give the magnitude of the release as a function of the time of year and the water surface elevation within the reservoir, and are established to most effectively operate the reservoir in accordance with the purposes for which it exists. For a given reservoir the release rule may be established to accommodate one or more of the following: hydro-electric power generation, flood control, recreation, downstream low-flow augmentation for water quality enhancement, wildlife propagation, irrigation, water supply, and other beneficial uses.

Where one or more reservoirs are known to significantly affect the flow regime in a given stream, the best approach to determine the flow magnitude of a certain frequency-duration is to utilize hydrologic routing techniques to route the reservoir releases, other artificial discharges, and natural base flows through the stream. Parameters in the routing model may be calibrated by comparing historical gaging station records with predictions for the corresponding conditions. The calibrated routing model may then be used to predict flow magnitudes at various points in the stream under the low flow regime of interest. While routing is desirable, it is also typically costly, both in expenditure of time and financial resources, and may not be practical in wide scaled modeling for planning purposes. In many cases it is necessary to rely on existing gaging station records, release rule curves, and judgement to predict the effects of releases on downstream flow.

Point Source Wastewater Discharges. For modeling for water quality management purposes the rate of wastewater discharges is normally required for both present and future conditions. Present discharge rates are utilized to predict present stream water quality conditions, in model calibration and verification, to identify critical water quality stream reaches, and to define "water quality limited" and "effluent limited" segments. For larger municipal and industrial dischargers, reasonably accurate determinations of discharge rates can normally be made from operating records. For smaller facilities, field observations and estimates must often be used.

Discharge rates must be projected for future conditions for modeling to determine stream water quality and to establish permissible effluent quality throughout the planning period. These projections are typically made based on historical trends, population forecasts, the expected degree of elimination of infiltration/inflow into collection systems (including the extent of elimination of combined sewers, if any), industrial expansion forecasts, and other factors.

For steady state modeling, the discharge rate for a particular source is considered as constant over the time period modeled. For non-steady state modeling, the time variation of the discharge rate may be taken into account.

4-4.2 Water Quality Considerations

Critical instream water quality conditions generally occur during low flow periods when the stream flow originates, essentially, only from groundwater (base flow) and point source discharges. An accurate characterization of the water quality associated with each is essential in modeling.

Background Stream Water Quality Characterization. Background water quality is the terminology often used to refer to the quality of water in a given stream under low flow conditions in reaches unaffected to any significant degree by point source discharges. It would be characteristic of the upper reaches of a stream above any point source discharges, or of a downstream reach at a distance sufficiently far below a point source discharge so that its effects have been minimized by natural assimilation processes in the stream and by base flow dilution. Background water quality is best determined by direct field observations. For wide scaled modeling, the results of research studies and monitoring records for the area of concern are commonly used sources. Background water quality can normally be characterized as good with regard to DO resources in that concentrations of potentially DO demanding substances are typically low.

Point Source Discharge Water Quality Characterization. As with the discharge rate discussed above, discharge water quality is normally required for modeling both present and future conditions. Present discharge water quality parameters are best determined from the treatment facility operation records and records of regulatory agencies. Records for small facilities are often inadequate and must be supplemented by field testing and estimates. Compliance monitoring required through the National Pollutant Discharge Elimination System (NPDES) permits should greatly increase the quantity and accuracy of information available on the quality of point source discharges.

Quality of point source discharges may take the form of either input parameters to or output parameters from modeling for future conditions through the planning period. Various effluent quality levels may be input to a model to ascertain and compare the corresponding responses to measured receiving water quality. Permissible effluent quality for non-violation of instream water quality standards is a typical output from modeling in water quality management studies. In this case, comparison of outputs for various combinations of model input parameter values gives a good indication of the sensitivity of the model to changes in parameter values.

For future conditions, in either setting effluent quality levels for model input or in programming the model to give permissible effluent qualities as output, available treatment technology should be considered. The case of modeling to determine the least stringent permissible effluent quality for a particular discharge for non-violation of instream standards may be used for illustration. One approach is to have the model programmed to theoretically select values for each effluent quality parameter of interest. Based on this result, the facilities planner may ultimately recommend a treatment process scheme which will ensure the desired effluent quality. A second approach is to program the model to evaluate a series of discrete treatment levels, each characterized by a unique set of quality parameters. This characterization is based on practical considerations as to the known performance of various treatment process schemes. The treatment levels are typically evaluated by beginning with the least stringent (or least costly) and proceeding through the treatment level train until a level is found which results in non-violation of standards. A somewhat abbreviated treatment level train is illustrated in Table 4-2. Typical municipal wastewater is assumed, each level is characterized by values for parameters of significance in DO modeling, and ranking is made by decreasing effluent strength factor, SF. The estimated ultimate oxygen demand, UOD, and SF are indicators of the overall oxygen demanding potential, and are calculated as:

$$UOD = 1.5 \times BOD_5 + 4.5 \times NH_3\text{-}N \qquad \text{(Eq. 4-35)}$$

$$SF = UOD - DO \qquad \text{(Eq. 4-36)}$$

where BOD_5 is the 5-day BOD concentration, NH_3-N is the ammonia nitrogen concentration expressed as N, and DO is the dissolved oxygen concentration. Estimated total costs include capital, operation, and maintenance for all unit processes required to provide the corresponding effluent quality levels. Pretreat-

TABLE 4-2 Typical Effluent Quality Parameter Values Resulting from Different Levels of Treatment of Typical Municipal Wastewater.

Treatment Level Designation	5-Day BOD (mg/l)	NH_3-N (mg/l as N)	DO (mg/l)	UOD (mg/l)	SF (mg/l)	Typical Process Scheme to Achieve Level	Estimated Total Treatment Cost for 10 MGD Facility (10^6 Mid-1976 $, Present Worth)
T1A	30	20	2	135	133	Trickling filter	9.8
T1B	30	20	7	135	128	T1A + Effluent aeration	10.1
T2A	25	18	2	119	117	High rate activated sludge	10.3
T2B	25	18	7	119	112	T2A + Effluent aeration	10.6
T3A	20	15	3	98	95	Conventional activated sludge	11.0
T3B	20	15	7	98	91	T3A + Effluent aeration	11.3
T4A	10	15	3	83	80	T3A + Mixed media filtration	13.8
T4B	10	15	7	83	76	T4A + Effluent aeration	14.1
T5A	15	1	3	27	24	T2A + Second stage bionitrification	12.8
T5B	15	1	7	27	20	T5A + Effluent aeration	13.2
T6A	8	1	3	17	14	T5A + Mixed media filtration	15.7
T6B	8	1	7	17	10	T6A + Effluent aeration	16.0
T7A	3	1	3	9	6	T6A + Activated carbon absorption	23.8
T7B	3	1	7	9	2	T7A + Effluent aeration	24.1

ment, primary treatment, disinfection, and sludge disposal costs are included in all cases (Ref. 4-51).

A permissible treatment level resulting from the treatment train approach gives the facilities planner considerable guidance in recommending treatment process schemes. However, either of the above two approaches should ultimately result in closely similar conclusions.

4-4.3 Rate Coefficients

For particular modeling applications, rate coefficients are best determined from direct field/laboratory observations for the site under consideration. For wide scaled modeling for planning purposes it is usually necessary to depend upon literature values and predictive expressions for determination of coefficient values.

Deoxygenation Rate Coefficients. The CBOD removal rate coefficient, k_r, may be determined for an existing stream situation using the following procedure. Samples are collected from the stream at points through a segment, and the time of travel between the collection points is determined by the method subsequently described. Each sample is incubated in the laboratory under constant temperature. CBOD concentrations are determined by a suitable technique. A semi-log plot is then made of the log of either the 5-day BOD or ultimate BOD vs. the arithmetic travel time from the most upstream sampling point. Values of k_r may be determined from the slope of the line between data points. Values so determined will have to be adjusted from the laboratory incubation temperature to the desired temperature for subsequent modeling. Values for the CBOD deoxygenation coefficient, k_d, may be determined at points in a stream by incubating a sample collected from each point and observing the BOD vs. time by a suitable technique. The rate coefficient, k_d, may then be determined from these data by one of the conventional methods such as Thomas Slope, Least Squares, etc. (Ref. 4-30). Either of the above two determinations may be made by incubation of BOD samples directly in the stream. However, the benefits of stream incubation are generally not commensurate with the effort required. The actual temperature and temperature variations in the stream are reflected, but a subsequent adjustment of the resulting coefficients to the temperature desired for modeling is usually necessary anyway.

One deficiency in the above methods of coefficient determination is that the results are depended upon the ambient conditions in the stream at the time of sampling which, in general, will not be the same as those desired for modeling. For example, the actual BOD and other pollutant concentrations in the stream may be different. It has been shown that for a treated wastewater effluent, k_d decreases with an increase in the degree of BOD stabilization during treatment.

Thus, in general, k_d would be expected to decrease with decreasing BOD concentrations in streams where the BOD has been contributed by point source discharges.

One method for k_d determination, which has been used successfully for modeling stream segments below point source discharges in Kentucky, is based on BOD observations of diluted effluent samples with the electrolysis BOD technique (Ref. 4-30). Effluent samples are diluted with stream water samples collected upstream from the outfall in the approximate proportion that would correspond to the condition to be modeled. The diluted samples are then incubated on the electrolysis BOD apparatus, producing data of BOD vs. time. Corresponding values of k_d may then be determined graphically, as illustrated in Figures 4-4 and 4-5 (Ref. 4-10). The resulting k_d values are then adjusted from the laboratory incubation temperature to the temperature desired for modeling. This method may, with a reasonable amount of effort, be used to test several streams. The electrolysis BOD technique has tremendous advantages over the conventional bottle technique for rate coefficient determinations in that BOD concentrations of up to about 500 mg/l may be measured without dilution, and BOD vs. time data may be determined for a sample in a single bottle.

The NBOD deoxygenation rate coefficient, k_n, may be determined from a semi-log plot of NH_3-N concentration vs. time of travel in a manner analogous to that described above for k_d, or it may be determined from an analysis of sec-

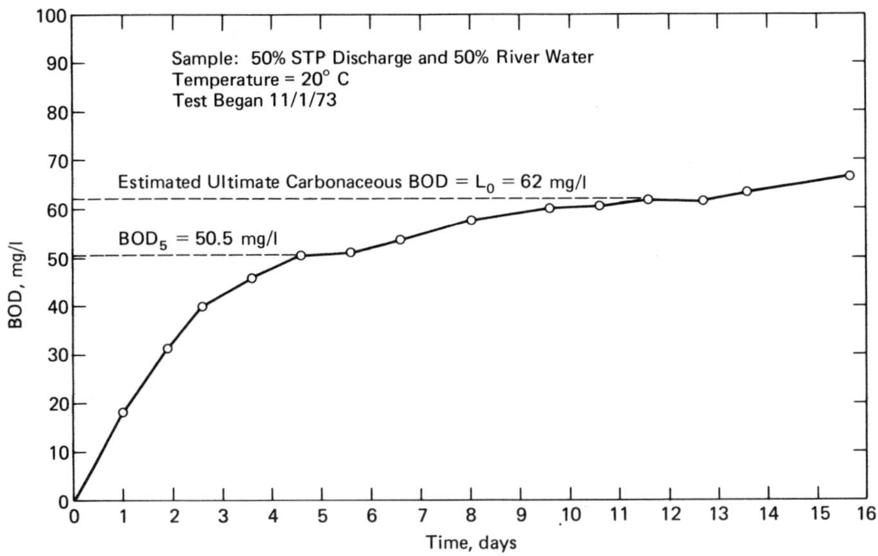

Fig. 4-4. Typical electrolysis BOD curve—Barren River—Bowling Green, Ky.

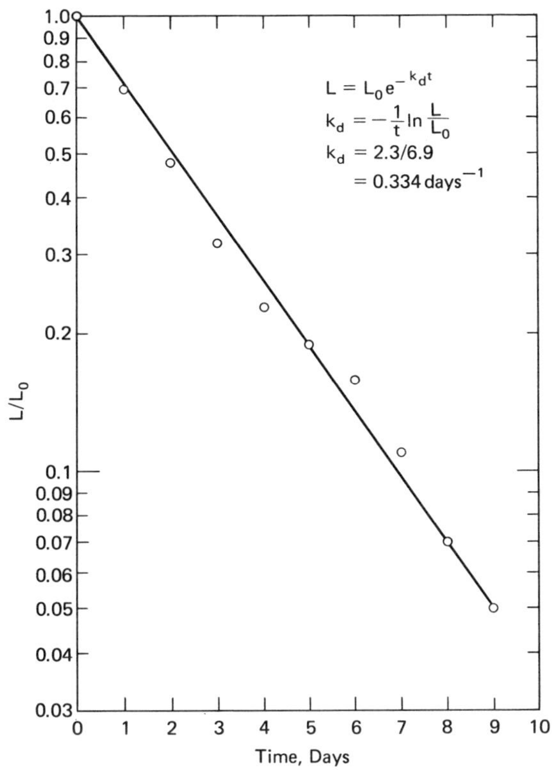

Fig. 4-5. Curve fitting for k_d determination–Barren River Bowling Green, Ky.

ond stage BOD data, such as that generated by an electrolysis BOD test as described above for CBOD. Less information is available from the literature on k_n values than on k_d values. However, it has been found that under otherwise similar conditions, k_n rates are greater in smaller free-flowing streams than in larger sluggish rivers. This is thought to be due to the relatively greater availability of fixed surfaces for attachment of nitrifying bacteria in the smaller, generally shallower streams.

Based on the above considerations, reviews of the available literature, and the results of research studies performed in the state, the following predictive expressions are currently (mid-1976) being used for modeling purposes in the State of Kentucky:

$$k_r = k_d = 0.018 \text{ BOD}_5 \quad \text{(Eq. 4-37)}$$

$$k_d, \text{ maximum} = 0.50 \quad \text{(Eq. 4-38)}$$

$$k_d, \text{ minimum} = 0.15 \quad \text{(Eq. 4-39)}$$

$$k_n = 0.030 + 0.019\ S^{1.2} \qquad \text{(Eq. 4-40)}$$

$$k_n,\ \text{maximum} = 0.30 \qquad \text{(Eq. 4-41)}$$

$$k_n,\ \text{minimum} = 0.10. \qquad \text{(Eq. 4-42)}$$

These expressions predict a set of constant coefficient values to be utilized for a particular stream segment. BOD_5 is the 5-day BOD concentration in mg/l at the beginning (upstream end) of the segment, and S is the average stream bed slope in ft/mi through the segment. All coefficient values are for 25°C and have dimensions of day^{-1}. These expressions are used for small to moderately sized streams. Certain constant values are used for some long uniform stretches of the major rivers. It should be emphasized that these are empirical expressions and are not necessarily of general applicability.

Atmospheric Reaeration Rate Coefficients. Much research has been devoted to determination of reaeration rates. Based on semi-theoretical considerations, the volumetric atmospheric reaeration rate coefficient, k_a, can be expressed as a function of the diffusivity coefficient for oxygen into water, D_l, the mean stream velocity, U, and the mean stream depth, H, as (Ref. 4-69):

$$k_a = \frac{(D_l U)^{1/2}}{H^{3/2}} \qquad \text{(Eq. 4-43)}$$

This expression is dimensionally correct, and any set of consistent units may be used. D_l has a value of about 8.1×10^{-5} ft^2/hr at 20°C.

Empirical expressions of the general form of Equation 4-43 have resulted from various studies over the years. The following are some which have been widely used for predictive purposes, and are commonly referred to by the names of the investigators by whom they were developed:

O'Connor-Dobbins: $k_a = 13.8\ U^{0.5} H^{-1.50}$ (Eq. 4-44)

Churchill, *et al.*: $k_a = 13.00\ U^{0.969} H^{-1.673}$ (Eq. 4-45)

Langbein-Durum: $k_a = 8.47\ U^{1.00} H^{-1.33}$ (Eq. 4-46)

Owens, *et al.*: $k_a = 24.5\ U^{0.97} H^{-1.85}$ (Eq. 4-47)

An additional expression which also included stream energy grade slope, S_e, and the acceleration of gravity, g, was developed by Thackson-Krenkel as:

$$k_a = 26.9\ [1 + U^{0.50}(gH)^{-0.25}]\ [(gS_e)^{0.50} H^{-0.50}] \qquad \text{(Eq. 4-48)}$$

These five expressions contain a total of only three hydraulic variables: U, H, and S_e. These expressions are presented as summarized and adjusted to a common temperature of 25°C by Tsivoglou (Ref. 4-49). (Refs. 4-5, 4-27, 4-28, 4-33, 4-35, and 4-42 may be consulted for the original developments.) Each expression gives k_a at 25°C in units of day^{-1}. In all cases, dimensions of hydraulic variables are U in ft/sec, H in ft, and S_e in ft/ft.

The difficulty in evaluating the validity of predictive expressions for k_a, such as those above, lies in the fact that k_a cannot be measured directly in natural stream situations. The traditional method of evaluation consists of observing a DO profile in a stream and fitting a predicted DO profile by adjustment of k_a values used in the DO profile predictive equation to obtain the best fit. Implicit in this methodology is that values for all other factors included in the predictive DO expression can be determined with reasonable accuracy. However, in practice this is normally a questionable assumption due to the complexities and multitude of factors affecting the DO concentration.

Over the last decade Tsivoglou has developed a tracer technique which he terms a direct method for k_a determination. Its theory, evaluation, and application have been described in detail in previous publications (Refs. 4-44, 4-45, 4-46, 4-47, 4-48, and 4-49). The field application of the tracer technique involves the use of three tracers: tritiated water, a dissolved radioactive gas (typically krypton-85), and a fluorescent dye (typically, DuPont Rhodamine WT-20% solution). For the stream reach covered, and at the prevailing flow, the tracer study provides direct evaluation of the gas exchange capacity, time of flow, and dispersion, both total and longitudinal.

The fluorescent dye yields time of flow and longitudinal dispersion information for the reach, and also indicates when to sample for the other two tracers. The tritiated water provides an accurate measure of the total dispersion. The dissolved tracer gas is dispersed to the same extent as the tritiated water, and, in addition, is lost to the atmosphere as the result of turbulent mixing and surface replacement. The change in concentration of the dissolved tracer gas between two sampling points, relative to the change in concentration of tritium, indicates the percentage of dissolved gas that has been lost within the reach. This is then converted to the exchange capacity for oxygen by use of the known krypton-85: oxygen ratio previously determined by Tsivoglou. The tracer method assumes, essentially, that the concentration of tritiated water changes within the reach only as the result of dispersion, that the krypton-85 concentration changes additionally only because of loss to the atmosphere, and that the ratio of exchange coefficient values for krypton-85 and oxygen is a constant (0.83) that is not significantly affected by temperature, turbulence, or usual concentrations of water pollutants, within the range of interest (Ref. 4-44).

The field tracer technique employs the device of simultaneity in order to avoid the many problems and the many sources of significant error associated with any attempt to independently evaluate turbulence or the rate of surface replacement. Thus, the three tracers are released as a single simultaneous dose, and, in addition, the concentrations of both tritium and krypton-85 are subsequently measured simultaneously in any single stream sample. As a result of these methods, the gas exchange capacity can be evaluated accurately by the use of an instantaneous point dose and by subsequent point sampling. A uniform cross-sectional dose (essentially impossible to achieve in a practical situation) is not required, and it is not necessary to attempt to perform accurate tracer mass balance. The device of simultaneity thus makes the task of evaluating stream reaeration capacity both much simpler and more accurate than is otherwise possible (Ref. 4-44).

For a particular reach under study, the reaeration rate coefficient is calculated from the following expression:

$$k_a = \frac{\ln (R_1/R_2)}{0.83(t_f)}, \qquad \text{(Eq. 4-49)}$$

where

k_a = atmospheric (oxygen) reaeration rate coefficient, time^{-1};
R_1 = measured ratio of krypton-85 activity to tritium activity at the upstream end of reach
R_2 = corresponding ratio at downstream end of reach
t_f = time of flow through reach, time, and
0.83 = ratio of exchange coefficient for krypton to exchange coefficient for oxygen.

The value of k_a is a function of temperature, and thus the value determined from Equation 4-49 is dependent upon the prevailing water temperature at the time of determination of R_1 and R_2.

Based on the results of his studies with the tracer technique, Tsivoglou has developed and evaluated the following predictive equation referred to as the "energy dissipation model;"

$$k_a = c\left(\frac{\Delta h}{t_f}\right), \qquad \text{(Eq. 4-50)}$$

where, Δh is the water surface elevation change, t_f is the time of flow through a stream reach, and c is termed the "escape coefficient." Tsivoglou (Ref. 4-48) has summarized the results of his numerous studies and thereby has tested the energy dissipation model against field data.

Relatively low values of c are usually associated with larger streams, or with relatively large, deep, unobstructed straight channels uniform in such properties as slope, cross-sectional geometry, etc. Relatively high values of c usually occur for smaller streams characterized by abrupt changes of slope, sharp changes of direction, frequent obstructions, rapids, shoals, and falls. These phenomena are reflected in the general dependency of c on the stream discharge Q. Mean values of c from 577 and 28 observations resulted in the following two expressions for the respective Q ranges:

$$k_a = 0.060 \left(\frac{\Delta h}{t_f}\right) \qquad 25 < Q < 300 \text{ cfs} \qquad \text{(Eq. 4-51)}$$

$$k_a = 0.123 \left(\frac{\Delta h}{t_f}\right) \qquad 1 < Q < 10 \text{ cfs}, \qquad \text{(Eq. 4-52)}$$

where k_a is at 25°C with units of day^{-1}, Δh is in ft, and t_f is in days. These are intended only to indicate the general dependency of c on Q, and Ref. 4-48 should be consulted for details of application for actual predictive purposes.

By defining the average water surface elevation slope through a stream segment of length, L, as $S = \Delta h/L$, the average velocity, U, through the segment as $U = L/t_f$, and making the appropriate substitutions, Eq. 4-50 may be rewritten as:

$$k_a = c\, S\, U. \qquad \text{(Eq. 4-53)}$$

Thus, the energy dissipation model also alternatively predicts k_a as directly proportional to the product of slope and velocity as defined here. By an analysis of his data, Tsivoglou (Ref. 4-48) developed the following expressions for prediction of k_a as a function of slope alone when velocity (or time of travel) data are unavailable:

$$k_a = 0.67\, S, \qquad \text{(Eq. 4-54)}$$

where, k_a is at 25°C with units of day^{-1} and S in in ft/mi. Tsivoglou recommends this expression for approximate predictive purposes for streams with slopes of less than about 20 ft/mi when time of passage (or mean velocity) information cannot be obtained for direct use in the energy dissipation model.

A series of field studies utilizing Tsivoglou's tracer technique for k_a determination was performed over a three year period (1973-75) on small to moderately sized streams in Kentucky (Ref. 4-11). A concentrated effort was made in these studies to perform the field experiments on streams under as low flow conditions as possible with the objective of producing results for subsequent use for predictive purposes in modeling of extreme low flow conditions. As a result of this work, the following empirical expression was developed to predict k_a as a function of stream bed slope (approximately equal to average water surface elevation slope as defined above):

$$k_a = 0.30 + 0.19\, S^{1.2}, \qquad \text{(Eq. 4-55)}$$

where, k_a is at 25°C with units of day^{-1} and S in ft/mi. This expression is best applied for free-flowing streams with associated drainage basin areas less than about 1000 sq mi with specific discharges of less than about 0.20 cfs/sq mi, and has produced good correlations between observed and predicted values of k_a under these conditions. As used here, specific discharge is defined as the actual discharge in a stream at a particular point divided by the drainage basin area upstream from that point, and serves as a useful general indicator of the flow regime.

As is readily apparent, Equations 4-54 and 4-55 will not necessarily produce the same value of k_a for identical values of S, although both were based on results from studies using the same experimental techniques. This apparent inconsistency results from the fact that Equation 4-54 reflects the "average" relationship between k_a and S for some 600 observations for streams of widely varying size under widely varying flow regimes; whereas Equation 4-55 was developed specifically from data for small to moderately sized streams under low flow conditions. Although he did not specifically identify specific discharge as an independent variable, Tsivoglou has discussed the effects of its two components, stream size and stream discharge, on k_a (Ref. 4-48). The results of all the available studies utilizing the tracer technique indicate that k_a decreases with decreasing specific discharge for a particular stream reach, but the relationship has not been completely formulated to date.*

Various studies have been conducted to evaluate the reaeration resulting from turbulence created in streams from abrupt changes in water surface elevation, such as that which occurs at dams and waterfalls. Studies in Kentucky have resulted in the following simple expression, which has produced good correlation between observed and predicted reaeration at instream dams and waterfalls under low flow conditions (Refs. 4-55, 4-11):

$$r = e^{0.16\,H}. \qquad \text{(Eq. 4-56)}$$

This expression is for a temperature of 25°C, where r is ratio of the upstream to the downstream DO deficit concentrations and H is the height of fall or difference in water surface elevation in feet. The temperature dependency of r is formulated below. In modeling applications these phenomena are generally considered as discontinuities. Using Equation 4-56 (or similar expressions) at a particular site, r is first computed. Then the downstream DO concentration, C_d, is predicted as a function of the upstream DO concentration, C_u, and the

*More data have become available and additional data analysis has been performed by the author since this original writing. Based on an evaluation of all the data from the Kentucky tracer studies plus data from tracer studies of ten streams in Wisconsin, the following expression, which is a modification of Eq. 4-55 and incorporates the specific discharge parameter, has been developed for k_a predication:

$$k_a = (0.63 + 0.40\,S^{1.15})q^{0.25}$$

Where q is the specific stream discharge in cfs/sq mi. This expression applies for $0.05 \leqslant q \leqslant 1.00$ and predicts the observed Kentucky and Wisconsin data with good accuracy.

appropriate DO saturation concentration, C_s, as:

$$C_d = C_s - \frac{1}{r}(C_s - C_u) \qquad \text{(Eq. 4-57)}$$

Gameson (Ref. 4-14), Holler (Ref. 4-19), and others have formulated expressions similar to Equation 4-56.

Temperature Adjustments. All the rate coefficients discussed above are temperature dependent. In general, coefficient values obtained either from the literature or from experimental studies will be based on a different temperature than that desired for a particular modeling application. Temperature adjustments are usually made with an expression of the following form:

$$k_{T_2} = k_{T_1}\, \theta^{T_2 - T_1}, \qquad \text{(Eq. 4-58)}$$

where, θ is a temperature adjustment coefficient to adjust a coefficient value, k, from temperature T_1 to T_2. Many results have been published in the literature as to values of θ for temperature adjustment for various rate coefficients. θ has also been found to be temperature dependent, but may usually be considered to be constant over a range of about 10°C. The predictive expressions presented above have all been normalized to 25°C. The following temperature adjustment expressions which have been used for modeling in Kentucky contain typical values of θ for the temperature range of interest in most surface water modeling applications. For temperatures outside the range 15–30°C, the literature should be consulted.

$$k_{d,T} = k_{d,25^\circ\mathrm{C}}\ 1.04^{(T-25)} \qquad \text{(Eq. 4-59)}$$

$$k_{n,T} = k_{n,25^\circ\mathrm{C}}\ 1.08^{(T-25)} \qquad \text{(Eq. 4-60)}$$

$$k_{a,T} = k_{a,25^\circ\mathrm{C}}\ 1.022^{(T-25)} \qquad \text{(Eq. 4-61)}$$

$$\ln r_T = (\ln r_{25})\ 1.022^{(T-25)} \qquad \text{(Eq. 4-62)}$$

In each of these expressions T is expressed in °C.

4-4.4 Time of Travel/Velocity

In stream modeling on a segment-by-segment basis, a constant velocity of flow is normally utilized for each segment. This parameter is typically referred to as either average velocity, time of travel, time of passage, or time of flow, where the time of travel is equal to the length of the segment divided by the average velocity. Field determinations of values for this parameter are made in two general ways. Direct determination of velocity at a particular point in a stream may be made by conventional gaging techniques in which velocity measurements are made with a current meter at various points in the cross section, the cross-sectional geometry is measured, and the mean velocity is calculated assuming a parabolic vertical velocity distribution. Direct determination of average time of travel through a stream reach may be made by field measurement of the time required for the passage of a slug dose of material, usually dye, through the reach, using either the peak or centroid of its concentration cloud as an indicator.

A method for accurate simulation of stream velocity is of utmost importance in stream DO modeling. Characteristics of a stream, such as its reaeration capacity, change from segment-to-segment. If the time of travel through the various segments is inaccurately simulated, then the relative effects of various segment-variable parameters are not properly reflected. In the case of multiple point source discharges in close proximity on a stream or stream network, accurate velocity simulation is important, as related to the relationship between the effects of the various discharges on the stream water quality and the allocation of the stream's assimilation capacity among the various discharges for water quality management purposes. For example, consider the simple case of two discharges located a fixed distance apart on a single stream with a limited assimilation capacity, the true time of travel through the reach between the two being just sufficient for stabilization of the upstream discharge to such an extent that the permissible loading from the downstream discharge is unaffected. If in modeling this situation the time of travel through the reach is underestimated, then the predicted extent of stabilization of the upstream discharge might be underpredicted by the model to such an extent that the permissible loading from the downstream discharger would be decreased.

The expression most commonly used for calculation of velocity of flow in open channels is the Manning equation:

$$U = \frac{1.486}{n} R^{2/3} S_e{}^{1/2} \qquad \text{(Eq. 4-63)}$$

where

U = velocity, in ft/sec
n = Manning roughness coefficient
R = hydraulic radius, in ft, and
S_e = slope of the energy grade line, in ft/ft.

Information has been published on selection of n values for natural streams of various characteristics (Ref. 4-2, 4-6). Utilization of Equation 4-63 for predictive purposes in wide scaled modeling of natural streams has several disadvantages: 1) The hydraulic properties n, R, and S_e, are typically reasonably constant only through very short reaches; 2) it is generally almost impossible to select appropriate values for n without direct field calibration through direct measurement of V, R, and S_e; 3) under low flow conditions to be modeled for most water quality management applications, small to moderately sized streams do not exhibit uniform open channel flow behavior as assumed for the equation, but are typically characterized by alternating pools (with nearly flat slope) and riffles or shoals (with moderate to steep slopes), both with highly variable cross-sectional geometry; and 4) the amount of effort required to gather the necessary data on the hydraulic and geometric properties is prohibitive.

Attempts at using the Manning equation for low flow velocity predictions in natural Kentucky streams were unsuccessful in that very poor correlation between predicted and observed values was obtained. This prompted a series of research studies in which a technique employing a fluorescent dye was used for field determinations of time of travel. Each of the radioactive tracer studies described above for k_a determination also yielded corresponding data on time of travel, as the dye was used as the indicator of when to sample for the radioactive tracer. These studies concentrated on small to moderately sized free flowing streams under low flow conditions to provide information needed for modeling under these conditions. As a result of these studies, the following empirical expressions were developed to predict velocity as a function of two variables—stream bed slope and specific discharge—for which values could be determined for modeling with a reasonable amount of effort (Refs. 4-11, 4-12):

$$\text{for } 0.30 \leqslant q < 0.90, \quad U = 0.40 + 5.6(q - 0.08)S^{0.5} \qquad \text{(Eq. 4-64)}$$

$$\text{for } q < 0.30, \quad U = 0.40 + 4.1qS^{0.5} \qquad \text{(Eq. 4-65)}$$

where

U = average velocity through a stream reach, in mi/day
S = average stream bed slope through the reach, in ft/mi, and
q = specific discharge = actual stream discharge/average drainage basin area associated with the reach, in cfs/sq mi.

Specific discharges for modeling of low flow conditions are almost always less than 0.9 cfs/sq mi, the range covered by these expressions. Good correlation has been obtained between values predicted by these expressions and values observed by the dye tracer technique in free flowing streams with drainage basin areas between 5 and 250 sq mi (Ref. 4-12). It should be emphasized that these expressions were developed from data collected from relatively small streams in a limited geographical area, and therefore may not be of general applicability.

For 303 and 201 water quality management studies in Kentucky, Equations 4-64 and 4-65 (or slight modifications thereof) have been used for velocity prediction for low flow conditions in streams with drainage basin areas of up to about 1000 sq mi. For larger streams, usually characterized by relatively constant geometric properties over relatively long distances, segment velocities were estimated by continuity from the known discharge under consideration and estimates of the cross sectional area characteristic of the segment ($U = Q/A$).

4-4.5 Other Required Model Inputs

Stream Longitudinal Profiles and Cross-Sectional Geometry. Profile and cross-section information is typically necessary for prediction of certain parameters required in the DO modeling discussed in the foregoing. Typically, a dilemma exists in obtaining this information for wide scaled modeling for planning purposes. It is characterized on the one hand by a need for a large quantity of accurate detailed information, and on the other hand by the limitation imposed by the scope of the modeling effort as to how much of this information can be gathered and assimilated. Thus, the model must generally be structured so that it has the capability and flexibility to produce the required simulation from a data base of somewhat limited scope.

Field determination of these paramcters may be accomplished by conventional surveying techniques through distance, leveling, and sounding measurements. Profile and cross-sectional geometry data for some major streams are available in published form, particularly from the U.S. Army Corps of Engineers. Although stream geometry is constantly changing, these previously published data may be utilized for approximation in modeling for water quality management. Published data on the geometry of small to moderately sized streams is extremely limited, while field determinations are costly in both time and financial resources.

Several of the foregoing expressions were developed to predict modeling parameters as a function of stream bed slope and drainage basin area. These are attractive variables for wide scaled modeling of small streams in that their values can be determined with an amount of effort generally commensurate with the scope of time and resources available. Both may be determined from U.S.G.S. topographic maps: area by planimetry and slope by plots of contour crossings vs. stream station. The drainage basin area is an indirect indicator of the cross-sectional properties. Use of drainage basin area as a variable for prediction implies that within a certain topographic region, streams with the same drainage basin area should have similar cross-sectional properties. Use of average stream bed slope for prediction implies that it is an indicator of hydraulic gradient. Determination of stream bed slope from topographic maps is a relatively crude method, but over a stream reach of substantial length it gives a good approximation of the average bed slope within the reach. This, in turn, is an approximation of the average slope of the water surface elevation under low flow conditions.

Methods of determination of parameter values for predictive purposes in modeling should be consistent with the methods used in developing the corresponding predictive expressions. For example, for modeling in Kentucky, values of k_a, experimentally determined by the radioactive tracer technique for representative streams, were correlated with average stream bed slope (as determined from topographic maps) to produce expressions to predict k_a as a function of slope. These expressions were then used for predicting k_a for other streams for which slope was determined from maps by the same method.

Stream Stationing System. In modeling stream networks, as is generally required for planning purposes, it is necessary to adopt a standard method for stationing of individual streams. Such a system usually involves stationing in horizontal distances, and provides a unique location identifier for any point on any stream in the total stream network under consideration. The system should be designed for optimum amenability to computer modeling and should be consistent as well as practical for the existing systems of the area. For example, the U.S.G.S. and U.S. Army Corps of Engineers have standard mile point systems for many major streams for which available data are correlated with stream mile point.

The system recently adopted for the state of Kentucky may be used for illustration. A single set of 7.5-minute U.S.G.S. topographic maps are utilized for measurement purposes and for permanent reference. Essentially all surface streams in the state are stationed by measurement with appropriate methods (marking stations are at 0.1 mi intervals on the topographic maps). For each stream, stationing begins at the mouth and proceeds upstream to the headwaters. All pertinent features, such as confluences, point source discharges, etc., are identified from the topographic maps and other sources, and are referenced as a function of station. This information is stored on computer tape for retrieval for subsequent modeling and other purposes. The flexibility of this system is such that the information may be continuously supplemented and corrected. The details of the system are beyond the scope of this text, but an example will serve to illustrate how any point on any stream in the state is characterized by a unique set of numbers.

With the exception of a few minor tributaries which discharge directly into the Mississippi River, the discharges from all streams in the state ultimately flow into the Ohio River, which is used as the reference basis for the system. The system can accommodate streams up to eight stream orders lower than the Ohio River, which is of sufficient scope to include all streams of any significance. The following set of numbers identifying the location of a point source discharge is typical:

435.60 51.90 17.80 34.00 10.30

This set uniquely locates the Lexington, Kentucky, Town Branch Wastewater Treatment Facility outfall on Town Branch Creek at 10.30 mi upstream from its confluence with South Elkhorn Creek, at 34.00 mi upstream from its confluence with Elkhorn Creek, at 17.80 mi upstream from its confluence with the Kentucky River, and at 51.90 mi upstream from its confluence with the Ohio River at MP 435.60. The system permits location of any point to the nearest 0.01 mi.

Temperature. Selection of the appropriate temperature for a particular DO modeling application is a particularly critical determination because several of the model parameters are temperature dependent, especially the rate coefficients and DO saturation concentration. Under otherwise identical conditions, the higher the temperature the more critical the DO conditions. Although generally not as extensive as discharge data, sufficient historical data on surface water temperature is generally available for most localities to permit selection of appropriate temperatures for modeling. In cases where long term temperature data are available, frequency-duration analyses may be performed in a manner analogous to that described above for stream discharge or flow. A value termed the 10-year, 7-day high temperature could conceivably come from such an analysis. In most cases, insufficient data are available to go to this level of detail in analysis and judgement must be exercised in selecting the appropriate temperature.

For determination of permissible point source effluent quality such as that required in 303 and 201 planning, DO modeling at critical conditions is performed. This generally means some combination of low flow and high temperature. If the low flow regime is established (10-year, 7-day low flow, for example), then the historical flow and temperature records should be examined to determine the time of the year of occurrence of this low flow and the corresponding water temperatures at this time of the year. The corresponding temperature used for modeling should be the highest temperature which historically occurs at the same time of year as the specified low flow and with the approximate same frequency and duration. For example, in Kentucky, extreme high temperatures historically occur in August when flows are typically considerably higher than the extreme low flows which historically occur in early October when the temperature is typically considerably less than the extreme August values. For 303 and 201 modeling in Kentucky a temperature of 25°C has generally been used in conjunction with 10-year, 7-day low flows. From the historical records, this is the approximate maximum temperature which can be expected for a 7-day duration in late September or early October when the low flows occur. Analysis for selected sites has also indicated that the extreme low flow, lower temperature (typical of early October) condition is usually more DO critical than the extreme high temperature, higher flow (typical of August) condition.

The extreme temperature, extreme flow relationship is obviously a function of geography and climate, but, in at least some localities such as Kentucky, the probability of occurrance of extreme low flow and extreme high temperature at the same time for any significant duration is small.

Special consideration must be given to the effects of thermal discharges on receiving water temperatures for DO modeling. Modeling of instream temperature is discussed briefly later in the section on modeling of parameters other than DO.

Discharge Inventories–Magnitude and Quality. In DO modeling for a stream network system, an inventory of all discharges to the streams in the system is necessary to provide input data. In modeling to simulate conditions for a particular date, each discharge must be characterized as to its magnitude and quality expected to occur at the time to be simulated. In modeling to determine permissible effluent quality for discharges under future conditions, only discharge magnitude is a required input in that discharge quality is the output in this case. All discharge points should be located in accordance with a standard stationing system, as described above.

Applicable Water Quality Standards. One of the primary objectives of DO modeling for water quality management planning purposes is the determination of the extent of water quality improvement necessary to meet instream water quality standards. Thus, these water quality standards should be quantified and incorporated into a planning model. In simulating a particular set of conditions, they serve to indicate the degree of violation or non-violation through a comparison of concentrations predicted by the model and the standard or permissible concentrations. In simulations to determine permissible effluent quality for discharges for non-violation of instream standards, they serve as a basis for comparison with predicted instream concentrations until non-violation is achieved. This is normally accomplished in the model by beginning with a low discharge effluent quality and increasing the effluent quality by reiterative techniques until non-violation of the specified instream standards is achieved. This technique also requires specification of the absolute minimum permissible effluent quality, or the poorest quality permissible for the effluent regardless of instream conditions. For a municipal discharge under current conditions this would typically be the effluent quality defined for conventional secondary treatment.

Water quality standards values for incorporation into a model must, of course, be specified in accordance with those established by the appropriate regulatory agency for the waters to be modeled. Instream standards for DO modeling for planning are generally relatively simple. For modeling in Kentucky in relation to 303 and 201 water quality management studies, the two significant standards have been, in the general case, a minimum permissible instream DO concentration of 5.0 mg/l and a maximum instream NH_3-N concentration ranging from 1 to 4 mg/l. The NH_3-N limitation is due to ammonia toxicity considerations. Various levels within the range have been applied for different instream pH conditions and in accordance with information on the accute vs. chronic toxicity effects of ammonia. Of the two forms comprising the total NH_3-N concentration—free or unionized ammonia (NH_3) and the ionized ammonium ion (NH_4^+)—the former is much more toxic to aquatic life. The proportion of free ammonia increases sharply with increasing pH in the pH range characteristic of natural streams. Thus, a greater concentration of total NH_3-N may be tolerated at a lower pH with the same potential toxicity as a lower total concentration at a higher pH. Also, a somewhat higher total concentration can be tolerated over a short time period with the same effects as a somewhat lower total concentration over an extended period of time. Different permissible concentration limits are sometimes established for these two cases, termed the acute toxicity limit for the former and the chronic toxicity limit for the latter. Since the extreme low flow is generally of shorter duration in a stream with a natural base flow, a higher concentration can be tolerated at the extreme low flow than can be tolerated in a stream which carries essentially only the effluent from a point source discharge for extended periods of time under dry weather conditions.

4-5 EXAMPLE APPLICATION OF A ONE DIMENSIONAL STEADY STATE DO MODEL FOR WATER QUALITY MANAGEMENT

Experience gained in performing modeling for water quality management planning in Kentucky will serve as a basis for an example to illustrate how the foregoing principles and methodologies may be synthesized into a workable model.

Prior to actual wide scaled modeling, research studies were performed to characterize the properties of a number of representative streams throughout the state. These results and information from the literature were then synthesized into a comprehensive model for application to all streams in the state. Accumulation of the necessary data base, development of a stream stationing system, and actual application of the model were accomplished by a team consisting of professors and graduate students from the University of Louisville and the University of Kentucky and personnel from state agencies. The same modeling techniques were applied for each of the state's ten river basins for the 303 River Basin Planning Studies. Discharge inventories were provided by consulting engineering firms performing the individual river basin studies, who in turn utilized the results of the modeling effort in preparing their respective plans. The model has also been used to supplement the results of the 303 studies in 201 Facilities Plans and in establishing discharge limitations for NPDES permits.

The overall model was designed to be utilized in a continuing planning process with the flexibility to accommodate changes resulting from new information gained from such sources as monitoring programs and model calibration studies.

A stream stationing system was developed as described above as a reference for both the model input and output. The data base consisted of a computerized inventory of all streams. As a function of stream station, this inventory included elevations; 10-year, 7-day low flows; drainage basin areas; and locations of all point source discharges, stream confluences, and other pertinent stream features such as dams and waterfalls. Mathematical formulations included in the model were mass balance equations, basic DO equations, and expressions to predict required model parameters as a function of input data. Output from the model was in various forms including DO profiles, assimilation capacities of stream segments, and permissible quality of point source discharges, all under various sets of conditions.

For the 303 studies, modeling proceeded along the following lines for each river basin: 1) Based on an inventory of existing flowrate and quality for all point source discharges, a single computer run produced predicted DO profiles for streams in the basin under low flow conditions. These results permitted the classification of "water quality limited" and "effluent limited" stream segments and provided an indication of the degree of severity of potential DO problems in the basin; 2) for a given set of future conditions, reiterative techniques were used to produce a set of output indicating the permissible effluent quality for each point source discharge for non-violation of stream water quality standards throughout the basin. Allocations of wastewater loads to the various dischargers for future conditions were made on the basis of these results; and 3) for several selected stream segments in the basin, computer runs were made to determine the stream waste assimilation capacity. These results were intended to serve as guidance for potential relocation of existing discharges or optimum location of future new dischargers.

4-5.1 Model Description

A detailed description of all of the mathemetical expressions and computer techniques involved in the model is beyond the scope of this text, but the

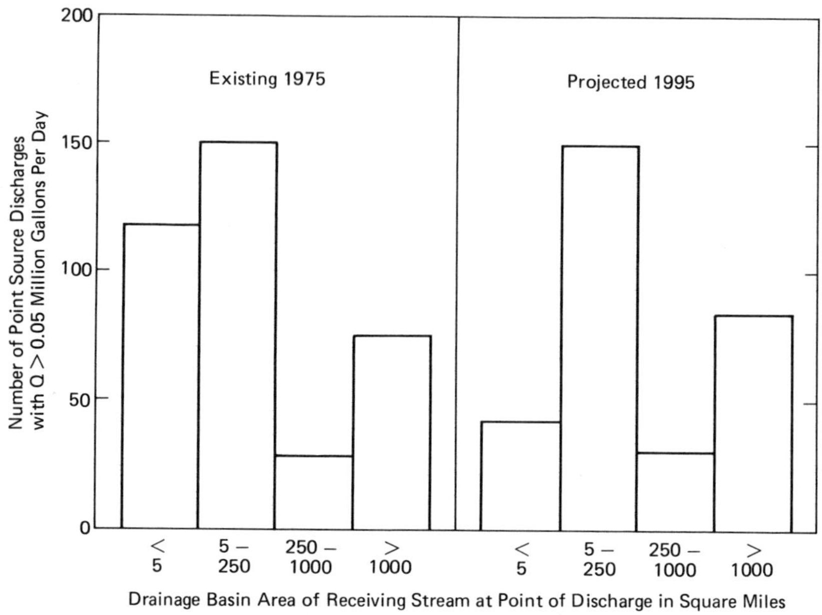

Fig. 4-6. Distribution of point source discharges in Kentucky as a function of receiving stream size.

essential features of the model will be presented as a basis for understanding its scope and application. The features of the model are described here as specifically applied to small to moderately size streams (drainage basin areas less than about 1000 sq mi). Based on the findings of the 303 planning studies and as shown in Figure 4-6, the majority of point source discharges in Kentucky are to streams in this category, both at present and as projected for the future. In some cases, different techniques were utilized in modeling for the large streams, particularly for reaeration and velocity predictions.

For a given stream, the basic DO equations were applied on a segment-by-segment basis. Segment boundaries were determined by the location of slope changes, point source discharges, confluences, dams, or in general, any feature which results in a change in value of any of the model parameters which are assumed constant within each segment. The basic DO equation in the model was Equation 4-33, which reflects the effects of several sources and sinks of DO. However, as generally applied for modeling for the planning studies, it was assumed that reaeration was the only DO source and CBOD and NBOD were the only DO sinks. For the purpose of applying the model to a particular segment, consider the segment to be defined as beginning at upstream station, X_o, and extending to the downstream station, X_1, with a general point, X, being located at a distance X downstream from X_o. Under these conditions and assumed DO sources and sinks, Equation 4-33 reduces to the following expression, which was the basic DO equation used in the model:

$$D = D_o\, e^{-k_a (X/U)} + \frac{k_d L_o}{k_a - k_d} \left(e^{-k_d (X/U)} - e^{-k_a (X/U)}\right) + \frac{k_n N_o}{k_a - k_n} \left(e^{-k_n (X/U)} - e^{-k_a (X/U)}\right) \qquad \text{(Eq. 4-66)}$$

Saturation DO concentrations were determined as a function of temperature and elevation as:*

for $T \lesseqgtr 27°C$

$$C_s = (14.65 - 0.406\, T + 0.00810\, T^2 - 0.000073\, T^3)\,(1.0 - \frac{0.027E}{760}) \qquad \text{(Eq. 4-67)}$$

for $T > 27°C$

$$C_s = (14.65 - 0.416\, T + 0.00813\, T^2 - 0.000069\, T^3)\,(1.0 - \frac{0.027\, E}{760}), \qquad \text{(Eq. 4-68)}$$

and by definition:

$$C = C_s - D, \qquad \text{(Eq. 4-69)}$$

where

D = DO deficit concentration at X, mg/l

*Equations 4-67 and 4-68 were developed to generate C_s vs. T values to correspond to those found in standard tables such as Table 422:I, p. 446–447, *Standard Methods for the Examination of Water and Wastewater*, 14th Edition, APHA-AWWA-WPCF, 1975.

The Committee on Sanitary Engineering Research of the Sanitary Engineering Division (now the Environmental Engineering Division) of the American Society of Civil Engineers conducted a study "to determine true saturation values throughout the normal range of natural stream water temperatures." The results were reported as "Solubility of Atmospheric Oxygen in Water," *J. Sanitary Eng. Div., A.S.C.E. 86,* No. SA 4, July, 1960, pp. 41–53. The following expression applicable for $0 \leq T \leq 30°C$ resulted from this study.

$$C_s = 14.652 - 0.41022T + 0.0079910T^2 - 0.000077774T^3$$

where C_s is the sea level DO saturation concentration. Both of the above sources seem to be in rather widespread use for determining C_s values. It should be noted that they produce significantly different C_s values in the temperature range of interest in most DO modeling applications. For example, for sea level at 25°C, the *Standard Methods* tables give C_s = 8.4 mg/l while the A.S.C.E. expression gives C_s = 8.18 mg/l.

D_o = DO deficit concentration at X_o, mg/l
L_o = ultimate CBOD concentration at X_o, mg/l
N_o = ultimate NBOD concentration at X_o, mg/l
k_a = atmospheric reaeration rate coefficient for segment, day^{-1}
k_d = CBOD deoxygenation rate coefficient for segment, day^{-1}
k_n = NBOD deoxygenation rate coefficient for segment, day^{-1}
U = mean velocity for reach = length of reach/time of travel for reach, mi/day
X/U = time of travel from X_o to X, days
C_s = DO saturation concentration for segment, mg/l
T = mean temperature of segment, °C
E = mean elevation of reach, ft above msl, and
C = DO concentration at X, mg/l.

Values for k_d and k_n were determined from Equations 4-37 through 4-42. Values for k_a were determined from Equation 4-55. Reaeration at dams and falls was determined from Equations 4-56 and 4-57. Temperature adjustments were made from Equations 4-59 through 4-62. Initial concentrations (at X_o) of DO deficit, CBOD, NBOD, and temperature (D_o, L_o, N_o, T_o) were determined by assuming instantaneous complete mixing of input and stream flow and computation by mass balance as follows:

$$C_o = \frac{RC_u + C_i}{1 + R} \qquad \text{(Eq. 4-70)}$$

$$R = \frac{Q_u}{Q_i} \qquad \text{(Eq. 4-71)}$$

where

C_o = instream concentration immediately below input
C_u = upstream concentration
C_i = input concentration
R = dilution ratio
Q_u = upstream flowrate, and
Q_i = input flowrate.

4-5.2 Example Applications

The general model as described in the foregoing has the flexibility to produce output of various forms depending upon the desired results. Output from the model from application to a typical stream network is too voluminous to be included here, but selected examples are presented below for illustration.

DO Profiles. Figure 4-7 schematically shows a simplified hypothetical stream network involving two streams and two point source discharges. The larger

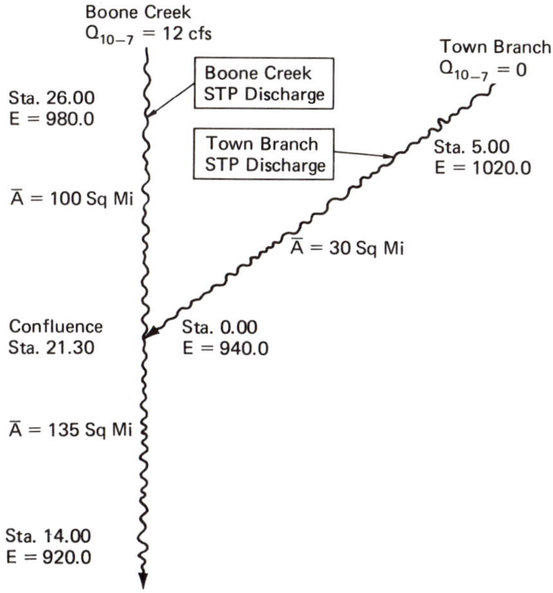

Fig. 4-7. Schematic of stream network for DO profile analysis shown in Exhibits 4-1A and 4-1B.

stream (Boone Creek) has a flatter slope and an upstream 10-year, 7-day low flow of 12.0 cfs while the smaller stream (Town Branch) has a steeper slope and an upstream 10-year, 7-day low flow of zero.

Exhibit 4-1A presents the results of a DO profile analysis to predict instream conditions under low flow conditions at 25°C with effluent quality and magnitude for the two discharges set at existing values. Asterisks associated with the last two columns indicate violations of the permissible instream NH_3-N and DO standards which were set at 2.0 mg/l maximum and 5.0 mg/l minimum, respectively.

Exhibit 4-1B presents the results of an analysis for the same network for a future condition. Here, the magnitudes of the two discharges have been set at their projected values, an allocation of permissible effluent quality has been made to each discharge for non-violation of the instream standards, and the resulting predicted DO profiles have been determined for the streams. It may be noted that the permissible effluent NH_3-N concentrations are primarily controlled by the instream NH_3-N concentrations in the two segments immediately below the two discharges, while the permissible effluent BOD concentrations are primarily controlled by the instream DO concentration in the segment downstream from the stream confluence.

STREAM DISSOLVED OXYGEN PROFILE ANALYSIS RESULTS – TOWN BRANCH–TRIBUTARY TO BOONE CRK

```
INPUT PARAMETERS IMMEDIATELY UPSTREAM FROM MILE POINT=   5.00

     Q=   0.00 CFS

     BOD-5=  0.00 MG/L

     NH3-N=  0.00 MG/L

     DO=  0.00 MG/L


INPUT PARAMETERS-SEGMENT   1
     STATION UPSTREAM END =    5.00
     STATION DOWNSTREAM END =    0.00
     ELEVATION UPSTREAM END = 1020.0
     ELEVATION DOWNSTREAM END =   940.0
     STREAM DRAINAGE BASIN AREA =    30. SQ. MI.
     CALCULATED SLOPE =     16.00 FT/MI
     INSTREAM SPECIFIC DISCHARGE = 0.103 CFS/SQ MI
     CALCULATED STREAM VELOCITY =   2.095 MI/DAY
     DISCHARGE IDENTIFICATION = TOWN BRANCH STP
     EFFLUENT DISCHARGE =    2.0000 MGD
     EFFLUENT 5-DAY BOD = 30.00 MG/L
     EFFLUENT AMMONIA N = 16.00 MG/L
     EFFLUENT DISSOLVED OXYGEN =   3.50 MG/L
     CALCULATED EFFLUENT STRENGTH FACTOR = 114. MG/L
     TEMPERATURE FOR SEGMENT = 25. DEGREES C
     CALCULATED DISSOLVED OXYGEN SATURATION =   8.12 MG/L
```

SEG NO	MILE PT	TT(DAYS)	Q(CFS)	KA(/DAY)	KD(/DAY)	KN(/DAY)	BOD5(MG/L)	NH3N(MG/L)	DO(MG/L)
1	5.00	0.00	3.100	5.593	0.500	0.300	30.00	16.00 *	3.50 *
1	4.48	0.25	3.100	5.593	0.500	0.300	26.47	14.84 *	1.39 *
1	3.95	0.50	3.100	5.593	0.500	0.300	23.36	13.77 *	1.40 *
1	3.43	0.75	3.100	5.593	0.500	0.300	20.62	12.78 *	1.88 *
1	2.91	1.00	3.100	5.593	0.371	0.300	18.79	11.85 *	2.34 *
1	2.38	1.25	3.100	5.593	0.371	0.300	17.13	11.00 *	2.78 *
1	1.86	1.50	3.100	5.593	0.371	0.300	15.61	10.20 *	3.19 *
1	1.33	1.75	3.100	5.593	0.371	0.300	14.23	9.46 *	3.58 *
1	0.81	2.00	3.100	5.593	0.256	0.300	13.34	8.78 *	3.51 *
1	0.29	2.25	3.100	5.593	0.256	0.300	12.52	8.15 *	3.74 *
1	0.00	2.39	3.100	5.593	0.256	0.300	12.08	7.82 *	3.89 *

Exhibit 4-1A

Determination of Permissible Effluent Quality/Sensitivity. The results from the application of a DO model to a wide range of conditions will usually give an indication as to the relative significance of the various model parameters. From determinations of permissible effluent qualities for point source discharge with this particular model, it was readily apparent that the results are most strongly affected by values of the dilution ratio (upstream flowrate/discharge flowrate) and receiving stream slope. By using these two parameters, each of which can be determined with a minimum amount of effort, as variables and assuming fixed values for other model parameters, an estimate of the permissible quality for a particular effluent may be made. Or, from a set of contributions of values of slope and dilution ratio with other parameters fixed, a corresponding set of permissible effluent qualities may be generated. Such output is useful in estimates for wasteload allocations in the preliminary phases of water quality planning studies.

Exhibit 4-2A presents the results of such an analysis. The assumed or fixed conditions are given in the heading and the results are self-explanatory. Ultimate

STREAM DISSOLVED OXYGEN PROFILE ANALYSIS RESULTS - BOONE CREEK

INPUT PARAMETERS IMMEDIATELY UPSTREAM FROM MILE POINT= 26.00

Q= 12.00 CFS

BOD-5= 2.00 MG/L

NH3-N= 0.22 MG/L

DO= 7.00 MG/L

INPUT PARAMETERS-SEGMENT 1
STATION UPSTREAM END = 26.00
STATION DOWNSTREAM END = 21.30
ELEVATION UPSTREAM END = 980.0
ELEVATION DOWNSTREAM END = 940.0
STREAM DRAINAGE BASIN AREA = 100. SQ. MI.
CALCULATED SLOPE = 8.51 FT/MI
INSTREAM SPECIFIC DISCHARGE = 0.244 CFS/SQ MI
CALCULATED STREAM VELOCITY = 3.318 MI/DAY
DISCHARGE IDENTIFICATION = BOONE CREEK STP
EFFLUENT DISCHARGE = 8.0000 MGD
EFFLUENT 5-DAY BOD = 30.00 MG/L
EFFLUENT AMMONIA N = 16.00 MG/L
EFFLUENT DISSOLVED OXYGEN = 3.50 MG/L
CALCULATED EFFLUENT STRENGTH FACTOR = 114. MG/L
TEMPERATURE FOR SEGMENT = 25. DEGREES C
CALCULATED DISSOLVED OXYGEN SATURATION = 8.13 MG/L

SEG NO	MILE PT	TT(DAYS)	Q(CFS)	KA(/DAY)	KD(/DAY)	KN(/DAY)	BOD5(MG/L)	NH3N(MG/L)	DO(MG/L)
1	26.00	0.00	24.400	2.781	0.292	0.278	16.23	8.24 *	5.22
1	25.17	0.25	24.400	2.781	0.292	0.278	15.09	7.69 *	3.66 *
1	24.34	0.50	24.400	2.781	0.292	0.278	14.02	7.17 *	3.09 *
1	23.51	0.75	24.400	2.781	0.292	0.278	13.04	6.69 *	3.00 *
1	22.68	1.00	24.400	2.781	0.292	0.278	12.12	6.24 *	3.13 *
1	21.85	1.25	24.400	2.781	0.218	0.278	11.47	5.82 *	3.15 *
1	21.30	1.42	24.400	2.781	0.218	0.278	11.07	5.56 *	3.26 *

INPUT PARAMETERS-SEGMENT 2
STATION UPSTREAM END = 21.30
STATION DOWNSTREAM END = 14.00
ELEVATION UPSTREAM END = 940.0
ELEVATION DOWNSTREAM END = 920.0
STREAM DRAINAGE BASIN AREA = 135. SQ. MI.
CALCULATED SLOPE = 2.74 FT/MI
INSTREAM SPECIFIC DISCHARGE = 0.204 CFS/SQ MI
CALCULATED STREAM VELOCITY = 1.782 MI/DAY
DISCHARGE IDENTIFICATION = DOWNSTREAM FROM CONF
EFFLUENT DISCHARGE = 2.0000 MGD
EFFLUENT 5-DAY BOD = 12.08 MG/L
EFFLUENT AMMONIA N = 7.82 MG/L
EFFLUENT DISSOLVED OXYGEN = 3.89 MG/L
CALCULATED EFFLUENT STRENGTH FACTOR = 49. MG/L
TEMPERATURE FOR SEGMENT = 25. DEGREES C
CALCULATED DISSOLVED OXYGEN SATURATION = 8.14 MG/L

SEG NO	MILE PT	TT(DAYS)	Q(CFS)	KA(/DAY)	KD(/DAY)	KN(/DAY)	BOD5(MG/L)	NH3N(MG/L)	DO(MG/L)
2	21.30	0.00	27.500	0.937	0.201	0.100	11.18	5.81 *	3.33 *
2	20.85	0.25	27.500	0.937	0.201	0.100	10.63	5.67 *	3.02 *
2	20.41	0.50	27.500	0.937	0.201	0.100	10.11	5.53 *	2.83 *
2	19.96	0.75	27.500	0.937	0.201	0.100	9.61	5.39 *	2.73 *
2	19.52	1.00	27.500	0.937	0.201	0.100	9.14	5.26 *	2.70 *
2	19.07	1.25	27.500	0.937	0.201	0.100	8.69	5.13 *	2.71 *
2	18.63	1.50	27.500	0.937	0.201	0.100	8.27	5.00 *	2.77 *
2	18.18	1.75	27.500	0.937	0.150	0.100	7.96	4.88 *	2.91 *
2	17.74	2.00	27.500	0.937	0.150	0.100	7.67	4.76 *	3.05 *
2	17.29	2.25	27.500	0.937	0.150	0.100	7.39	4.64 *	3.19 *
2	16.84	2.50	27.500	0.937	0.150	0.100	7.12	4.53 *	3.32 *
2	16.40	2.75	27.500	0.937	0.150	0.100	6.85	4.41 *	3.46 *
2	15.95	3.00	27.500	0.937	0.150	0.100	6.60	4.31 *	3.59 *
2	15.51	3.25	27.500	0.937	0.150	0.100	6.36	4.20 *	3.72 *
2	15.06	3.50	27.500	0.937	0.150	0.100	6.12	4.10 *	3.85 *
2	14.62	3.75	27.500	0.937	0.150	0.100	5.90	3.99 *	3.87 *
2	14.17	4.00	27.500	0.937	0.150	0.100	5.68	3.90 *	3.91 *
2	14.00	4.10	27.500	0.937	0.150	0.100	5.60	3.86 *	3.93 *

Exhibit 4-1A (continued)

```
STREAM DISSOLVED OXYGEN PROFILE ANALYSIS RESULTS - TOWN BRANCH-TRIBUTARY TO BOONE CRK

  INPUT PARAMETERS IMMEDIATELY UPSTREAM FROM MILE POINT=   5.00

        Q=   0.00 CFS

        BOD-5=  0.00 MG/L

        NH3-N=  0.00 MG/L

        DO=  0.00 MG/L


  INPUT PARAMETERS-SEGMENT    1
        STATION UPSTREAM END =    5.00
        STATION DOWNSTREAM END =    0.00
        ELEVATION UPSTREAM END = 1020.0
        ELEVATION DOWNSTREAM END =   940.0
        STREAM DRAINAGE BASIN AREA =    30. SQ. MI.
        CALCULATED SLOPE =    16.00 FT/MI
        INSTREAM SPECIFIC DISCHARGE = 0.124 CFS/SQ MI
        CALCULATED STREAM VELOCITY =  2.434 MI/DAY
        DISCHARGE IDENTIFICATION = TOWN BRANCH STP
        EFFLUENT DISCHARGE =   2.4000 MGD
        EFFLUENT 5-DAY BOD = 25.00 MG/L
        EFFLUENT AMMONIA N =  2.00 MG/L
        EFFLUENT DISSOLVED OXYGEN =  7.00 MG/L
        CALCULATED EFFLUENT STRENGTH FACTOR =  40. MG/L
        TEMPERATURE FOR SEGMENT = 25. DEGREES C
        CALCULATED DISSOLVED OXYGEN SATURATION =  8.12 MG/L
```

SEG NO	MILE PT	TT(DAYS)	Q(CFS)	KA(/DAY)	KD(/DAY)	KN(/DAY)	BOD5(MG/L)	NH3N(MG/L)	DO(MG/L)
1	5.00	0.00	3.720	5.593	0.450	0.300	25.00	2.00	7.00
1	4.39	0.25	3.720	5.593	0.450	0.300	22.34	1.86	5.37
1	3.78	0.50	3.720	5.593	0.450	0.300	19.96	1.72	5.22
1	3.17	0.75	3.720	5.593	0.450	0.300	17.84	1.60	5.41
1	2.57	1.00	3.720	5.593	0.321	0.300	16.46	1.48	6.00
1	1.96	1.25	3.720	5.593	0.321	0.300	15.19	1.37	6.25
1	1.35	1.50	3.720	5.593	0.321	0.300	14.02	1.28	6.42
1	0.74	1.75	3.720	5.593	0.321	0.300	12.94	1.18	6.55
1	0.13	2.00	3.720	5.593	0.233	0.300	12.21	1.10	6.80
1	0.00	2.05	3.720	5.593	0.233	0.300	12.05	1.08	6.83

Exhibit 4-1B

oxygen demand, UOD, and strength factor, SF, were calculated as defined in Equations 4-35 and 4-36, respectively, as indicators of overall permissible effluent strength. As an indicator of relative cost, the present worth cost for a 10 MGD capacity treatment facility was calculated from a formulation of cost vs. SF, using the values in Table 4-2. In each case, the value given for "Crit Time" represents the time of travel downstream below the discharge to the point where the corresponding "minimum DO" occurs. For this set of assumed conditions, permissible effluent quality is not further decreased for dilutions of greater than about 12:1, regardless of the stream slope. For a given dilution ratio, permissible effluent quality is not further decreased for stream slopes greater than about 25 ft/mi.

The accuracy with which a given situation can be simulated by a model is obviously a function of the accuracy with which the input parameters can be determined. For DO modeling of a given situation, there is generally a good probability that one or more of the parameter values used will differ significantly from the corresponding true values. Thus, especially for water quality planning and management, it is desirable to perform an analysis on the model being used to

STREAM DISSOLVED OXYGEN PROFILE ANALYSIS RESULTS - BOONE CREEK

INPUT PARAMETERS IMMEDIATELY UPSTREAM FROM MILE POINT= 26.00

Q= 12.00 CFS

BOD-5= 2.00 MG/L

NH3-N= 0.22 MG/L

DO= 7.00 MG/L

INPUT PARAMETERS-SEGMENT 1
STATION UPSTREAM END = 26.00
STATION DOWNSTREAM END = 21.30
ELEVATION UPSTREAM END = 980.0
ELEVATION DOWNSTREAM END = 940.0
STREAM DRAINAGE BASIN AREA = 100. SQ. MI.
CALCULATED SLOPE = 8.51 FT/MI
INSTREAM SPECIFIC DISCHARGE = 0.275 CFS/SQ MI
CALCULATED STREAM VELOCITY = 3.689 MI/DAY
DISCHARGE IDENTIFICATION = BOONE CREEK STP
EFFLUENT DISCHARGE = 10.0000 MGD
EFFLUENT 5-DAY BOD = 25.00 MG/L
EFFLUENT AMMONIA N = 3.30 MG/L
EFFLUENT DISSOLVED OXYGEN = 5.00 MG/L
CALCULATED EFFLUENT STRENGTH FACTOR = 47. MG/L
TEMPERATURE FOR SEGMENT = 25. DEGREES C
CALCULATED DISSOLVED OXYGEN SATURATION = 8.13 MG/L

SEG NO	MILE PT	TT(DAYS)	Q(CFS)	KA(/DAY)	KD(/DAY)	KN(/DAY)	BOD5(MG/L)	NH3N(MG/L)	DO(MG/L)
1	26.00	0.00	27.500	2.781	0.269	0.278	14.96	1.96	5.87
1	25.08	0.25	27.500	2.781	0.269	0.278	13.99	1.82	5.53
1	24.16	0.50	27.500	2.781	0.269	0.278	13.08	1.70	5.46
1	23.23	0.75	27.500	2.781	0.269	0.278	12.23	1.59	5.51
1	22.31	1.00	27.500	2.781	0.269	0.278	11.43	1.48	5.62
1	21.39	1.25	27.500	2.781	0.269	0.278	10.69	1.38	5.75
1	21.30	1.27	27.500	2.781	0.192	0.278	10.64	1.37	5.78

INPUT PARAMETERS-SEGMENT 2
STATION UPSTREAM END = 21.30
STATION DOWNSTREAM END = 14.00
ELEVATION UPSTREAM END = 940.0
ELEVATION DOWNSTREAM END = 920.0
STREAM DRAINAGE BASIN AREA = 135. SQ. MI.
CALCULATED SLOPE = 2.74 FT/MI
INSTREAM SPECIFIC DISCHARGE = 0.231 CFS/SQ MI
CALCULATED STREAM VELOCITY = 1.969 MI/DAY
DISCHARGE IDENTIFICATION = DOWNSTREAM FROM CONF
EFFLUENT DISCHARGE = 2.4000 MGD
EFFLUENT 5-DAY BOD = 12.05 MG/L
EFFLUENT AMMONIA N = 1.08 MG/L
EFFLUENT DISSOLVED OXYGEN = 6.83 MG/L
CALCULATED EFFLUENT STRENGTH FACTOR = 16. MG/L
TEMPERATURE FOR SEGMENT = 25. DEGREES C
CALCULATED DISSOLVED OXYGEN SATURATION = 8.14 MG/L

SEG NO	MILE PT	TT(DAYS)	Q(CFS)	KA(/DAY)	KD(/DAY)	KN(/DAY)	BOD5(MG/L)	NH3N(MG/L)	DO(MG/L)
2	21.30	0.00	31.220	0.937	0.194	0.100	10.81	1.34	5.90
2	20.81	0.25	31.220	0.937	0.194	0.100	10.29	1.30	5.55
2	20.32	0.50	31.220	0.937	0.194	0.100	9.80	1.27	5.31
2	19.82	0.75	31.220	0.937	0.194	0.100	9.34	1.24	5.16
2	19.33	1.00	31.220	0.937	0.194	0.100	8.90	1.21	5.06
2	18.84	1.25	31.220	0.937	0.194	0.100	8.47	1.18	5.02
2	18.35	1.50	31.220	0.937	0.194	0.100	8.07	1.15	5.02
2	17.85	1.75	31.220	0.937	0.150	0.100	7.77	1.12	5.14
2	17.36	2.00	31.220	0.937	0.150	0.100	7.49	1.10	5.26
2	16.87	2.25	31.220	0.937	0.150	0.100	7.21	1.07	5.36
2	16.38	2.50	31.220	0.937	0.150	0.100	6.95	1.04	5.47
2	15.88	2.75	31.220	0.937	0.150	0.100	6.69	1.02	5.56
2	15.39	3.00	31.220	0.937	0.150	0.100	6.44	0.99	5.66
2	14.90	3.25	31.220	0.937	0.150	0.100	6.21	0.97	5.74
2	14.41	3.50	31.220	0.937	0.150	0.100	5.98	0.94	5.83
2	14.00	3.71	31.220	0.937	0.150	0.100	5.80	0.92	5.87

Exhibit 4-1B (continued)

RESULTS OF ANALYSIS TO PREDICT PERMISSIBLE EFFLUENT PARAMETERS AS A FUNCTION OF DILUTION RATIO AND SLOPE

ASSUMED UPSTREAM WATER QUALITY: BOD-5 = 2.0 MG/L, NH3-N = 0.22 MG/L, DO = 7.0 MG/L.

ASSUMED DISSOLVED OXYGEN SATURATION CONCENTRATION = 8.2 MG/L. ASSUMED TEMPERATURE = 25 DEGREES C.

VARIABLE DEOXYGENATION COEFFICIENTS.

MAXIMUM PERMISSIBLE IN-STREAM NH3-N = 2.0 MG/L.

THE PERMISSIBLE EFFLUENT PARAMETERS ARE DETERMINED AS FOLLOWS:

DO = 2.0 MG/L GREATER THAN THE MINIMUM REQUIRED TO GIVE AN INITIAL IN-STREAM CONCENTRATION OF 5.0 MG/L. ABSOLUTE MINIMUM = 2.0 MG/L.

NH3-N = MAXIMUM CONCENTRATION PERMISSIBLE TO GIVE A MAXIMUM IN-STREAM CONCENTRATION AS INDICATED ABOVE BUT NOT TO EXCEED 2/3 OF THE PERMISSIBLE BOD-5 CONCENTRATION, BOTH IN MG/L. ABSOLUTE MAXIMUM = 20.0 MG/L.

BOD-5 = MAXIMUM CONCENTRATION PERMISSIBLE TO GIVE A MINIMUM IN-STREAM DO CONCENTRATION OF 5.0 MG/L WITH PERMISSIBLE DO AND NH3-N SET AS ABOVE. ABSOLUTE MAXIMUM = 30.0 MG/L.

COST ESTIMATED FOR A 10 MGD TREATMENT FACILITY TO PROVIDE REQUIRED EFFLUENT QUALITY.

DILUTION RATIO	SLOPE (FT/MI)	KD (/DAY)	KN (/DAY)	KA (/DAY)	MINIMUM DO (MG/L)	CRIT TIME (DAYS)	MAX NH3N (MG/L)	PERMISSIBLE EFFLUENT QUALITY PARAMETERS BOD-5 MG/L	NH3-N MG/L	DO MG/L	UOD MG/L	SF MG/L	COST M$PW
0.0	0.0	0.150	0.100	0.300	5.06	4.25	2.00	3.0	2.0	7.0	14.	7.	20.8
0.0	2.0	0.162	0.100	0.737	5.19	2.25	2.00	9.0	2.0	7.0	23.	16.	14.8
0.0	5.0	0.252	0.161	1.611	5.03	1.25	2.00	14.0	2.0	7.0	30.	23.	13.2
0.0	10.0	0.342	0.300	3.311	5.19	0.75	2.00	19.0	2.0	7.0	38.	31.	12.9
0.0	25.0	0.500	0.300	9.342	5.84	0.25	2.00	30.0	2.0	7.0	54.	47.	12.5
0.5	0.0	0.150	0.100	0.300	5.45	3.25	1.41	3.0	2.0	6.0	14.	8.	19.7
0.5	2.0	0.156	0.100	0.737	5.18	2.00	2.00	12.0	2.9	6.0	31.	25.	13.1
0.5	5.0	0.240	0.161	1.611	5.12	1.00	2.00	19.0	2.9	6.0	41.	35.	12.8
0.5	10.0	0.348	0.300	3.311	5.03	0.50	2.00	28.0	2.9	6.0	55.	49.	12.4
0.5	25.0	0.372	0.300	9.342	6.33	0.00	2.00	30.0	2.9	6.0	58.	52.	12.3
1.0	0.0	0.150	0.100	0.300	5.18	3.00	1.44	4.0	2.7	5.0	18.	13.	15.8
1.0	2.0	0.153	0.100	0.737	5.16	1.75	2.00	15.0	3.8	5.0	39.	34.	12.8
1.0	5.0	0.243	0.161	1.611	5.01	1.00	2.00	25.0	3.8	5.0	54.	49.	12.4
1.0	10.0	0.288	0.300	3.311	5.64	0.50	2.00	30.0	3.8	5.0	62.	57.	12.2
1.0	25.0	0.288	0.300	9.342	6.00	0.00	2.00	30.0	3.8	5.0	62.	57.	12.2
2.0	0.0	0.150	0.100	0.300	5.17	2.50	1.26	5.0	3.3	3.0	22.	19.	13.4
2.0	2.0	0.156	0.100	0.737	5.00	1.50	2.00	22.0	5.6	3.0	58.	55.	12.3
2.0	5.0	0.204	0.161	1.611	5.47	0.50	2.00	30.0	5.6	3.0	70.	67.	11.9
2.0	10.0	0.204	0.300	3.311	5.67	0.00	2.00	30.0	5.6	3.0	70.	67.	11.9
2.0	25.0	0.204	0.300	9.342	5.67	0.00	2.00	30.0	5.6	3.0	70.	67.	11.9
5.0	0.0	0.150	0.100	0.300	5.10	3.25	1.41	11.0	7.3	2.0	49.	47.	12.5
5.0	2.0	0.150	0.100	0.737	5.58	1.75	2.00	30.0	10.9	2.0	94.	92.	11.2
5.0	5.0	0.150	0.161	1.611	6.17	0.00	2.00	30.0	10.9	2.0	94.	92.	11.2
5.0	10.0	0.150	0.300	3.311	6.17	0.00	2.00	30.0	10.9	2.0	94.	92.	11.2
5.0	25.0	0.150	0.300	9.342	6.17	0.00	2.00	30.0	10.9	2.0	94.	92.	11.2
10.0	0.0	0.150	0.100	0.300	5.01	3.75	1.54	22.0	14.7	2.0	99.	97.	11.1
10.0	2.0	0.150	0.100	0.737	6.09	1.75	2.00	30.0	19.8	2.0	134.	132.	10.1
10.0	5.0	0.150	0.161	1.611	6.55	0.00	2.00	30.0	19.8	2.0	134.	132.	10.1
15.0	0.0	0.150	0.100	0.300	5.14	3.75	1.46	30.0	20.0	2.0	135.	133.	10.1
15.0	2.0	0.150	0.100	0.737	6.48	1.25	1.46	30.0	20.0	2.0	135.	133.	10.1

**** AVERAGE STRENGTH FACTOR FOR TREATMENT LEVELS REQUIRED IN THIS RUN = 61. MG/L ****

**** AVERAGE COST FOR TREATMENT LEVELS REQUIRED IN THIS RUN = 12.7 MILLION DOLLARS PRESENT WORTH ****

Exhibit 4-2A

test its sensitivity to changes in parameter values or assumed conditions. Sensitivity analyses may be of several types as applied to a DO model. It may be a comparison of DO profiles or a comparison of permissible stream loadings for different sets of input parameter values. One or more parameters may be varied while others are held constant. Typical or expected ranges of input parameters are needed. A probability distribution of values for a given parameter as determined by research or from the literature is desirable, but frequently impossible to obtain.

The output presented in Exhibits 4-2B through 4-2F illustrates the effects of variation of input parameter values on the output permissible effluent quality, and may be considered as a form of sensitivity analysis. It may be used as a general guideline for estimation of permissible effluent quality under the various specified conditions. The SF and cost values for a specific dilution and slope combination may be compared for the various conditions. The overall averages of SF and cost for each exhibit give a rough indication of the relative effects over the range of dilution ratio and slope combinations evaluated. Exhibit 4-2A represents the "normal" model output and may be used as a basis for comparison with the results of Exhibits 4-2B through 4-2F.

Exhibits 4-2B and 4-2C illustrate the effects of variation of maximum permissible instream NH_3-N concentration. Exhibit 4-2B presents results for a maximum permissible instream NH_3-N concentration of 4.0 m/gl as compared with 2.0 mg/l for Exhibit 4-2A, while all other conditions are the same as for Exhibit 4-2A. Exhibit 4-2C presents results for the case of no maximum permissible instream NH_3-N concentration *per se*, that is, only the DO standard determines the permissible effluent quality while other conditions are the same as for Exhibit 4-2A.

Exhibits 4-2D through 4-2F illustrate the effects of variation in coefficient values. Exhibit 4-2D presents results for assumed constant deoxygenation coefficients (k_d = 0.40/day and k_n = 0.30/day), while other conditions are the same as for Exhibit 4-2A. Exhibit 4-2E presents results for the case when all three coefficients have been adjusted in a direction which will produce favorable effects on instream DO. For this case, k_a = 2.0 times normal k_a, k_d = 0.67 times normal k_d, and k_n = 0.67 times normal k_n, where the normal coefficient values are those presented in Exhibit 4-2A and all other conditions are the same as for Exhibit 4-2A. Exhibit 4-2F presents results for the case when all three coefficients have been adjusted in a direction which will produce unfavorable effects on instream DO. For this case, k_a = 0.5 times normal k_a, k_d = 1.5 times normal k_d, and k_n = 1.5 times normal k_n.

Numerous comparisons and evaluations may be made from Exhibits 4-2A through 4-2F. They, in general, serve to illustrate that variations within the accuracy of determination or prediction of input values for a parameter or combination of parameters may have a significant effect on the results. This is an im-

RESULTS OF ANALYSIS TO PREDICT PERMISSIBLE EFFLUENT PARAMETERS AS A FUNCTION OF DILUTION RATIO AND SLOPE

ASSUMED UPSTREAM WATER QUALITY: BOD-5 = 2.0 MG/L, NH3-N = 0.22 MG/L, DO = 7.0 MG/L.

ASSUMED DISSOLVED OXYGEN SATURATION CONCENTRATION = 8.2 MG/L. ASSUMED TEMPERATURE = 25 DEGREES C.

VARIABLE DEOXYGENATION COEFFICIENTS.

MAXIMUM PERMISSIBLE IN-STREAM NH3-N = 4.0 MG/L.

THE PERMISSIBLE EFFLUENT PARAMETERS ARE DETERMINED AS FOLLOWS:

DO = 2.0 MG/L GREATER THAN THE MINIMUM REQUIRED TO GIVE AN INITIAL IN-STREAM CONCENTRATION OF 5.0 MG/L. ABSOLUTE MINIMUM = 2.0 MG/L.

NH3-N = MAXIMUM CONCENTRATION PERMISSIBLE TO GIVE A MAXIMUM IN-STREAM CONCENTRATION AS INDICATED ABOVE BUT NOT TO EXCEED 2/3 OF THE PERMISSIBLE BOD-5 CONCENTRATION, BOTH IN MG/L. ABSOLUTE MAXIMUM = 20.0 MG/L.

BOD-5 = MAXIMUM CONCENTRATION PERMISSIBLE TO GIVE A MINIMUM IN-STREAM DO CONCENTRATION OF 5.0 MG/L WITH PERMISSIBLE DO AND NH3-N SET AS ABOVE. ABSOLUTE MAXIMUM = 30.0 MG/L.

COST ESTIMATED FOR A 10 MGD TREATMENT FACILITY TO PROVIDE REQUIRED EFFLUENT QUALITY.

DILUTION RATIO	SLOPE (FT/MI)	KD (/DAY)	KN (/DAY)	KA (/DAY)	MINIMUM DO (MG/L)	CRIT TIME (DAYS)	MAX NH3N (MG/L)	PERMISSIBLE EFFLUENT QUALITY PARAMETERS BOD-5 MG/L	NH3-N MG/L	DO MG/L	UOD MG/L	SF MG/L	COST M$PW
0.0	0.0	0.150	0.100	0.300	5.06	4.25	2.00	3.0	2.0	7.0	14.	7.	20.8
0.0	2.0	0.150	0.100	0.737	5.04	2.50	4.00	6.0	4.0	7.0	27.	20.	13.2
0.0	5.0	0.198	0.161	1.611	5.17	1.25	4.00	11.0	4.0	7.0	35.	28.	13.0
0.0	10.0	0.288	0.300	3.311	5.17	0.75	4.00	16.0	4.0	7.0	42.	35.	12.8
0.0	25.0	0.500	0.300	9.342	5.59	0.25	4.00	30.0	4.0	7.0	63.	56.	12.2
0.5	0.0	0.150	0.100	0.300	5.45	3.25	1.41	3.0	2.0	6.0	14.	8.	19.7
0.5	2.0	0.150	0.100	0.737	5.08	2.25	3.63	8.0	5.3	6.0	36.	30.	13.0
0.5	5.0	0.204	0.161	1.611	5.00	1.25	4.00	16.0	5.9	6.0	50.	44.	12.6
0.5	10.0	0.288	0.300	3.311	5.10	0.50	4.00	23.0	5.9	6.0	61.	55.	12.3
0.5	25.0	0.372	0.300	9.342	6.33	0.00	4.00	30.0	5.9	6.0	71.	65.	12.0
1.0	0.0	0.150	0.100	0.300	5.18	3.00	1.44	4.0	2.7	5.0	18.	13.	15.8
1.0	2.0	0.150	0.100	0.737	5.08	2.00	3.44	10.0	6.7	5.0	45.	40.	12.7
1.0	5.0	0.198	0.161	1.611	5.02	1.00	4.00	20.0	7.8	5.0	65.	60.	12.1
1.0	10.0	0.288	0.300	3.311	5.04	0.50	4.00	30.0	7.8	5.0	80.	75.	11.7
1.0	25.0	0.288	0.300	9.342	6.00	0.00	4.00	30.0	7.8	5.0	80.	75.	11.7
2.0	0.0	0.150	0.100	0.300	5.17	2.50	1.26	5.0	3.3	3.0	22.	19.	13.4
2.0	2.0	0.150	0.100	0.737	5.07	1.50	3.26	14.0	9.3	3.0	63.	60.	12.1
2.0	5.0	0.192	0.161	1.611	5.04	1.00	4.00	28.0	11.6	3.0	94.	91.	11.3
2.0	10.0	0.204	0.300	3.311	5.64	0.25	4.00	30.0	11.6	3.0	97.	94.	11.2
2.0	25.0	0.204	0.300	9.342	5.67	0.00	4.00	30.0	11.6	3.0	97.	94.	11.2
5.0	0.0	0.150	0.100	0.300	5.10	3.25	1.41	11.0	7.3	2.0	49.	47.	12.5
5.0	2.0	0.150	0.100	0.737	5.04	2.00	3.41	29.0	19.3	2.0	130.	128.	10.2
5.0	5.0	0.150	0.161	1.611	5.95	0.75	3.52	30.0	20.0	2.0	135.	133.	10.1
5.0	10.0	0.150	0.300	3.311	6.17	0.00	3.52	30.0	20.0	2.0	135.	133.	10.1
5.0	25.0	0.150	0.300	9.342	6.17	0.00	3.52	30.0	20.0	2.0	135.	133.	10.1
10.0	0.0	0.150	0.100	0.300	5.01	3.75	1.54	22.0	14.7	2.0	99.	97.	11.1
10.0	2.0	0.150	0.100	0.737	6.08	1.75	2.02	30.0	20.0	2.0	135.	133.	10.1
10.0	5.0	0.150	0.161	1.611	6.55	0.00	2.02	30.0	20.0	2.0	135.	133.	10.1
15.0	0.0	0.150	0.100	0.300	5.14	3.75	1.46	30.0	20.0	2.0	135.	133.	10.1
15.0	2.0	0.150	0.100	0.737	6.48	1.25	1.46	30.0	20.0	2.0	135.	133.	10.1

**** AVERAGE STRENGTH FACTOR FOR TREATMENT LEVELS REQUIRED IN THIS RUN = 72. MG/L ****

**** AVERAGE COST FOR TREATMENT LEVELS REQUIRED IN THIS RUN = 12.3 MILLION DOLLARS PRESENT WORTH ****

Exhibit 4-2B

RESULTS OF ANALYSIS TO PREDICT PERMISSIBLE EFFLUENT PARAMETERS AS A FUNCTION OF DILUTION RATIO AND SLOPE

ASSUMED UPSTREAM WATER QUALITY: BOD-5 = 2.0 MG/L, NH3-N = 0.22 MG/L, DO = 7.0 MG/L.

ASSUMED DISSOLVED OXYGEN SATURATION CONCENTRATION = 8.2 MG/L. ASSUMED TEMPERATURE = 25 DEGREES C.

VARIABLE DEOXYGENATION COEFFICIENTS.

NO MAXIMUM PERMISSIBLE IN-STREAM NH3-N CRITERION PER SE, ONLY THE DO STANDARD DETERMINES TREATMENT REQUIREMENTS.

THE PERMISSIBLE EFFLUENT PARAMETERS ARE DETERMINED AS FOLLOWS:

DO = 2.0 MG/L GREATER THAN THE MINIMUM REQUIRED TO GIVE AN INITIAL IN-STREAM CONCENTRATION OF 5.0 MG/L. ABSOLUTE MINIMUM = 2.0 MG/L.

NH3-N = MAXIMUM CONCENTRATION PERMISSIBLE TO GIVE A MAXIMUM IN-STREAM CONCENTRATION AS INDICATED ABOVE BUT NOT TO EXCEED 2/3 OF THE PERMISSIBLE BOD-5 CONCENTRATION, BOTH IN MG/L. ABSOLUTE MAXIMUM = 20.0 MG/L.

BOD-5 = MAXIMUM CONCENTRATION PERMISSIBLE TO GIVE A MINIMUM IN-STREAM DO CONCENTRATION OF 5.0 MG/L WITH PERMISSIBLE DO AND NH3-N SET AS ABOVE. ABSOLUTE MAXIMUM = 30.0 MG/L.

COST ESTIMATED FOR A 10 MGD TREATMENT FACILITY TO PROVIDE REQUIRED EFFLUENT QUALITY.

DILUTION RATIO	SLOPE (FT/MI)	KD (/DAY)	KN (/DAY)	KA (/DAY)	MINIMUM DO (MG/L)	CRIT TIME (DAYS)	MAX NH3N (MG/L)	PERMISSIBLE EFFLUENT QUALITY PARAMETERS BOD-5 MG/L	NH3-N MG/L	DO MG/L	UOD MG/L	SF MG/L	COST M$PW
0.0	0.0	0.150	0.100	0.300	5.06	4.25	2.00	3.0	2.0	7.0	14.	7.	20.8
0.0	2.0	0.150	0.100	0.737	5.04	2.50	4.00	6.0	4.0	7.0	27.	20.	13.2
0.0	5.0	0.150	0.161	1.611	5.36	1.25	5.33	8.0	5.3	7.0	36.	29.	13.0
0.0	10.0	0.180	0.300	3.311	5.28	0.75	6.67	10.0	6.7	7.0	45.	38.	12.7
0.0	25.0	0.396	0.300	9.342	5.08	0.25	14.67	22.0	14.7	7.0	99.	92.	11.2
0.5	0.0	0.150	0.100	0.300	5.45	3.25	1.41	3.0	2.0	6.0	14.	8.	19.7
0.5	2.0	0.150	0.100	0.737	5.08	2.25	3.63	8.0	5.3	6.0	36.	30.	13.0
0.5	5.0	0.156	0.161	1.611	5.14	1.25	5.41	12.0	8.0	6.0	54.	48.	12.5
0.5	10.0	0.192	0.300	3.311	5.11	0.75	6.74	15.0	10.0	6.0	68.	62.	12.1
0.5	25.0	0.372	0.300	9.342	5.31	0.25	13.41	30.0	20.0	6.0	135.	129.	10.2
1.0	0.0	0.150	0.100	0.300	5.18	3.00	1.44	4.0	2.7	5.0	18.	13.	15.8
1.0	2.0	0.150	0.100	0.737	5.08	2.00	3.44	10.0	6.7	5.0	45.	40.	12.7
1.0	5.0	0.162	0.161	1.611	5.00	1.00	5.44	16.0	10.7	5.0	72.	67.	11.9
1.0	10.0	0.189	0.300	3.311	5.16	0.50	6.44	19.0	12.7	5.0	85.	80.	11.5
1.0	25.0	0.288	0.300	9.342	6.00	0.00	10.11	30.0	20.0	5.0	135.	130.	10.2
2.0	0.0	0.150	0.100	0.300	5.17	2.50	1.26	5.0	3.3	3.0	22.	19.	13.4
2.0	2.0	0.150	0.100	0.737	5.07	1.50	3.26	14.0	9.3	3.0	63.	60.	12.1
2.0	5.0	0.156	0.161	1.611	5.14	1.00	5.04	22.0	14.7	3.0	99.	96.	11.1
2.0	10.0	0.192	0.300	3.311	5.10	0.50	6.37	28.0	18.7	3.0	126.	123.	10.4
2.0	25.0	0.204	0.300	9.342	5.67	0.00	6.81	30.0	20.0	3.0	135.	132.	10.1
5.0	0.0	0.150	0.100	0.300	5.10	3.25	1.41	11.0	7.3	2.0	49.	47.	12.5
5.0	2.0	0.150	0.100	0.737	5.04	2.00	3.41	29.0	19.3	2.0	130.	128.	10.2
5.0	5.0	0.150	0.161	1.611	5.95	0.75	3.52	30.0	20.0	2.0	135.	133.	10.1
5.0	10.0	0.150	0.300	3.311	6.17	0.00	3.52	30.0	20.0	2.0	135.	133.	10.1
5.0	25.0	0.150	0.300	9.342	6.17	0.00	3.52	30.0	20.0	2.0	135.	133.	10.1
10.0	0.0	0.150	0.100	0.300	5.01	3.75	1.54	22.0	14.7	2.0	99.	97.	11.1
10.0	2.0	0.150	0.100	0.737	6.08	1.75	2.02	30.0	20.0	2.0	135.	133.	10.1
10.0	5.0	0.150	0.161	1.611	6.55	0.00	2.02	30.0	20.0	2.0	135.	133.	10.1
15.0	0.0	0.150	0.100	0.300	5.14	3.75	1.46	30.0	20.0	2.0	135.	133.	10.1
15.0	2.0	0.150	0.100	0.737	6.48	1.25	1.46	30.0	20.0	2.0	135.	133.	10.1

**** AVERAGE STRENGTH FACTOR FOR TREATMENT LEVELS REQUIRED IN THIS RUN = 81. MG/L ****

**** AVERAGE COST FOR TREATMENT LEVELS REQUIRED IN THIS RUN = 12.1 MILLION DOLLARS PRESENT WORTH ****

Exhibit 4-2C

RESULTS OF ANALYSIS TO PREDICT PERMISSIBLE EFFLUENT PARAMETERS AS A FUNCTION OF DILUTION RATIO AND SLOPE

ASSUMED UPSTREAM WATER QUALITY: BOD-5 = 2.0 MG/L, NH3-N = 0.22 MG/L, DO = 7.0 MG/L.

ASSUMED DISSOLVED OXYGEN SATURATION CONCENTRATION = 8.2 MG/L. ASSUMED TEMPERATURE = 25 DEGREES C.

ASSUMED CONSTANT DEOXYGENATION COEFFICIENTS: KD=0.4/DAY AND KN=0.3/DAY.

MAXIMUM PERMISSIBLE IN-STREAM NH3-N = 2.0 MG/L.

THE PERMISSIBLE EFFLUENT PARAMETERS ARE DETERMINED AS FOLLOWS:

DO = 2.0 MG/L GREATER THAN THE MINIMUM REQUIRED TO GIVE AN INITIAL IN-STREAM CONCENTRATION OF 5.0 MG/L. ABSOLUTE MINIMUM = 2.0 MG/L.

NH3-N = MAXIMUM CONCENTRATION PERMISSIBLE TO GIVE A MAXIMUM IN-STREAM CONCENTRATION AS INDICATED ABOVE BUT NOT TO EXCEED 2/3 OF THE PERMISSIBLE BOD-5 CONCENTRATION, BOTH IN MG/L. ABSOLUTE MAXIMUM = 20.0 MG/L.

BOD-5 = MAXIMUM CONCENTRATION PERMISSIBLE TO GIVE A MINIMUM IN-STREAM DO CONCENTRATION OF 5.0 MG/L WITH PERMISSIBLE DO AND NH3-N SET AS ABOVE. ABSOLUTE MAXIMUM = 30.0 MG/L.

COST ESTIMATED FOR A 10 MGD TREATMENT FACILITY TO PROVIDE REQUIRED EFFLUENT QUALITY.

								PERMISSIBLE EFFLUENT QUALITY PARAMETERS					
DILUTION RATIO	SLOPE (FT/MI)	KD (/DAY)	KN (/DAY)	KA (/DAY)	MINIMUM DO (MG/L)	CRIT TIME (DAYS)	MAX NH3N (MG/L)	BOD-5 MG/L	NH3-N MG/L	DO MG/L	UOD MG/L	SF MG/L	COST M$PW
0.0	0.0	0.400	0.300	0.301	5.95	2.25	0.67	1.0	0.7	7.0	4.	-3.	30.8
0.0	2.0	0.400	0.300	0.737	5.76	1.50	1.33	2.0	1.3	7.0	9.	2.	25.8
0.0	5.0	0.400	0.300	1.611	5.21	1.00	2.00	7.0	2.0	7.0	20.	13.	15.9
0.0	10.0	0.400	0.300	3.311	5.02	0.75	2.00	18.0	2.0	7.0	36.	29.	13.0
0.0	25.0	0.400	0.300	9.342	6.21	0.25	2.00	30.0	2.0	7.0	54.	47.	12.5
0.5	0.0	0.400	0.300	0.301	5.61	1.75	0.52	1.0	0.7	6.0	4.	-2.	29.6
0.5	2.0	0.400	0.300	0.737	5.21	1.25	1.41	3.0	2.0	6.0	14.	8.	19.7
0.5	5.0	0.400	0.300	1.611	5.00	1.00	2.00	10.0	2.9	6.0	28.	22.	13.2
0.5	10.0	0.400	0.300	3.311	5.00	0.50	2.00	25.0	2.9	6.0	50.	44.	12.6
0.5	25.0	0.400	0.300	9.342	6.33	0.00	2.00	30.0	2.9	6.0	58.	52.	12.3
1.0	0.0	0.400	0.300	0.301	5.43	1.75	0.44	1.0	0.7	5.0	4.	-1.	28.5
1.0	2.0	0.400	0.300	0.737	5.39	1.00	1.11	3.0	2.0	5.0	14.	9.	18.6
1.0	5.0	0.400	0.300	1.611	5.09	0.75	2.00	11.0	3.8	5.0	33.	28.	13.0
1.0	10.0	0.400	0.300	3.311	5.11	0.50	2.00	30.0	3.8	5.0	62.	57.	12.2
1.0	25.0	0.400	0.300	9.342	6.00	0.00	2.00	30.0	3.8	5.0	62.	57.	12.2
2.0	0.0	0.400	0.300	0.301	5.22	1.50	0.37	1.0	0.7	3.0	4.	1.	26.3
2.0	2.0	0.400	0.300	0.737	5.24	1.00	1.04	4.0	2.7	3.0	18.	15.	15.0
2.0	5.0	0.400	0.300	1.611	5.03	0.75	2.00	15.0	5.6	3.0	47.	44.	12.6
2.0	10.0	0.400	0.300	3.311	5.56	0.25	2.00	30.0	5.6	3.0	70.	67.	11.9
2.0	25.0	0.400	0.300	9.342	5.67	0.00	2.00	30.0	5.6	3.0	70.	67.	11.9
5.0	0.0	0.400	0.300	0.301	5.03	2.00	0.52	3.0	2.0	2.0	14.	12.	16.3
5.0	2.0	0.400	0.300	0.737	5.15	1.25	1.19	9.0	6.0	2.0	41.	39.	12.7
5.0	5.0	0.400	0.300	1.611	5.10	0.75	2.00	30.0	10.9	2.0	94.	92.	11.2
5.0	10.0	0.400	0.300	3.311	6.17	0.00	2.00	30.0	10.9	2.0	94.	92.	11.2
5.0	25.0	0.400	0.300	9.342	6.17	0.00	2.00	30.0	10.9	2.0	94.	92.	11.2
10.0	0.0	0.400	0.300	0.301	5.03	2.25	0.57	6.0	4.0	2.0	27.	25.	13.1
10.0	2.0	0.400	0.300	0.737	5.09	1.50	1.29	18.0	12.0	2.0	81.	79.	11.6
10.0	5.0	0.400	0.300	1.611	5.68	0.75	2.00	30.0	19.8	2.0	134.	132.	10.1
15.0	0.0	0.400	0.300	0.301	5.03	2.25	0.58	9.0	6.0	2.0	41.	39.	12.7
15.0	2.0	0.400	0.300	0.737	5.05	1.50	1.33	27.0	18.0	2.0	122.	120.	10.5

**** AVERAGE STRENGTH FACTOR FOR TREATMENT LEVELS REQUIRED IN THIS RUN = 43. MG/L ****

**** AVERAGE COST FOR TREATMENT LEVELS REQUIRED IN THIS RUN = 15.6 MILLION DOLLARS PRESENT WORTH ****

Exhibit 4-2D

RESULTS OF ANALYSIS TO PREDICT PERMISSIBLE EFFLUENT PARAMETERS AS A FUNCTION OF DILUTION RATIO AND SLOPE

ASSUMED UPSTREAM WATER QUALITY: BOD-5 = 2.0 MG/L, NH3-N = 0.22 MG/L, DO = 7.0 MG/L.

ASSUMED DISSOLVED OXYGEN SATURATION CONCENTRATION = 8.2 MG/L. ASSUMED TEMPERATURE = 25 DEGREES C.

VARIABLE DEOXYGENATION COEFFICIENTS. KA, KD, AND KN ALL MODIFIED TO GIVE FAVORABLE EFFECTS ON DO.

MAXIMUM PERMISSIBLE IN-STREAM NH3-N = 2.0 MG/L.

THE PERMISSIBLE EFFLUENT PARAMETERS ARE DETERMINED AS FOLLOWS:

DO = 2.0 MG/L GREATER THAN THE MINIMUM REQUIRED TO GIVE AN INITIAL IN-STREAM CONCENTRATION OF 5.0 MG/L. ABSOLUTE MINIMUM = 2.0 MG/L.

NH3-N = MAXIMUM CONCENTRATION PERMISSIBLE TO GIVE A MAXIMUM IN-STREAM CONCENTRATION AS INDICATED ABOVE BUT NOT TO EXCEED 2/3 OF THE PERMISSIBLE BOD-5 CONCENTRATION, BOTH IN MG/L. ABSOLUTE MAXIMUM = 20.0 MG/L.

BOD-5 = MAXIMUM CONCENTRATION PERMISSIBLE TO GIVE A MINIMUM IN-STREAM DO CONCENTRATION OF 5.0 MG/L WITH PERMISSIBLE DO AND NH3-N SET AS ABOVE. ABSOLUTE MAXIMUM = 30.0 MG/L.

COST ESTIMATED FOR A 10 MGD TREATMENT FACILITY TO PROVIDE REQUIRED EFFLUENT QUALITY.

DILUTION RATIO	SLOPE (FT/MI)	KD (/DAY)	KN (/DAY)	KA (/DAY)	MINIMUM DO (MG/L)	CRIT TIME (DAYS)	MAX NH3N (MG/L)	PERMISSIBLE EFFLUENT QUALITY PARAMETERS BOD-5 MG/L	NH3-N MG/L	DO MG/L	UOD MG/L	SF MG/L	COST M$PW
0.0	0.0	0.121	0.067	0.600	5.26	3.00	2.00	10.0	2.0	7.0	24.	17.	14.3
0.0	2.0	0.205	0.067	1.473	5.15	1.25	2.00	17.0	2.0	7.0	35.	28.	13.0
0.0	5.0	0.301	0.108	3.221	5.09	0.75	2.00	25.0	2.0	7.0	47.	40.	12.7
0.0	10.0	0.335	0.201	6.623	5.97	0.50	2.00	30.0	2.0	7.0	54.	47.	12.5
0.0	25.0	0.335	0.201	18.685	7.00	0.00	2.00	30.0	2.0	7.0	54.	47.	12.5
0.5	0.0	0.121	0.067	0.600	5.13	2.50	2.00	14.0	2.9	6.0	34.	28.	13.0
0.5	2.0	0.201	0.067	1.473	5.13	1.25	2.00	24.0	2.9	6.0	49.	43.	12.6
0.5	5.0	0.249	0.108	3.221	5.82	0.50	2.00	30.0	2.9	6.0	58.	52.	12.3
0.5	10.0	0.249	0.201	6.623	6.33	0.00	2.00	30.0	2.9	6.0	58.	52.	12.3
0.5	25.0	0.249	0.201	18.685	6.33	0.00	2.00	30.0	2.9	6.0	58.	52.	12.3
1.0	0.0	0.121	0.067	0.600	5.05	2.25	2.00	18.0	3.8	5.0	44.	39.	12.7
1.0	2.0	0.193	0.067	1.473	5.24	1.00	2.00	30.0	3.8	5.0	62.	57.	12.2
1.0	5.0	0.193	0.108	3.221	6.00	0.00	2.00	30.0	3.8	5.0	62.	57.	12.2
1.0	10.0	0.193	0.201	6.623	6.00	0.00	2.00	30.0	3.8	5.0	62.	57.	12.2
1.0	25.0	0.193	0.201	18.685	6.00	0.00	2.00	30.0	3.8	5.0	62.	57.	12.2
2.0	0.0	0.117	0.067	0.600	5.07	2.00	2.00	25.0	5.6	3.0	62.	59.	12.1
2.0	2.0	0.137	0.067	1.473	5.67	0.00	2.00	30.0	5.6	3.0	70.	67.	11.9
2.0	5.0	0.137	0.108	3.221	5.67	0.00	2.00	30.0	5.6	3.0	70.	67.	11.9
2.0	10.0	0.137	0.201	6.623	5.67	0.00	2.00	30.0	5.6	3.0	70.	67.	11.9
2.0	25.0	0.137	0.201	18.685	5.67	0.00	2.00	30.0	5.6	3.0	70.	67.	11.9
5.0	0.0	0.100	0.067	0.600	5.89	1.75	2.00	30.0	10.9	2.0	94.	92.	11.2
5.0	2.0	0.100	0.067	1.473	6.17	0.00	2.00	30.0	10.9	2.0	94.	92.	11.2
5.0	5.0	0.100	0.108	3.221	6.17	0.00	2.00	30.0	10.9	2.0	94.	92.	11.2
5.0	10.0	0.100	0.201	6.623	6.17	0.00	2.00	30.0	10.9	2.0	94.	92.	11.2
5.0	25.0	0.100	0.201	18.685	6.17	0.00	2.00	30.0	10.9	2.0	94.	92.	11.2
10.0	0.0	0.100	0.067	0.600	6.34	1.75	2.00	30.0	19.8	2.0	134.	132.	10.1
10.0	2.0	0.100	0.067	1.473	6.55	0.00	2.00	30.0	19.8	2.0	134.	132.	10.1
10.0	5.0	0.100	0.108	3.221	6.55	0.00	2.00	30.0	19.8	2.0	134.	132.	10.1
15.0	0.0	0.100	0.067	0.600	6.65	1.00	1.46	30.0	20.0	2.0	135.	133.	10.1
15.0	2.0	0.100	0.067	1.473	6.69	0.00	1.46	30.0	20.0	2.0	135.	133.	10.1

**** AVERAGE STRENGTH FACTOR FOR TREATMENT LEVELS REQUIRED IN THIS RUN = 71. MG/L ****

**** AVERAGE COST FOR TREATMENT LEVELS REQUIRED IN THIS RUN = 11.9 MILLION DOLLARS PRESENT WORTH ****

Exhibit 4-2E

RESULTS OF ANALYSIS TO PREDICT PERMISSIBLE EFFLUENT PARAMETERS AS A FUNCTION OF DILUTION RATIO AND SLOPE

ASSUMED UPSTREAM WATER QUALITY: BOD-5 = 2.0 MG/L, NH3-N = 0.22 MG/L, DO = 7.0 MG/L.

ASSUMED DISSOLVED OXYGEN SATURATION CONCENTRATION = 8.2 MG/L. ASSUMED TEMPERATURE = 25 DEGREES C.

VARIABLE DEOXYGENATION COEFFICIENTS. KA, KD, AND KN ALL MODIFIED TO GIVE UNFAVORABLE EFFECTS ON DO.

MAXIMUM PERMISSIBLE IN-STREAM NH3-N = 2.0 MG/L.

THE PERMISSIBLE EFFLUENT PARAMETERS ARE DETERMINED AS FOLLOWS:

DO = 2.0 MG/L GREATER THAN THE MINIMUM REQUIRED TO GIVE AN INITIAL IN-STREAM CONCENTRATION OF 5.0 MG/L. ABSOLUTE MINIMUM = 2.0 MG/L.

NH3-N = MAXIMUM CONCENTRATION PERMISSIBLE TO GIVE A MAXIMUM IN-STREAM CONCENTRATION AS INDICATED ABOVE BUT NOT TO EXCEED 2/3 OF THE PERMISSIBLE BOD-5 CONCENTRATION, BOTH IN MG/L. ABSOLUTE MAXIMUM = 20.0 MG/L.

BOD-5 = MAXIMUM CONCENTRATION PERMISSIBLE TO GIVE A MINIMUM IN-STREAM DO CONCENTRATION OF 5.0 MG/L WITH PERMISSIBLE DO AND NH3-N SET AS ABOVE. ABSOLUTE MAXIMUM = 30.0 MG/L.

COST ESTIMATED FOR A 10 MGD TREATMENT FACILITY TO PROVIDE REQUIRED EFFLUENT QUALITY.

DILUTION RATIO	SLOPE (FT/MI)	KD (/DAY)	KN (/DAY)	KA (/DAY)	MINIMUM DO (MG/L)	CRIT TIME (DAYS)	MAX NH3N (MG/L)	PERMISSIBLE EFFLUENT QUALITY PARAMETERS BOD-5 MG/L	NH3-N MG/L	DO MG/L	UOD MG/L	SF MG/L	COST M$PW
0.0	0.0	0.225	0.150	0.151	5.91	4.50	0.67	1.0	0.7	7.0	4.	-3.	30.8
0.0	2.0	0.225	0.150	0.368	5.70	3.00	1.33	2.0	1.3	7.0	9.	2.	25.8
0.0	5.0	0.225	0.242	0.805	5.07	1.75	2.00	5.0	2.0	7.0	17.	10.	17.5
0.0	10.0	0.243	0.450	1.656	5.09	1.00	2.00	9.0	2.0	7.0	23.	16.	14.8
0.0	25.0	0.513	0.450	4.671	5.03	0.50	2.00	19.0	2.0	7.0	38.	31.	12.9
0.5	0.0	0.225	0.150	0.151	5.56	3.75	0.52	1.0	0.7	6.0	4.	-2.	29.6
0.5	2.0	0.225	0.150	0.368	5.12	2.75	1.41	3.0	2.0	6.0	14.	8.	19.7
0.5	5.0	0.225	0.242	0.805	5.15	1.50	2.00	5.0	2.9	6.0	20.	14.	15.2
0.5	10.0	0.234	0.450	1.656	5.06	0.75	2.00	12.0	2.9	6.0	31.	25.	13.1
0.5	25.0	0.504	0.450	4.671	5.04	0.50	2.00	27.0	2.9	6.0	53.	47.	12.5
1.0	0.0	0.225	0.150	0.151	5.36	3.25	0.44	1.0	0.7	5.0	4.	-1.	28.5
1.0	2.0	0.225	0.150	0.368	5.31	2.25	1.11	3.0	2.0	5.0	14.	9.	18.6
1.0	5.0	0.225	0.242	0.805	5.05	1.50	2.00	6.0	3.8	5.0	26.	21.	13.2
1.0	10.0	0.229	0.450	1.656	5.00	0.75	2.00	15.0	3.8	5.0	39.	34.	12.8
1.0	25.0	0.432	0.450	4.671	5.52	0.25	2.00	30.0	3.8	5.0	62.	57.	12.2
2.0	0.0	0.225	0.150	0.151	5.15	2.75	0.37	1.0	0.7	3.0	4.	1.	26.3
2.0	2.0	0.225	0.150	0.368	5.15	1.75	1.04	4.0	2.7	3.0	18.	15.	15.0
2.0	5.0	0.225	0.242	0.805	5.21	1.00	1.70	7.0	4.7	3.0	31.	28.	13.0
2.0	10.0	0.225	0.450	1.656	5.04	0.75	2.00	19.0	5.6	3.0	53.	50.	12.4
2.0	25.0	0.306	0.450	4.671	5.67	0.00	2.00	30.0	5.6	3.0	70.	67.	11.9
5.0	0.0	0.225	0.150	0.151	5.20	3.75	0.41	2.0	1.3	2.0	9.	7.	20.3
5.0	2.0	0.225	0.150	0.368	5.05	2.50	1.19	9.0	6.0	2.0	41.	39.	12.7
5.0	5.0	0.225	0.242	0.805	5.04	1.50	1.96	16.0	10.7	2.0	72.	70.	11.8
5.0	10.0	0.225	0.450	1.656	5.31	0.75	2.00	30.0	10.9	2.0	94.	92.	11.2
5.0	25.0	0.225	0.450	4.671	6.17	0.00	2.00	30.0	10.9	2.0	94.	92.	11.2
10.0	0.0	0.225	0.150	0.151	5.09	4.25	0.51	5.0	3.3	2.0	22.	20.	13.2
10.0	2.0	0.225	0.150	0.368	5.07	2.75	1.23	17.0	11.3	2.0	76.	74.	11.7
10.0	5.0	0.225	0.242	0.805	5.07	1.75	2.00	30.0	19.8	2.0	134.	132.	10.1
15.0	0.0	0.225	0.150	0.151	5.04	4.50	0.54	8.0	5.3	2.0	36.	34.	12.8
15.0	2.0	0.225	0.150	0.368	5.01	3.00	1.29	26.0	17.3	2.0	117.	115.	10.6

**** AVERAGE STRENGTH FACTOR FOR TREATMENT LEVELS REQUIRED IN THIS RUN = 37. MG/L ****

**** AVERAGE COST FOR TREATMENT LEVELS REQUIRED IN THIS RUN = 16.1 MILLION DOLLARS PRESENT WORTH ****

Exhibit 4-2F

portant point for consideration in the interpretation of modeling results for water quality management.

4-6 MODELING OF WATER QUALITY PARAMETERS OTHER THAN DO

Experience in model calibration and verification for water quality parameters other than DO is limited. However, those discussed below and others will play an increasingly significant role in modeling for water quality management as standards and criteria become more stringent, modeling becomes more sophisticated, and nonpoint source modeling becomes extensive.

The behavior of truely conservative substances is relatively easily simulated, their concentrations being primarily controlled by dilution and determined by mass balance calculations. Few chemical constituents are truly conservative in natural waters because their concentrations may be affected by biological synthesis, absorption or ion exchange on sediments, and other processes. However, in many cases these effects may be of negligible significance. Examples of substances normally considered to be conservative for modeling purposes are inorganics such as sodium, potassium, chloride, and sulfate salts.

Modeling the behavior of substances serving as biological nutrients, such as nitrogen and phosphorus compounds, requires expressions describing the chemical behavior of the substances; expressions describing growth, synthesis, decay, and other biological phenomena; and expressions describing physical phenomena such as movement of the organisms. For example, some factors affecting the concentration of various forms of nitrogen in a lake are the extent of stratification, oxygen availability, bacterial nitrification, algal photosynthesis and respiration, settling rate of microbes, regeneration from bottom sediments, etc. In addition to its role as a biological nutrient, reduced inorganic nitrogen represents a potential oxygen demanding material and must generally be included as a DO sink in DO modeling as discussed in detail in the foregoing.

Modeling of pH is difficult. Its behavior is most strongly dependent upon the ionic strength and buffering capacity of the water under consideration, and it is affected by many chemical and biochemical reactions characteristic of natural waters. Algal activity is particularly important, e.g., CO_2 depletion during photosynthesis resulting in a pH increase and CO_2 generation during respiration resulting in a pH decrease. For a particular situation, the factors significantly affecting pH must be identified and quantified in the model.

Temperature is a necessary and important parameter for DO modeling due to its affect on several of the other variables (as discussed in preceding sections of this chapter). For this purpose, estimates of the appropriate temperature for a given application can generally be made with sufficient accuracy from historical and other data. Testing of DO models for sensitivity to changes in temperature is also fairly routine. However, temperature limits per se are specified in most water quality standards and in some instances represent the controlling parameter for permissible discharge limitations. Probably the most important example is represented by discharges of heated effluents from multiple electric generating installations on major rivers and estuaries. Accurate model simulation of temperature behavior in these situations is essential to proper water quality management. A considerable amount of work has been performed on thermal pollution and associated modeling and the published literature should be consulted for detailed applications (Refs. 4-34, 4-16). In the simplist case of a one dimensional nondispersive river, the behavior of temperature below a point source input of excess heat may generally be described by downstream exponential decay of the initial temperature excess (actual river temperature minus ambient river temperature).

The behavior of either total or fecal coliforms has usually been described by first order kinetics where the die-off or decay rate is proportional to the number present. In some cases the phenomenon termed "aftergrowth" has been demonstrated below point source discharges of disinfected effluents, in which case first order decay is not applicable. Other factors being equal, coliform die-off is more rapid in saline than in fresh waters.

Sediment is a relatively unimportant parameter in most modeling applications for low flow regimes, such as is typically performed for 303 and 201 planning studies. However, it becomes an essential parameter for nonpoint source modeling typical of 208 water quality management studies. Most nonpoint source models predict sediment contribution as a function of slope, soil type, cover type, and other factors associated with the contributing land area, and intensity, duration, and other factors associated with the storm event. In a stream, the behavior of sediment particles is governed by their physical and chemical characteristics. Erosion, sedimentation, and resuspension is governed by the hydraulic properties of the stream and by the size, shape, specific gravity, and other properties of the particles. Suspended sediment can indirectly affect the concentration of other constituents in several ways. Ion exchange properties of certain clays may result in changes in concentrations of soluble inorganic ions. Positive charged ammonium ions may become bound to negatively charged sediment particles and thus be removed from solution. Sediment induced turbidity may result in decreased light penetration which may in turn result in reduced photosynthesis. Chemical reactions from sediment originating in areas characterized by certain mining activities may result in changes in iron, sulfate, and pH concentrations.

4-7 REPRESENTATIVE MODELS AVAILABLE FOR WATER QUALITY MANAGEMENT STUDIES

In response to the water quality management studies under the various sections of Public Law 92-500, numerous models of various types have been developed.

Since the basic concepts behind models of a particular type are similar, and they differ primarily only in their approach to a specific problem, only selected models of the types considered essential for water quality management applications will be reviewed here. Examples of some of the better known models of each type are presented with their capabilities in terms of particular constituents, dimensions, and general scope. Examples of model types which may be applicable in water quality management studies, but are not discussed here, are models for population projection, sewerage system design, wastewater treatment facility design, and sewerage system-treatment facility optimization.

The available models to be presented have been divided into five categories. The first three are stream, estuary, and lake/reservoir, and are for simulating the instream effects of constituents affecting water quality. The fourth category is called nonpoint source/urban runoff, and is used to describe the generation of pollutants washed off a given watershed into a receiving water system. The fifth category is multiple function models, and is used to describe models which overlap in category, e.g., by changing certain coefficients some models can be used to simulate pollutant transport in several types of receiving water systems. Another example would be models which have various blocks which can simulate nonpoint source/urban runoff pollutant generation as well as simulate instream effects of the pollutant generation. For all the models described in this section, only a general description will be given. For a more complete discussion the references cited should be consulted.

4-7.1 Stream Models

This category is used to describe models which have equations describing transport and reaction kinetics for stream and river systems which can be simulated as behaving without tidal or lake/reservoir effects. These models were constructed primarily to simulate dissolved oxygen, but some have the capability of simulating other constituents.

SNSIM. This model, developed by Braster, *et al.* (Ref. 4-3), is typical of those designed to solve one dimensional steady state problems with insignificant dispersion relating to sources and sinks of dissolved oxygen. The model solves the integrated form of the Streeter-Phelps equation and is similar to the one described in the foregoing Section 4-5. The model expands the Streeter-Phelps equation to account for inputs of oxygen-demanding constituents from point and distributed sources of both carbonaceous and nitrogenous origin and distributed oxygen demand from benthic deposits and photosynthetic activity. Withdrawals from the system can also be simulated. SNSIM gives the option of inputting stream depth and velocity directly or calculating them based on exponential correlations with flow. The reaeration rate can be directly inputted or calculated as an exponential function of velocity and depth or as a function of the stream slope and time of flow. The system to be simulated using SNSIM is represented as a system of reaches which lie between stream confluences. Each reach is divided into sections which are characterized by constant physical, chemical, biological, and hydrological characteristics. Point source waste inputs and withdrawals can only occur at the beginning or end of a section. The equations used in SNSIM are similar to those described previously in this chapter.

DOSAG-I. This model was originally developed by the Texas Water Development Board (Ref. 4-40) for use in simulating point and distributed sources of carbonaceous and nitrogenous oxygen demand and their impacts on the dissolved oxygen of a stream. The model solves the integrated form of the Streeter-Phelps equation and is applicable to systems which can be simulated as one dimensional steady state with no dispersion. Unlike the previous models described, DOSAG-I does not have the ability to simulate the oxygen demand from benthic deposits or net oxygen production from photosynthetic activity. The velocity and depth used in DOSAG-I are based on exponential relationships with flow, and the reaeration rate can be calculated as described above for SNSIM, as an exponential relationship with flow, or by an equation developed by Thackston and Krenkel (see foregoing Section 4-4.3). The system to be simulated is divided into stream stretches which are composed of stream reaches. For a given stream reach all constants of the dissolved oxygen sag equation and the velocity and depth coefficients are held constant. All point inputs and withdrawals must occur at the beginning of a reach. Therefore, a new reach must be designated at any point where there is a significant change in the hydraulic, biological, or physical characteristics of a channel, including the addition of a point waste source or withdrawal. DOSAG-I through an iteration scheme has the ability to study the potential of increasing the dissolved oxygen through flow augmentation. The equations used in DOSAG-I are also similar to those previously described.

DOSAG-III. The original DOSAG-I was modified for the Environmental Protection Agency by Water Resources Engineers, and the resulting product was called DOSAG-III (Ref. 4-58). DOSAG-III is identical to DOSAG-I except that several additional constituents and the ability to use the Tsivoglou technique (see foregoing Section 4-4.3) for calculating the reaeration rate have been added. The model has the ability to simulate the steady state behavior of chlorophyll A; ammonia, nitrite, and nitrate nitrogen; phosphorus; carbonaceous biochemical oxygen demand; benthic oxygen demand; dissolved oxygen; coliforms; and three conservative materials. Growth of algae in DOSAG-III is governed by nutrient and light limitation kinetics. The transformation of ammonia nitrogen to nitrite nitrogen, the transformation of nitrite nitrogen to nitrate nitrogen, and the oxygen consumed in the reaction are modeled. A coupling exists between nitrate nitrogen and the production of algae to close the loop. Phosphorus in the model is represented as orthophosphate, and the link between phosphorus and algal production closes the phosphorus loop. Carbonaceous biochemical oxygen demand is modeled, as in DOSAG-I, by first order kinetics, benthic oxygen demand is simulated as a fixed demand, coliforms are simulated by first order kinetics. Dissolved oxygen, as simulated in DOSAG-III, is a function of: the concentration of dissolved oxygen, the saturation concentration of dissolved oxygen at the local temperature and pressure, algal photosynthesis, algal respiration, oxygen uptake by ammonia nitrogen oxidation, oxygen uptake by nitrite nitrogen oxidation, and atmospheric reaeration.

SNISM, DOSAG-I, and DOSAG-III use an explicit solution technique, i.e., they solve the integrated forms of the differential equations that describe the instream reactions. They all also use a Lagrangian approach, i.e., they trace the reactions in a single parcel of water as it traverses a stream.

QUAL-I. This model was also developed by the Texas Water Development Board (Ref. 4-41) and uses a finite difference solution technique to solve the differential equation representing one dimensional transport due to longitudinal dispersion and advection and the sources and sinks for nonconservative constituents. The dispersion coefficient is calculated in the model in terms of velocity, depth, and the Manning coefficient. The model calculates depth and velocity using exponential relationships with flow. QUAL-I has the ability to calculate the heat exchange between a water body and the atmosphere, using net shortwave solar radiation, longwave atmospheric radiation, water surface back radiation, evaporation, conduction, and assumptions related to the energy budget and internal mixing. The model calculates the reaeration rate using the same formulations in DOSAG-1, and does not simulate the nitrogenous component of oxygen demand. An approximation is included in the model to calculate the flow augmentation necessary to increase the dissolved oxygen concentration to some specified level. Routing of conservative constituents through mass balance can also be accomplished with this model. The output from QUAL-I can be an intermediate summary which gives a time history (as the system approaches steady state) and the associated spatial distribution of selected water quality constituents or a final summary of stream conditions after steady state has been reached (including concentrations of the water quality constituents in every reach of the system being simulated). Unlike the previously described models which utilize a reach as a minimum element for computation and allow only discharges to be inputted at the beginning and end of a reach, QUAL-I uses a standard computational element of a set length throughout the system. Computational elements with similar hydrologic and physical properties are grouped together as reaches. Reaches are also established between confluences of streams to be simulated. This computational element scheme is flexible, in that additional inputs and withdrawals can be put into any computational element between confluences without the necessity to resegment the system. Like the previously described models, wastewater from point sources can be inputted only at a constant rate and concentration.

QUAL-II. This model is a modification of QUAL-I and was developed by Water Resources Engineers for the Environmental Protection Agency (Ref. 4-57). QUAL-II has all the capabilities of QUAL-I plus the ability to directly attain a steady state solution for every constituent except temperature. In addition, QUAL-II can simulate chlorophyll A, the nitrogen series, phosphorus, carbonaceous biochemical oxygen demand, benthic oxygen demand, dissolved oxygen, coliforms, and one conservative constiuent. Both QUAL-I and QUAL-II use an implicit solution technique; i.e., they solve the differential equations directly using a finite difference approach. Rather than using a LaGrangian approach to track a parcel of water as it traverses a stream, QUAL-I and QUAL-II use a Eulerian approach in which the reactions are traced for all reaches simultaneously throughout time. The interactions among the nonconservative constituents in QUAL-II are simulated by the same equations as in DOSAG-III, except that the differential equations are solved directly in QUAL-II.

4-7.2 Estuary Models

Estuary models are utilized to simulate systems that are tidal influenced. In these systems, the effects of dispersion become much more significant than in most freshwater systems. Because of the effects of tidal action on the hydraulics of the estuarine system, the estuary models become more mathematically complex than freshwater models, with both advection and dispersion generally being significant transport mechanisms. Two approaches to the simulation of a system over time are generally used—the steady state and the time varying. The steady state approach averages constituents over the tidal cycle, while the time varying approach simulates events in a real time fashion over specified time steps. To simulate the effects of tidal action on the advection and dispersion of an estuarine system, models are generally segmented in one, two, or three dimensions as each individual system may dictate. One and two dimensional segmentations are common and three dimensional analyses have been conducted in rare cases.

Steady State Segmented Estuary Model. This model contains the concepts discussed by Thomann (Ref. 4-43), and was documented by Bunce and Hetling (Ref. 4-4). This estuary model was documented for the one dimensional steady state case of organic pollution. To use the model, an estuary is divided into sections. Dissolved oxygen in a given section can be gained or lost by: advection or dispersion of oxygen from or into an adjacent section, reaeration into the section from the atmosphere, production of oxygen in the section of photosynthesis, or use of oxygen by benthic deposits or other immediate oxygen demands. The model works in terms of ultimate oxygen demand and does not separate the carbonaceous and nitrogenous fractions. An implicit matrix inversion approach is used to solve the one dimensional advection-dispersion equations.

Mississippi State University Estuary Model. This model is a two dimensional, vertically well mixed, steady state intertidal model which was developed by Shindala, *et al.* (Ref. 4-37). The Mississippi State Estuary Model is similar to the Segmented Steady State Estuary Model except that it is two dimensional and the carbonaceous and nitrogenous oxygen demand fractions are separated. The model can simulate the conservative constituent chloride or the effects, in the lateral and longitudinal direction, of dissolved oxygen sources and sinks. Dissolved oxygen sources included in the model are atmospheric reaeration and the oxygen produced by algal photosynthesis. Carbonaceous demanding materials, nitrogenous demanding materials, algal respiration, and demand from benthic deposits are all dissolved oxygen sinks. The model uses the O'Connor-Dobbins formulation (see foregoing Section 4-14.3) for calculating the reaeration coefficient in a given segment. An implicit solution technique is used. Time variations in constituent concentrations are averaged over a tidal cycle. To use this model in two dimensions an estuary or tidal river must be subdivided into segments in both the longitudinal and lateral directions.

Dynamic Estuary Model (DEM). This model, described by Feigner and Harris (Ref. 4-9), is an example of a model which can be used to simulate in two dimensions on a time varying (real time) basis. Unlike the Mississippi State Model for which an estuary must be divided into segments, the DEM uses a channel and node approach. Inputs and withdrawals to the system occur at the nodes and transport between nodes occurs along the channels. The model assumes that the estuarine system is well mixed in the vertical direction and thus, the node-channel representation is in the longitudinal and lateral directions. The model can simulate both nonconservative and conservative constituents, including the coupled effects of biochemical oxygen demand and dissolved oxygen. A single coefficient for the decay of a nonconservative constituent and a single reaeration coefficient must be used for the entire system under simulation. In addition, a constant value of the dissolved oxygen saturation concentration is used. For each simulation, constant inputs at each node are assumed. The model has the capability of including a constant input from precipitation and a constant output from evaporation, although this must be done at a given node. The model is divided into a hydraulic model and a quality model. The hydraulic model is usually run at time steps in the order of ½ to 5 minutes. Once a dynamic steady state equilibrium in the hydraulic model is attained, the summary values are stored on tape for subsequent use in the quality model, which can be run in time steps of about an hour against the summary hydraulic information. The hydraulic run consumes a relatively large amount of computer time. For a given set of hydraulics, numerous quality runs, which require considerably less computer time, can be made.

4-7.3 Lake/Reservoir Models

Although in many cases the two words, lake and reservoir, are used interchangeably, a distinction is made for this discussion. A reservoir is generally considered to be an artificial body of water, in the sense that it is man made for a specific purpose, while a lake is generally considered to be a creation of nature. From a modeling standpoint, the key difference between a lake and a reservoir lies in the hydraulic retention time. In general, lakes have long hydraulic retention times, on the order of years, while reservoirs have retention times generally on the order of months. Sophisticated mathematical representations of biological processes have generally been applied to reservoirs, while more generalized "black box" type models have been applied to lakes. These generalized lake models are most often called input-output models because they attempt to relate the input of a constituent to a lake or the concentration of the constituent in the lake with the physical properties of the lake, such as mean depth and hydraulic retention time. The constituent most frequently considered in these models is phosphorus. The input-output approach has evolved based primarily on the work of Vollenweider (Ref. 4-53), and is sometimes referred to as the Vollenweider Approach.

The complex biological process representation, because of the relatively long hydraulic retention time, is seldom used in lakes. However, the input-output formulation may in some cases be used with success for reservoirs. The Vollenweider formulation was developed on large lakes in the Northern United States and Canada, and therefore its use on reservoirs is sometimes questioned. A study by Tapp (Ref. 4-39) and a study by Gakstatter, *et al.* (Ref. 4-13), indicated that the input-output formulations have merit when applied to reservoirs. For this reason, the models to be described will not be subdivided between lake and reservoir models. Many of the representations and transport mechanisms used in the complex reservoir models have been described earlier in this chapter and will not be repeated under the description of a specific model.

An additional phenomenon which is present in many reservoirs is the density stratification that occurs due to temperature differences between the upper and lower layers in the reservoir. This density difference, coupled with the design of the reservoir outlet structure, greatly influences the transport of pollutants through the reservoir. Since no theory behind the input-output models has been previously presented, a description of the theory will accompany each input-output model. When considering reservoirs, the question arises as to whether the additional time and expense for a complex process model formulation justifies the decision that must be made in terms of water quality management. This question has been addressed in a study by Tapp (Ref. 4-39), which compared the impact on water quality management decisions from using a complex reservoir model and input-output models. Although this study addressed only one case, it did give insight into the question.

Vollenweider Model. This model is the forerunner of most input-output models in use today and was developed by Vollenweider (Ref. 4-53). The Vollenweider Model is a nutrient budget model for phosphorus derived from the following mass balance equation:

$$\frac{dm_w}{dt} = J - \sigma m_w - \frac{Q}{V} m_w, \qquad \text{(Eq. 4-72)}$$

where,

m_w = mass of substance in the lake, in mg
J = yearly supply of substance to the lake, in mg/yr
σ = a sedimentation rate coefficient, in yr^{-1}
Q = outflow rate, in m^3/hr
V = lake volume, in m^3, and
t = time, in yr.

As indicated in Fig. 4-8, Vollenweider plotted phosphorus loading in $g/m^2/yr$ vs. the mean depth divided by the retention time of a lake. Vollenweider then empirically defined a basic loading tolerance for the case where the mean depth divided by the retention time was much less than one. Using the solution to the equation, the loading tolerance lines were projected throughout the commonly encountered ranges of mean depth divided by retention time. The lower line was termed the permissible limit and the upper line was termed the dangerous limit and was defined as twice the permissible limit. The permissible limit was said to separate oligotrophic and mesotrophic lakes and the dangerous limit was to separate mesotrophic and eutrophic lakes. To find the allowable loading based on the Vollenweider curve, a mean depth and hydraulic retention time are used to enter the abscissa (Figure 4-8). Depending upon whether the goal is to be in the mesotrophic (dangerous) or oligotrophic (permissible) range, a loading can then be obtained on the ordinate. The model considers the lake as a whole and does not separate the hypolimnion from the epilimnion when stratification exists. Vollenweider's model was originally calibrated on large lakes in the Northern United States and Canada.

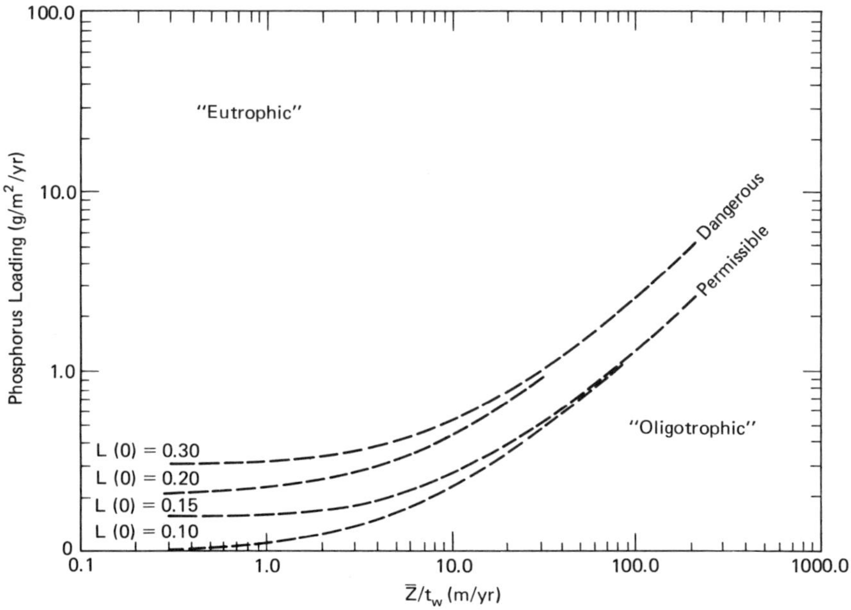

Fig. 4-8. Graphical representation of the Vollenweider Model.

Dillon Model. This model is a furthering of the Vollenweider approach and was developed by Vollenweider and Dillon (Ref. 4-54). The Dillon Model uses the steady state solution to the Vollenweider approach, which in terms of the total phosphorus concentration in a water body, was formulated as:

$$[P] = \frac{L}{\overline{Z}\sigma + (\overline{Z}/\tau_w)}, \quad \text{(Eq. 4-73)}$$

where

$[P]$ = total phosphorus concentration in the lake, in g/m^3
L = phosphorus loading to the lake, in $g/m^2/yr$
$\overline{Z}$ = mean depth, in m
σ = sedimentation rate coefficient, in yr^{-1}, and
τ_w = hydraulic retention time, in yr.

Vollenweider and Dillon then expressed the retention coefficient as:

$$R = \frac{\sigma}{\sigma + \rho_w}, \quad \text{(Eq. 4-74)}$$

where ρ_w = hydraulic washout coefficient or flushing rate = $1/\tau_w$, in yr^{-1}. From the above equations:

$$\sigma = \frac{R\rho_w}{1 - R} \qquad \text{(Eq. 4-75)}$$

$$[P] = \frac{L(1 - R_{exp})}{\overline{Z}\,\rho_w}, \qquad \text{(Eq. 4-76)}$$

where R_{exp} = retention coefficient independently calculated as:

$$R_{exp} = 1 - \frac{\sum q_a [P]_a}{\sum q_i [P]_i}, \qquad \text{(Eq. 4-77)}$$

where

q_a = outflow rate, in m^3/yr
$[P]_a$ = outflow concentration, in g/m^3
q_i = inflow rate, in m^3/hr, and
$[P]_i$ = inflow concentration, in g/m^3.

Therefore, if dangerous and critical lines are drawn on a plot of $L(1 - R)/\rho_w$ vs. $\overline{Z}$, the trends represent equal predictive phosphorus concentrations. This indicates that the prediction of the trophic state of the lake is based on a measure of the predictive phosphorus concentration in the lake, rather than on the phosphorus loading. The resulting curves developed for the Dillon Model are shown in Figure 4-9.

To compute an allowable phosphorus loading based on the Dillon Model, the abscissa is entered with the mean depth of the water body. Depending on whether the interest is in the oligotrophic state (permissible limit) or the mesotrophic state (dangerous limit), a value of $L(1 - R)/\rho$ is obtained. The value of R must either be estimated based on a comparison with other water bodies or calculated from actual data on the water body under study. Thus, the Dillon Model requires one piece of data more than the Vollenweider Model: the value of the phosphorus retention coefficient. Like the Vollenweider Model, the Dillon Model does not separately analyze the various layers in stratified lakes and was originally calibrated on large lakes in the Northern United States and Canada.

Larsen-Mercier Model. Working independently, Larsen and Mercier (Ref. 4-29) expressed Vollenweider's mass balance model in terms of concentration. This formulation is called the Larsen-Mercier Model and is expressed as:

$$\frac{d[P]}{dt} = \frac{\sum v_j [P_j]}{V} - \rho_w [P] - \sigma[P], \qquad \text{(Eq. 4-78)}$$

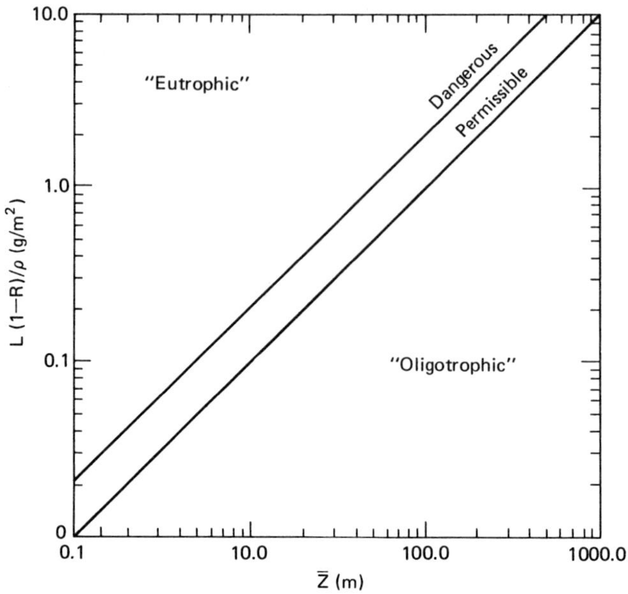

Fig. 4-9. Graphical representation of the Dillon Model.

where

$[P]$ = P concentration in the lake, in g/m^3
v_j = flowrate of the jth tributary, in m^3/yr
$[P_j]$ = P concentration in the jth tributary, in g/m^3
V = lake volume in m^3
σ = sedimentation rate coefficient, in yr^{-1}
ρ_w = theoretical washout coefficient, in yr^{-1}, and
t = time, in yr.

Utilizing the retention coefficient, R_{exp}, as defined for the Dillon Model in Equation 4-77, Equation 4-78 may be solved to yield:

$$[\overline{P}] = \frac{P_\infty}{1 - R_{exp}}, \qquad \text{(Eq. 4-79)}$$

where

$[\overline{P}]$ = mean influent P concentration, in g/m^3, and
P_∞ = steady state P concentration in the lake, in g/m^3.

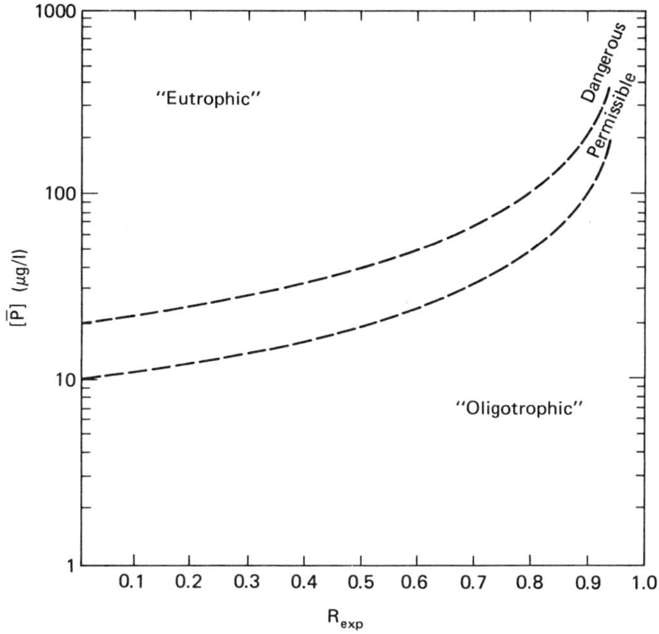

Fig. 4-10. Graphical representation of the Larsen-Mercier Model.

This equation describes the relationship between the steady state lake and mean input phosphorus concentrations. Larsen and Mercier selected phosphorus concentrations of 10 and 20 μg/l to delineate the oligotropic, mesotrophic, and eutrophic states. These values were selected, based on results from the literature suggesting that springtime concentrations of total phosphorus, in excess of 20 μg/l, were likely to produce average summer chlorophyll A concentrations of 10 μg/l or greater. Larsen and Mercier's curves are shown in Figure 4-10. To compute an allowable mean influent phosphorus concentration, the abscissa is entered with the phosphorus retention coefficient, which is calculated as described for the Dillon Model. From the abscissa, a mean influent phosphorus concentration is obtained. Knowing the annual inflow to the reservoir, a yearly loading can be calculated. The Larsen-Mercier Model also considers the water body as a whole rather than separating it into the layers which occur during stratification.

Snodgrass and O'Melia Model. The Vollenweider type formulations may be considered one-box models because they do not consider both the epilimnion and the hypolimnion separately, but consider the water body as a whole. A two-box model, which considers the epilimnion and hypolimnion as two distinct regions during summer stratification, with some interchange across the thermocline and a winter circulation scheme, has been developed by Snodgrass and O'Melia (Ref. 4-38). This model also considers orthophosphate and particulate phosphorus separately. A schematic of the model is shown in Figure 4-11. For the summer model, when stratification is prominent, four interdependent linear differential equations were obtained as a result of mass balances for orthophosphate and particulate phosphorous in each layer. For orthophosphate in the epilimnion:

$$V_e \frac{d[OP]_e}{dt} = \sum Q_j [OP]_j - Q[OP]_e - P_e V_e [OP]_e + \frac{k_{th}}{\overline{\overline{Z}}_{th}} A_{th} [OP]_h - \frac{k_{th}}{\overline{\overline{Z}}_{th}} A_{th} [OP]_e. \quad \text{(Eq. 4-80)}$$

For particulate phosphorus in the epilimnion:

$$V_e \frac{d[PP]_e}{dt} = \sum Q_j [PP]_e + p_e V_e [OP]_e - g_e A_{th} [PP]_e + \frac{k_{th}}{\overline{\overline{Z}}_{th}} A_{th} [PP]_h - \frac{k_{th}}{\overline{\overline{Z}}_{th}} A_{th} [PP]_e. \quad \text{(Eq. 4-81)}$$

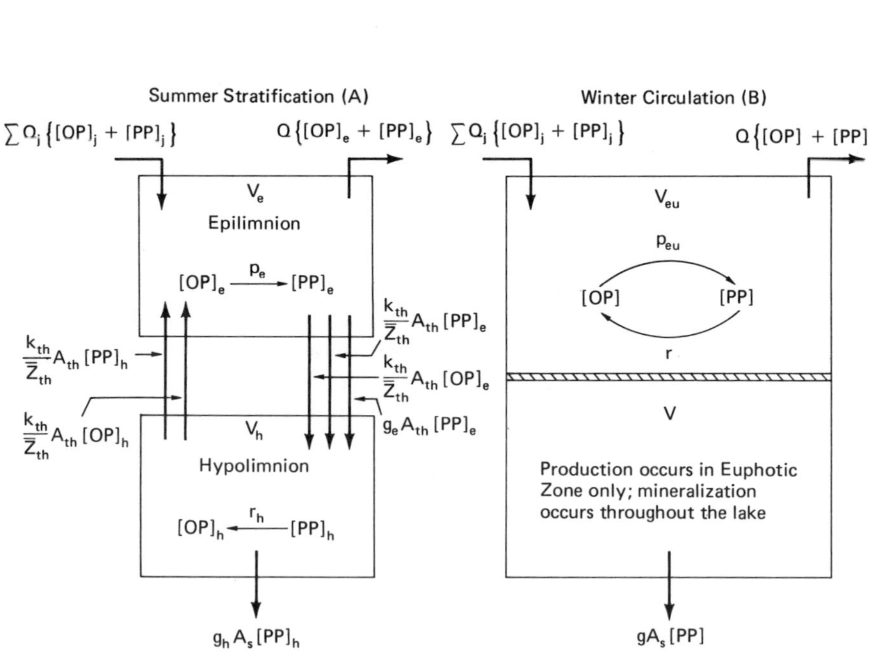

Fig. 4-11. Schematic of Snodgrass-O'Melia Model (from Ref. 4-38).

For orthophosphate in the hypolimnion:

$$V_h \frac{d[OP]_h}{dt} = r_h V_h [PP]_h + \frac{k_{th}}{\overline{Z}_{th}} A_{th} [OP]_e - \frac{k_{th}}{\overline{Z}_{th}} A_{th} [OP]_h. \quad \text{(Eq. 4-82)}$$

For particulate phosphorus in the hypolimnion:

$$V_h \frac{d[PP]_h}{dt} = g_e A_{th} [PP]_e - g_h A_s [PP]_h - r_h V_h [PP]_h + \frac{k_{th}}{\overline{Z}_{th}} A_{th} [PP]_e - \frac{k_{th}}{\overline{Z}_{th}} A_{th} [PP]_h. \quad \text{(Eq. 4-83)}$$

The subscripts e and h refer to the epilimnion and hypolimnion, and th and s denote the thermocline region and the sediment-water interface; $[OP]$ and $[PP]$ refer to the concentrations of orthophosphate and particulate phosphorus; p and r are rate coefficients for net production and decomposition; k is a vertical exchange coefficient (L^2T^{-1}), which includes the effects of molecular and turbulent diffusion, internal waves, erosion of the hypolimnion and other fluid processes on the transfer of heat and materials across the thermocline; $\overline{Z}$ is a mean depth; V is the volume of a layer or box; A is an interfacial area; Q_j is a land-based volumetric rate of inflow of water; Q is the volumetric rate of lake discharge; and g is an effective settling velocity. For the winter model, when stratification does not occur, two linear differential equations were obtained from mass balances of phosphorus between the euphotic and noneuphotic zones. For orthophosphate throughout the lake:

$$V \frac{d[OP]}{dt} = Q_j [OP]_j - Q[OP] - p_{eu} V_{eu} [OP] + rV [PP]. \quad \text{(Eq. 4-84)}$$

For particulate phosphorus throughout the lake:

$$V \frac{d[PP]}{dt} = Q_j [PP]_j - Q[PP] + p_{eu} V_{eu} [OP] - rV[PP] - gA_s [PP]. \quad \text{(Eq. 4-85)}$$

The subscript eu denotes the euphotic zone.

These models were joined by boundary conditions at the fall overturn and the spring stratification. At the fall overturn:

$$[OP] \text{ just after mixing} = \frac{[OP]_e V_e + [OP]_h V_h}{V} \quad \text{(Eq. 4-86)}$$

$$[PP] \text{ just after mixing} = \frac{[PP]_e V_e + [PP]_h V_h}{V}. \quad \text{(Eq. 4-87)}$$

At the beginning of spring stratification, the concentrations of orthophosphate and particulate phosphorus in the epilimnion and hypolimnion were set equal to the concentrations throughout the lake at the end of the winter circulation period. The coefficients used in the model were assumed to be constant over the appropriate summer or winter period.

The model has been calibrated for data from Lake Ontario. Based on this calibration, the model predictions of the combined effects of the areal phosphorus loading, the outflow from the water body, and the mean depth of the water body are presented in Figure 4-12. A spring concentration of total phosphorus of 20 μg/l or less was used to define the range for oligotrophic waters. The formulation of the Snodgrass-O'Melia Model indicates that lakes with large depths and large area hydraulic loadings can tolerate large phosphorus loadings. Based on the application of the model, a given lake nutrient loading should be reduced to a level at or below the three dimensional surface given in Figure 4-12 to allow the lake to remain oligotrophic. The Snodgrass-O'Melia Model is a much more theoretical formulation (depending on several coefficients) than the Vollenweider type approach, and therefore should be more applicable to a specific case where data are available. Additional calibration of the Snodgrass-O'Melia formulation may show it to be the necessary compromise between the semi-empirical models

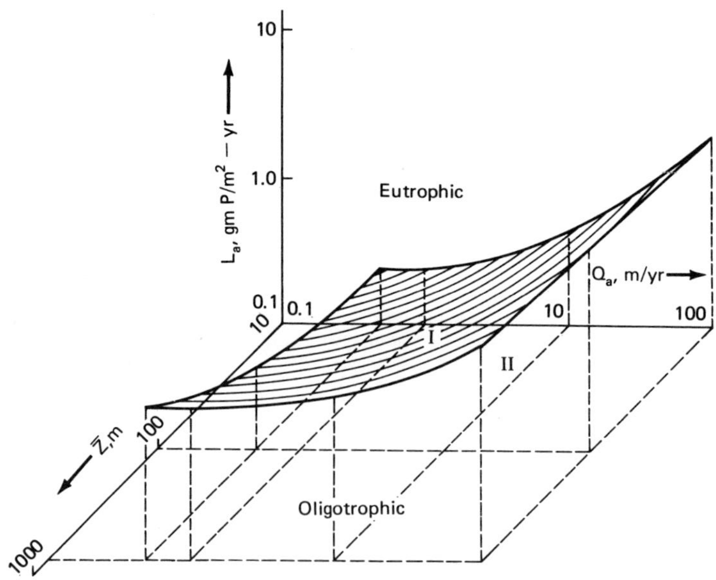

Fig. 4-12. Interrelationships among areal P loading (L_a), areal hydraulic loading (Q_a), and mean lake depth (Z), ([TP] = 20 μg/l) for Snodgrass-O'Melia Model (from Ref. 4-38).

of the Vollenweider type and the mathematically complex models formulated to describe several processes occurring simultaneously in the water body.

EPARES. This reservoir model was developed for EPA, by Water Resources Engineers (Ref. 4-59), from the original deep reservoir model which had the capability to simulate temperature only. The geometric representation used in EPARES assumes all mass and energy transfers only occur along the vertical axis of the reservoir. To facilitate the application of the necessary mass and energy balance equations, a reservoir is represented as a one dimensional system of horizontal elements with uniform thickness as illustrated in Figure 4-13. Each hydraulic element is treated as a continuously stirred tank reactor with completely mixed and uniform properties. The implicit assumption of this geometric structuring of the problem is that mass concentration and thermal gradients in the horizontal plane are insignificant in determining the water quality responses and thermal behavior of the reservoir along the vertical axis. Simulated results are, therefore, interpreted as being average conditions across the reservoir at a particular elevation. Water and water quality constituents are transported from element to element across the horizontal bounding planes by advection and diffusion processes. Since water quality variation within an element cannot exist, the simulated vertical concentration profile of any water quality constituent consists of a series of discrete steps with discontinuities at the boundaries between two elements. This discrete approximation of a reservoir profile should improve as the number of elements used to represent a reservoir is increased.

EPARES allocates tributary inflows among the various reservoir elements according to the density of the inflow and its relation to the density structure of the lake. The model assumes that density is only a function of temperature and the effects of suspended sediment and dissolved solids on density are negligible. The treatment of the various hydraulic elements as continuously stirred tank reactors provides that all constituents advected to the reservoir by inflow are completely distributed throughout the horizontal elements they enter. The hydraulic elements that contribute to a reservoir's outflow are similarly determined by density criteria, and the water quality of reservoir releases are calculated by mass balance among those elements that contribute to the outflow. The model sequentially solves, at each time step, the equations that represent the various biotic and chemical constituents. Temperatures are computed first, since the hydrodynamics of the system are completely controlled by temperature via its effect on water density. In addition, kinetic reaction rates and biological growth and respiration rates are temperature dependent. A tri-diagonal matrix is formulated that summarizes the rates of concentration changes in each element and the interactions of the two adjacent elements. The constituents simulated by EPARES are temperature, carbonaceous biochemical oxygen demand, benthic oxygen demand, phosphorus, chlorophyll A, ammonia, nitrite, nitrate, dissolved

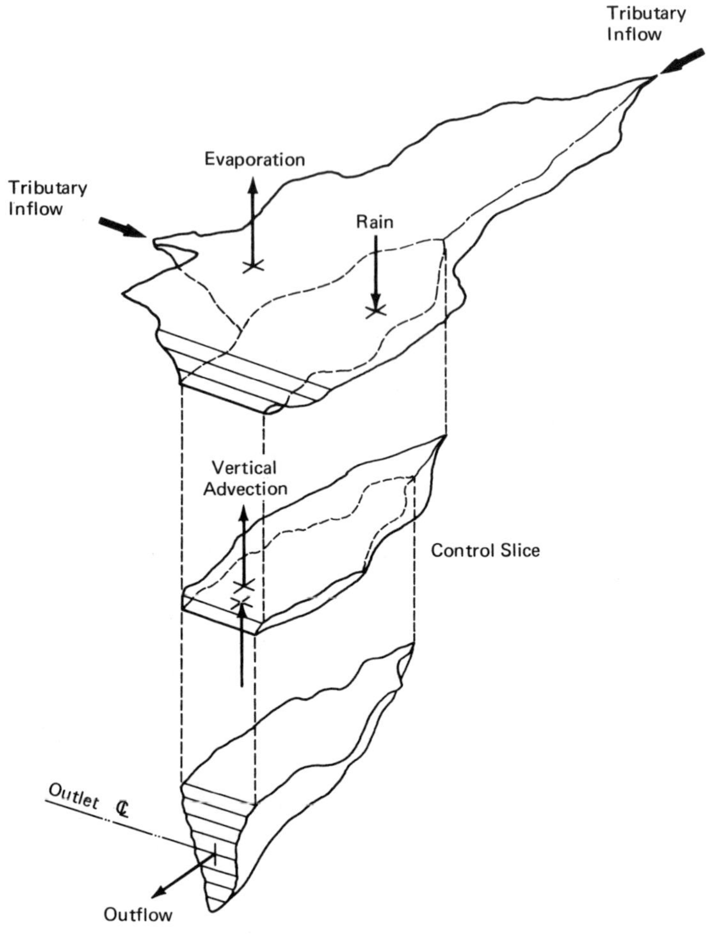

Fig. 4-13. Schematic of a stratified reservoir represented as a one-dimensional system of horizontal elements with uniform thickness.

oxygen, coliforms, and conservative materials. A data preparation program called MFIP must be used to prepare the meterological, hydraulic, and tributary water quality data for input to EPARES. The EPARES model can be used to simulate over no more than a one calendar year period.

EPAECO. This model was also developed for the Environmental Protection Agency by Water Resources Engineers (Ref. 4-60), and is identical to EPARES in relation to the temperature simulation routines. However, no data preparation routine is necessary as these routines are contained in the EPAECO coding di-

TABLE 4-3 Constituents in EPAECO and Their Order of Computation.

1. Temperature	8. Organic sediment	16. Total dissolved solids
2. Carbonaceous biochemical oxygen demand	9. Phosphorus	17. Alkalinity
3. Fish	10. Total inorganic carbon	18. pH
4. Benthos	11. Ammonia nitrogen	19. Carbon dioxide
5. Zooplankton	12. Nitrite nitrogen	
6. Algae	13. Nitrate nitrogen	
7. Detritus	14. Dissolved oxygen	
	15. Coliforms	

rectly. Where EPARES simulates only a few water quality constituents, EPAECO attempts to represent the major interactions in the ecological cycle. Like EPARES, the temperatures in EPAECO are simulated first. The other constituents are ordered such that current time step predation is incorporated in each computation. The constituents simulated in the model and their order of computation are shown in Table 4-3.

With the exception of the three classes of fish, the distribution of the biotic constituent concentrations with depth is determined by simultaneous solving the system of equations describing each constituent in all elements. As in EPARES, a tri-diagonal matrix is formulated that summarized the rates of concentration change in each element and the interactions of the two adjacent elements. The distribution of fish biomass with depth is not determined, but the biomass of the three classes of fish (coldwater, warmwater, and bottom feeders) is individually determined in terms of surface area units. The growth of the various biotic constituents in the model is determined by growth rate, respiration, mortality, and predation in the system. The growth rates utilized in EPAECO are a function of limiting factors based on light, nutrients, temperature, and dissolved oxygen. The respiration rates are divided into standard or passive rates, which are a function of temperature. Mortality is calculated as a function of temperature and dissolved oxygen, while predation in the model is calculated as a function of the predators growth rate and efficiency of predation. The growth limiting factors for the various biotic constituents in EPACEO are shown in Table 4-4. Like ESPARES, EPAECO can only be used to simulate a one-year period.

Battelle Model. This model was developed by Baca and Arnett (Ref. 4-1) and, according to these authors, can be used to predict short term changes in several water quality parameters in a reservoir as a function of nutrient loading, climatic conditions, lake morphometry, hydrologic features, and thermal and ecological regimes. Although the model is designed for short term predictions of up to 10 years, calibration, as described in the above reference, was only for less than a one year period. The temperature simulation routine is similar to that used in

TABLE 4-4 Growth Limiting Factors for Biotic Constituents in EPAECO.

Biotic Constituent	Food Source
Fish type 1 (coldwater)	Zooplankton
Fish type 2 (warmwater)	Zooplankton
Fish type 3 (bottom feeders)	Benthic animals
Benthos	Organic sediments
Zooplankton	Total algae
Algae type 1 (green algae)	Nitrogen, phosphorus, carbon, and light
Algae type 2 (blue green algae)	Nitrogen, phosphorus, carbon, and light

EPARES and EPAECO except that, with the Battelle Model, the reservoir can be segmented along the longitudinal axis as well as layered in the vertical direction. Unlike EPARES and EPAECO, the model is programmed to handle the anoxic reactions that may occur at the bottom of a reservoir. If the anoxic reactions were not programmed, inorganic constituents could accumulate to unreasonable levels at the reservoir bottom. The solution procedure used in the model is to simulate the heat transport, mixing, and fluid flow for the entire period in one day time steps. These results are then inputted to the limnological or water quality model for a finite difference technique solution to the equations for constituent transport for each of the constituents. The solution time steps range from 4 to 12 hours. The constituents simulated in the Battelle Model are temperature, dissolved oxygen, biochemical oxygen demand, phosphorus, ammonia, nitrate and nitrite, zooplankton, and phytoplankton.

4-7.4 Nonpoint Source/Urban Runoff Models

In many instances, the terms nonpoint source and urban runoff are used interchangeably. However, as these fields evolve, the term nonpoint source is commonly used to describe pollutants generated soley from runoff over agricultural, silvicultural, and mined lands. Urban runoff is used to describe the runoff from urban and suburban lands which may or may not flow in pipes. Models used in urban areas generally have routines similar to those used in rural areas for simulating pollutant generations from the open or rural part of the urban area. Urban runoff modeling is a relatively new field as is nonpoint source modeling. To date, most nonpoint source modeling for agricultural and silvicultural land centers around the use of the Universal Soil Loss Equation (USLE) to predict sediment yield from a given area. Other pollutants are then generally estimated as a fraction of the sediment.

Various agricultural agencies have conducted research over the years to define the parameters for use in the empirical (USLE) equation. They have finally reached a point where long term sediment loss can be adequately defined. However, as the time frame becomes shorter, the ability of the Universal Soil Loss Equation to predict sediment loss is diminished. The other approach commonly used, in an attempt to shorten the time frame of analysis, is to define pollutant availability and the transport by runoff based on a more detailed representation of the actual physical process occurring. Various hydrologic models, such as the Stanford Watershed Model (Ref. 4-7), are used to generate either on a continuous simulation or single storm event basis, the runoff necessary to wash the available pollutants from the land. Pollutant availability is related to numerous factors, such as land cover and use, soil type, and geology, and time between runoff events.

The actual transport of these pollutants is governed by complex physical and chemical relationships that must generally be simplified for operational reasons. These complex physical-chemical relationships have not been studied in as much depth as the relationships in the USLE, and in many cases empirical techniques must be used with the physical process relationships. These relationships have been used mainly on agricultural and urban lands. Nonpoint source pollutant simulation from mined lands is normally accomplished by the use of empirical or statistical techniques. However, in some instances, use of the USLE or the more physical process oriented techniques may be warranted.

Runoff hydrographs and pollutographs are integral components of most nonpoint source runoff models. While the hydrograph is a familiar traditional hydrologic concept, the pollutograph is a relatively new concept, stemming primarily from studies to define nonpoint source and urban runoff pollution generation. The runoff hydrograph is simply a record of flowrate or discharge vs. time at a particular point in the runoff receiving stream. The term pollutograph has been used to describe both the record of runoff pollutant concentration vs. time and the record of mass rate of pollutant generation vs. time, but is more widely applied in the latter sense. The mass rate pollutograph is produced from the product of the hydrograph and concentration pollutograph as illustrated schematically in Figure 4-14.

Universal Soil Loss Equation (USLE). The USLE is generally used to predict average annual soil losses from sheet and rill erosion by rainfall and its ensuing runoff. The application of the USLE requires knowledge of several local characteristics of the area of interest and for most areas can give reasonable estimates of the annual soil loss. The equation is described by the U.S. Department of Agriculture, Agricultural Research Service (Ref. 4-52) as:

$$A = (R)\,(K)\,(LS)\,(C)\,(P), \qquad \text{(Eq. 4-88)}$$

where A is the estimated average annual soil loss in tons per acre; R is the rainfall and runoff erosivity index; K is the soil-erodibility factor, which is the average soil loss per unit of R under arbitrarily selected "basic" conditions (it depends

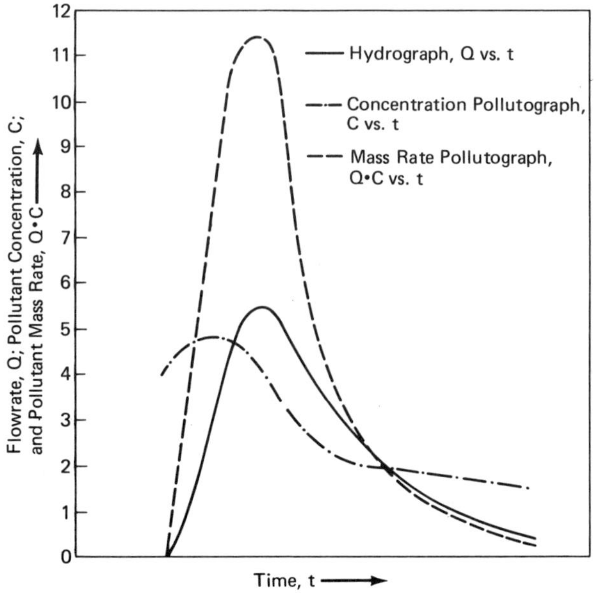

Fig. 4-14. Schematic illustration of relationship between runoff hydrograph and pollutograph.

on soil properties); *LS* is a dimensionless topographic factor that represents the combined effects of slope length and steepness; *C* is the cover and management factor; and *P* is the factor for supporting practices. Various charts and tables for using the USLE are given by the Agricultural Research Service (Ref. 4-52) and by the Midwest Research Institute (Ref. 4-32). The USLE was developed to give only the soil loss from a given area for agricultural production purposes. The amount of sediment that actually reaches a stream is termed the sediment yield, and is the gross erosion or soil loss minus the amount deposited enroute to the place of measurement. Sediment yield is generally calculated as a factor (which is a function of the drainage area) times the gross erosion. To date, very little generalized information has been developed to relate other pollutants to the sediment yield. Those relationships that have been developed are usually site specific and generalization is extremely difficult.

Agricultural Runoff Management Model (ARM). This model was developed for the Environmental Protection Agency by Hydrocomp (Ref. 4-21), and essentially coupled the Pesticide Transport and Runoff Model (PTR) (Ref. 4-20) with a model for other agricultural chemicals. The ARM uses a modification to the Stanford Watershed Model to simulate the hydrology of the area being modeled.

The model simulates the processes of precipitation, interception, runoff, infiltration, percolation, storage, evaporation and transportation as part of the hydrologic or water cycle. This is a continuous simulation model and instead of using the USLE for simulating erosion, it uses an equation developed by Negev to simulate the pickup of soil fines. In terms of pollutant generation, the model uses the mechanisms of adsorption-desorption, volatilization and degradation to establish a pollutant balance for the watershed being studied. This model gives sediment yield and other constituents but does not include channel routing. In addition, the model has only been calibrated for watersheds of two square miles or less. Therefore, pollutant yield formulations or a method of channel routing would have to be used in conjunction with the ARM if pollutant routing is desired over large watersheds.

Storage Treatment Overflow Runoff Model (STORM). An example of a model which has great utility in water quality management planning for urban runofff is STORM, developed by Water Resources Engineers and the Corps of Engineers (Ref. 4-23). The model was designed to estimate the quantity and quality of runoff from small, primarily urban, watersheds. Runoff is considered to be a constant fraction of hourly rainfall (less the available depression storage). The model does not route flow and pollutants but gives the output of each at the outlet of a basin in the form of hydrographs and pollutographs. If needed, storage and overflow routines can also be used to determine optimum storage capacity. The hydrological data used by STORM are a continuous record of hourly rainfall data, evaporation data, a rainfall factor, and depression storage. The land use data needed by STORM is the percent of the area under study in each of six land use categories (single family residential, multi-family residential, commercial, industrial, open, and nonurban), the percent of imperviousness of each area, the number of feet of gutter in the area, and the street sweeping interval. Information required by STORM to calculate water quality constituents are the dust and dirt buildup rate and the mass of BOD, suspended solids, settleable solids, orthophosphate, and total nitrogen per unit mass of dust and dirt. For nonurban areas, the mass of pollutants per unit area must be specified. The model uses the USLE for calculating erosion. It does not have the ability to utilize base flow. Since the model does not contain routing, its use is recommended for areas of less than twenty-five square miles.

Other Stormwater Models. Many other stormwater models have been developed. One of the most widely used is the Environmental Protection Agency Stormwater Management Model (SWMM), which will be described in the next section titled Multiple Function Models. Other stormwater models and their basic characteristics have been described by Heaney and Huber (Ref. 4-17), and are shown in Table 4-5.

TABLE 4-5 Comparison of Urban Runoff Models, March 1974.

Model	Surface Routing	Sewer Routing	Quality Routing	Degree of Sophistication of Surface Flow Routing	Degree of Sophistication of Sewer Flow Routing	Accurate Modeling of Surcharging	Flexibility of Modeling of Sewer Components	Explicit Modeling of Insystem Storage
Rational Method	Peak flows only	Peak flows only	No	Low	Low	No	Low	No
Chicago	Yes	No	No	Moderate	NA	NA	NA	NA
Unit hydrograph	Yes	In combination with surface	No	Low	Low	No	Low	No
Unit pulse	Yes	In combination with surface	No	Low–Moderate	Low–Moderate	No	Low	No
STORM	Yes	In combination with surface	Yes	Low	Low	No	Low	No
RRL	Yes	Yes	No	Moderate	Low–Moderate	No	Low	No
MIT	Yes	No	No	High	NA	NA	NA	NA
Battelle	Yes	Yes	Yes	Low	Moderate	No	Moderate	No
EPA-SWMM	Yes	Yes	Yes	High	Moderate	No	High	Yes
WRE-SWMM	Yes	Yes	Yes	High	High	Yes	High	No
Cincinnati (UCUR)	Yes	Yes	Yes	High	Low	No	Low	No
Dorsch (HVM)	Yes	Yes	No	High	High	Yes	High	?
SOGREAH	Yes	Yes	?	High	High	Yes	High	?
Hydrocomp	Yes	Yes	Yes	Moderate	Moderate	No	Low	No
Illinois (ISS)	Yes	Yes	No	Moderate	High	No	Low	No

Source: Heaney and Huber (Ref. 3-17).

4-7.5 Multiple Function Models

Many models used in water quality management planning can serve more than one function and fit into more than one of the previously described types of models. This section will describe four of the more commonly used of these multiple function models.

TABLE 4-5 *(Continued)*

Treatment Modeling	Receiving Model Available	Degree of Calibration/ Verification Required	Simulation Period	Availability	Documentation	Data Requirements
NA	No	Usually not verified	Individual storms	Non-proprietary	Good	Low
NA	No	Moderate	Individual storms	Non-proprietary	Fair	Moderate
NA	No	High	Individual storms	Non-proprietary	Fair	Moderate
NA	No	High	Individual storms	Non-proprietary	Fair	Moderate
Yes	No	Low	Long term	Non-proprietary	Good	Low
NA	No	Moderate	Individual storms	Non-proprietary	Good	Moderate
NA	No	Moderate	Individual storms	Proprietary	Fair	Moderate
No	No	Moderate	Individual storms	Proprietary	Poor	Moderate
Yes	Yes	Moderate	Individual storms	Non-proprietary	Good	Extensive
Yes	Yes	Moderate	Individual storms	Proprietary	Poor	Extensive
No	No	Moderate	Individual storms	Non-proprietary	Fair	Extensive
NA	Yes	Moderate	Individual storms or (separate) long term	Proprietary	Poor	Extensive
?	Yes	Moderate	Individual storms	Proprietary	Poor	Extensive
No	Yes	High	Individual storms or long term	Proprietary	Fair	Extensive
NA	No	Moderate	Individual storms	Non-proprietary	Good	Extensive

Simplified Guide. During the early 1970s, when water quality management planning began to emerge as a distinct field, a need existed for a simplified technique to quickly assess the discharge requirements necessary to protect instream water quality. To fulfill this need, Hydroscience (Ref. 4-25) developed for the Environmental Protection Agency a simplified technique for estimating allowable loadings to maintain instream water quality standards. The basic advection-dispersion equations were solved for their most simple cases. Tables, figures, and nomographs were presented for obtaining solutions for both streams and estuaries. The parameters modeled in the Simplified Guide include certain dissolved solids, coliforms, nutrients, oxidation of carbonaceous and nitrogenous compounds, and dissolved oxygen. Because of the many simplifying assumptions, the Simplified Guide should be used only where no data exists in order to get a rough estimate of what should be the effluent limitation for a particular discharge-receiving water situation. Some of the more significant simplifying assumptions are steady state conditions, similar carbonaceous and nitrogenous compounds oxidation rates, and a one dimensional system representation. If more than just a rough estimate is desired, caution should be exercised to assure that these, as well as other simplifying assumptions used in the Simplified Guide, are applicable for the system being studied.

Water Quality for River-Reservoir Systems (WQRRS) Model. This model is an ecological water quality model and was originally developed for the Corps of Engineers (Ref. 4-22) to simulate together river-reservoir systems. As originally developed, the model used steady state hydraulics in the river for input to the reservoir. However, the WQRRS model was modified by the Corps of Engineers (Ref. 4-24) so that the dynamic storm water inputs can be simulated in the river-reservoir system. The reservoir part of the WQRRS Model is almost identical to that described earlier for EPAECO. The constituents simulated in both the river and the reservoir are identical to those described for EPAECO. The WQRRS Model can also be used in a steady flow mode, using either a backwater analysis or a stage-flow relationship. Water quality and ecological relationships are simulated in a dynamic mode for both steady and nonsteady state hydraulics.

Stormwater Management Model (SWMM). This model was developed by a consortium of contractors (Ref. 4-31) for the Environmental Protection Agency to simulate the water cycle from precipitation to discharge and reaction in a receiving stream. The model is continually updated and Huber, *et al.* (Ref. 4-26) described significant changes made to SWMM. The model is divided into six different blocks, each with a separate function. The Executive Block is the main control block and assigns logical units and determines the sequence of blocks to be executed; it does no calculations. The Combine Block allows data sets to be aggregated for input to certain blocks, allowing complex drainage systems to be partitioned for simulation in smaller segments and subsequent combining. The Runoff Block computes the quality and quantity of stormwater runoff for a given precipitation event and stores the results in the form of hydrographs and pollutographs at inlets to the main sewer system. The Transport Block sets up conditions in the conveyance system by computing dry weather flow and infiltration throughout the sewer system. The primary function of the Transport Block is to route the flow and quality, picking up results from the Runoff Block and producing hydrographs and pollutographs for the entire system or at intermediate points. The Storage Block takes the results from the Runoff Block and modifies the quality and quantity based on some predefined level of treatment and treatment capacity. Costs of construction and operation and maintenance for a given type of treatment and treatment capacity are calculated.

The Receiving Water Block can accept the output from the Runoff Block, the Transport Block, or the Storage Block and calculate the resulting receiving water quality and quantity in a river, lake, estuary, or bay. Any of the four major computational blocks can be run independently of the others with proper input data. The SWMM generates runoff for a single storm event based on a rainfall hyetograph, street sweeping interval, antecedent number of dry days, five types of land use, and topography. The quality of runoff is based on the amount of dust and dirt picked up by the runoff, and the mass of various pollutants per unit mass of dust and dirt. The quality constituents simulated in the SWMM are 5-day biochemical oxygen demand, total suspended solids, total coliforms as a conservative, and dissolved oxygen. The suspended solids contribution from urban erosion is calculated by the Universal Soil Loss Equation.

The Runoff Block alone can also simulate chemical oxygen demand, settleable solids, total nitrogen, phosphate, and grease. The routing through the system by the Transport Block is done using a modified kinematic wave approach in accordance with the Manning equation and continuity. Complete mixing is assumed at all input points. The receiving water system is simulated by dividing the system into a series of nodes connected by links or channels. For each time step the equations of motion and continuity are applied to all nodes to calculate the hydrodynamics of the system. The equations for conservation of mass are then coupled with the hydrodynamics to solve for the quality of the system. The Receiving Water Block can be used to simulate the hydrodynamics without simulating quality. Care must be exercised when simulating quality in tidally affected systems with the Receiving Water Block of SWMM because dispersion is not included in the model formulation.

RECEIV-II. This model is a modification of the Receiving Water Block of SWMM and was developed for the Environmental Protection Agency by Raytheon (Ref. 4-36). The model can simulate phosphorus, coliforms, ammonia nitrogen, nitrite nitrogen, nitrate nitrogen, total nitrogen, carbonaceous biochemical oxygen demand, chlorophyll A, dissolved oxygen, salinity, and one metal ion for lake, river, and estuary receiving water systems. Like the Receiving Water Block of SWMM, RECEIV-II can be used to simulate on a time varying basis in a pseudo-two dimensional system of links and nodes. RECEIV-II can also simulate systems with multiple tidal inlets, and multiple shallow dams. Like the Receiving Water Block of SWMM, RECEIV-II does not include dispersion. The model was developed to be fully compatible with the other blocks of SWMM.

4-8. A SYSTEMS APPROACH TO WATER QUALITY MANAGEMENT

Water quality management planning requires a vast amount of data. To determine the necessary size of wastewater facilities, existing and projected populations and land use patterns must be known. When dealing with nonpoint source or urban runoff, drainage patterns must be defined. Data dealing with natural stream and overland flow networks should be readily available. Other information which may be useful is soil type or soil association, depth to bedrock, housing patterns, and employment information. The efficiency with which water quality management planning can be conducted depends on the ability of the engineer or planner to rapidly access and manipulate these data. If all data could be stored in a useable format and accessed in a format compatible with necessary models, the planning process could proceed at a relatively rapid rate. The major effort would be at the outset of the study in compiling and storing the vast amount of data required. Since most water quality management studies have budget constraints, a system for storing, accessing, manipulating, and displaying the data must be available at a relatively low cost.

Several systems have been developed toward fulfilling the needs described above; one was described by Grayman, *et al.* (Ref. 4-15). The type of data and necessary models, as well as potential applications of such systems are described below.

4-8.1 Geographical Data

The geographical data to be stored in a data management system depends on the type of problem being studied. For a typical water quality management study dealing with point and nonpoint sources of wastewater the following data would be desirable:

- Land Use—both present and projected.
- Population—both present and projected.
- Elevation.
- Stream boundaries.
- Drainage basin and subbasin boundaries.
- Location of municipal and industrial wastewater treatment plants.
- Location of water supply intakes.
- Location of water quality and effluent limited stream segments.
- Soils data (type and/or association).
- Depth to bedrock and depth to water table.
- Facility planning (201) boundaries.
- Stream flow information (high and low flow).
- Sewer service boundaries.

In addition, data on housing and employment, both present and projected, might prove useful.

4-8.2 Models

The data in the management system must be manipulated to solve a given problem. This is most commonly accomplished by the use of models which describe certain physical phenomena of interest. The system must be designed such that the data can be outputted in a format compatible with the models to be used, or the models must be designed to accept data in a format that the system can output. Models of the following types may be useful for the different types of water quality management studies:

- Models to check accuracy of the data base.
- Models to project and distribute population.
- Models for preliminary design and costing of alternative sewer systems.
- Models to design wastewater treatment plants based on flow and influent quality.
- Water quality models to establish effluent limitations for point sources.
- Models for estimating nonpoint source and urban runoff.
- Statistical models.

4-8.3 Potential Applications of Systems Approach

The potential applications of the system approach lie in the data/model linkages and subsequent output. If the problem to be solved were the design of alternative sewer locations, data such as soil type, depth to bedrock and groundwater, population data, land use data, and elevation would be taken from the system for input to the model for a given route. Several routes could be evaluated with relative ease until the least cost or most feasible route were determined. If the problem to be solved were the impacts on water quality from nonpoint sources of pollution, land use patterns and their pollution generation potential could be easily varied for input to the proper model to give the decision maker an idea of the pollution impact of a decision he might make. The list of possible applications of the systems approach could go on, but the numerous possibilities should be obvious in light of the foregoing discussions. The point is that if large amounts of data must be manipulated and displayed in a water quality management study, the systems approach can simplify the process and can easily provide adequate information to the decision maker.

4-9 SUMMARY

The requirements of PL 92-500 have resulted in a great acceleration in water quality management planning studies. Water quality modeling or simulation represents a key element of these studies, and many management decisions will be made based on the modeling results. These efforts have also resulted in a proliferation of technical publications on the subject. Those considered to be of particular significance to the objectives of this chapter are cited in the following References and Bibliography sections. However, others may be found by a thorough search of the available literature.

This chapter has covered the general concepts of water quality modeling, a review of models available for various applications, the structuring and application of a particular model, and a description of a systems approach to water quality management. Emphasis has been placed on practical application of the material for water quality management purposes, and the most recently available technical information has been included.

The ability of the engineer or planner conducting a water quality management study to interface with the decision maker will ultimately determine the success or failure of the study. This often depends upon whether the associated models have been structured and applied in such a manner so the results present the decision maker with a rational and understandable basis for his decisions. Unfortunately, this is not always the case. In their recently published study on the formulation and use of practical models for river quality assessment, Hines, *et al.*, (Ref. 4-18) concluded that "the failure to accept and trust models stems from the fact that many river quality models have not been formulated on the basis of sound data nor effectively applied to planning and management situations." Their findings relative to model deficiencies and suggestions for improvement are presented in Table 4-6. This is not to imply that all modeling efforts are characterized by any or all of these deficiencies; but the suggestions presented in Table 4-6 serve as a good set of guidelines for structuring and applying a model for water quality management purposes.

TABLE 4-6 Common Deficiencies of Applied River Quality Models and Suggestions for Improvement as given by Hines, et al. (Ref. 4-18).

Deficiencies	Suggestions for Improvement
Application to a variable or process that is too complex for formulation of a practical model.	Model only those variables and processes for which major parameters and assumptions may be quantified or verified through field and laboratory study. Define model limitations, bias, predictive accuracy, relative importance of each parameter, and applications commensurate with the level of understanding surrounding each variable or process. Use semiquantitative descriptive studies when knowledge is inadequate to formulate a practical, applied model.

TABLE 4-6 *(Continued)*

Deficiencies	Suggestions for Improvement
Application of a sophisticated, general case model without adequate understanding of the particular river in question.	Conduct premodeling studies to develop a better understanding of river hydrology and important river quality processes, and tailor the model to reflect these determined characteristics. Consult with independent mathematicians and computer scientists concerning model configuration, solution techniques, and other technical model questions before selection or use of any particular model.
Failure to recognize the importance of basin hydrology in defining the critical planning and management decision periods for simulation by the model.	Conduct a detailed assessment of natural and man-influenced basin hydrology to determine the critical recurring conditions(s) with respect to river quality. Establish recurrence probabilities for these conditions. Base production runs of the model (that is, predictions) within the framework of these recurrence probabilities.
Misapplication of model calibration and verification procedures.	Use separate, independent sets of data for calibration and verification. Minimize reliance on mathematical "curve fitting" techniques by independently establishing reasonable ranges for model parameters and assumptions through field and laboratory investigation.
Use of a poor data base for interpretation, model calibration, and model verification.	Collect a statistically reliable set of data specifically suited for analysis of the river quality variable or process being modeled. Concentrate data collection and analysis effort within periods when critical hydrologic and river quality conditions occur.
Failure to format results for easy understanding.	Consult continually with intended model users in order to update needs and instill familiarity with the model and its capabilities. Write user reports and products in nontechnical language; include discussion of model limitations and application. Display results of model production runs in graphical form whenever possible.

REFERENCES

4-1. Baca, Robert G. and Arnett, R. C., "A Limnological Model for Eutrophic Lakes and Impoundments," paper presented at the U.S. Environmental Protection Agency Conference on Modeling and Simulation, Cincinnati, Ohio, April 20–22, 1976.

4-2. Barnes, H. H., Jr., "Roughness Characteristics of Natural Channels," *Water Supply Paper 1849*, U.S. Geological Survey, Washington, D.C., 1976.

4-3. Braster, R. E., Chapra, S. C., and Nossa, G. A., "Documentation for SNSIM1/2," U.S. Environmental Protection Agency, Region II, New York, New York, February, 1975.

4-4. Bunce, R. E. and Hetling, L. J., "A Steady State Segmented Estuary Model," *Technical Paper Number 11*, Federal Water Pollution Control Administration, Charlottesville, Virginia.

4-5. Churchill, M. A., *et al.*, "Prediction of Stream Reaeration Rates," *J. Sanitary Eng. Div., A. S. C. E.* 88, No. SA4, July, 1962, p. 1.

4-6. Committee on Hydromechanics, "Friction Factors in Open Channels," *J. Hydraul. Div., A. S. C. E.* 89, No. HY2, Part 1, March, 1963, pp. 97–143.

4-7. Crawford, N. H. and Linsley, R. K., "Digital Simulation in Hydrology: Stanford Watershed Model IV," *Technical Report No. 39*, Department of Civil Engineering, Stanford University, Stanford, California, July, 1966.

4-8. Federal Water Pollution Control Administration, "Water Quality Criteria," Washington, D. C., April, 1968.

4-9. Feigner, K. D. and Harris, H. S., "Documentation Report FWQA Dynamic Estuary Model," Federal Water Quality Administration, July, 1970.

4-10. Foree, E. G., "Wastewater Assimilation Capacity of Water Quality Limited Stream Segments in Kentucky," *Technical Report UKY TR80-74-CE20*, University of Kentucky College of Engineering, Lexington, Kentucky, February, 1974.

4-11. Foree, E. G., "Summary Report: Low-Flow Reaeration and Velocity Characteristics of Kentucky Streams," *Technical Report UKY TR99-76-CE27*, University of Kentucky College of Engineering, Lexington, Kentucky, March, 1976.

4-12. Foree, E. G., "Reaeration and Velocity Prediction for Small Streams," *J. Environmental Div., A. S. C. E.* 102, No. EE5, October, 1976.

4-13. Gakstatter, J. H., *et al.*, "Lake Eutrophication: Results from the National Eutrophication Survey," paper presented at the 26th Annual AIBS Meeting, Oregon State University, Corvallis, Oregon, August 17–22, 1975.

4-14. Gameson, A. L. H., "Aeration at Weirs," *Notes on Water Pollution, No. 61*, Water Pollution Research Laboratory, Stevenage, Herts, England, June, 1973.

4-15. Grayman, W. M., *et al.*, "A Land Based Modeling System for Water Quality Management Studies," paper presented at the American Society of Civil Engineers Speciality Conference on Planning for Water Quality Management, Cornell University, Ithaca, New York, June, 1974.

4-16. Harleman, D. R. F., "Heat Disposal in Water Environment," *J. Hydraul. Div., A. S. C. E.* **101**, No. HY9, September, 1975, pp. 1120–1138.

4-17. Heaney, J. P. and Huber, W. C., "EPA Stormwater Management Model," paper presented at the University of Massachusetts Short Course—Applications of Stormwater Management Models, Amherst, Massachusetts, August 19–23, 1974.

4-18. Hines, W. G., *et al.*, "Formulation and Use of Practical Models for River Quality Assessment," *J. Water Pollution Control Federation* **47**, No. 10, October, 1975, pp. 2357–2370.

4-19. Holler, A. G., Jr., "Reaeration of Discharge through Hydraulic Structures," thesis presented to the University of Cincinnati, Cincinnati, Ohio, in partial fulfillment of the requirements for the degree Doctor of Philosophy, 1970.

4-20. Hydrocomp, Inc., "Pesticide Transport and Runoff Model for Agricultural Lands," *Report No. EPA-660/2-74-013*, U.S. Environmental Protection Agency, Washington, D.C., December, 1973.

4-21. Hydrocomp, Inc., "Modeling Pesticides and Nutrients on Agricultural Lands," *Report No. EPA-600/2-76-043*, U.S. Environmental Protection Agency, Washington, D.C.

4-22. Hydrologic Engineering Center, "Water Quality for River-Reservoir Systems," preliminary draft, U.S. Army Corps of Engineers, Davis, California, September, 1973.

4-23. Hydrologic Engineering Center, "Urban Stormwater Runoff-STORM," U.S. Army Corps of Engineers, Davis, California, October, 1974.

4-24. Hydrologic Engineering Center, "Chattahoochee River Water Quality Analysis," preliminary draft, U.S. Army Corps of Engineers, Davis, California, December, 1975.

4-25. Hydroscience, Inc., "Simplified Mathematical Modeling of Water Quality," U.S. Environmental Protection Agency, Washington, D.C., March, 1971.

4-26. Huber, W. C., *et al.*, "Storm Water Management Model Users Manual Version II," *Report No. EPA-670/2-75-017*, U.S. Environmental Protection Agency, Washington, D.C., March 1975.

4-27. Krenkel, P. S. and Orlob, G. T., "Turbulent Diffusion and the Reaeration Coefficient," *J. Sanitary Eng. Div., A. S. C. E.* **88**, No. SA2, March, 1962, pp. 53–83.

4-28. Langbein, W. B. and Durum, W. H., "The Aeration Capacity of Streams," *Geological Survey Circular 542*, U.S. Geological Survey, Washington, D.C., 1967.

4-29. Larsen, D. P. and Mercier, H. T., "Lake Phosphorus Loading Graphs: An Alternative," *National Eutrophication Survey Working Paper No. 174*, U.S. Environmental Protection Agency, July, 1975.

4-30. Metcalf & Eddy, Inc., *Wastewater Engineering: Collection, Treatment, Disposal*, McGraw-Hill, New York, New York, 1972.

4-31. Metcalf & Eddy, Inc., University of Florida, and Water Resources Engineers, Inc., "Storm Water Management Model," *Volumes I–IV*, U.S. Environmental Protection Agency, Washington, D.C., July–October, 1971.

4-32. Midwest Research Institute, "Loading Functions for Assessment of Water Pollution from Nonpoint Sources," *Report No. EPA-600/2-76-151*, U.S. Environmental Protection Agency, Washington, D.C., May, 1976.

4-33. O'Connor, D. J. and Dobbins, W. E., "The Mechanism of Reaeration in Natural Streams," *J. Sanitary Eng. Div., A. S. C. E.* **82**, No. SA6, December, 1956, p. 115.

4-34. O'Connor, D. J., *et al.*, "Notes on Stream and Estuarine Analysis," Manhattan College Summer Institute in Water Pollution Control, New York, New York, May, 1971.

4-35. Owens, M., *et al.*, "Some Reaeration Studies in Streams," *Inter. J. Air and Water Pollution* **8**, 1964, pp. 469–486.

4-36. Raytheon, Inc., "RECIEV-II, Water Quantity and Quality Model," reprinted from the New England River Basins Modeling Project for the U.S. Environmental Protection Agency, Portsmouth, Rhode Island, March, 1975.

4-37. Shindala, A., Zitta, V. L., and Corey, M. W., "Water Quality Models for the Pascagoula River Basin-II. Tidal Estuaries," *Report No. EIRS-CE-73-2*, Engineering Industrial Research Station, Mississippi State University, Mississippi State, Mississippi, May, 1973.

4-38. Snodgrass, W. J. and O'Melia, C. R., "Predictive Model for Phosphorus in Lakes," *Environmental Science and Technology* **9**, No. 10, 1975, pp. 937–934.

4-39. Tapp, J. S., "Comparison of Eutrophication Models," paper presented at the U.S. Environmental Protection Agency Conference on Modeling and Simulation, Cincinnati, Ohio, April 20–22, 1976.

4-40. Texas Water Development Board, "Simulation of Water Quality in Streams and Canals DOSAG-I," Systems Engineering Division, Austin, Texas, September, 1970.

4-41. Texas Water Development Board, "Simulation of Water Quality in Streams and Canals," *Report No. 128*, Austin, Texas, May, 1971.

4-42. Thackson, E. L. and Krenkel, P. A., "Reaeration Prediction in Natural Streams," *J. Sanitary Eng. Div., A. S. C. E.* **95**, No. SA1, February, 1969, pp. 65–94.

4-43. Thomann, R. V., *Systems Analysis and Water Quality Management*, McGraw-Hill, New York, New York, 1972.

4-44. Tsivoglou, E. C., "Tracer Measurement of Stream Reaeration," Federal Pollution Control Administration Report, Washington, D.C., June, 1967.

4-45. Tsivoglou, E. C., "Direct Tracer Measurement of the Reaeration Capacity of Streams and Estuaries," *Water Pollution Control Research Series Report 16050*, U.S. Environmental Protection Agency, January, 1972.

4-46. Tsivoglou, E. C., *et al.*, "Tracer Measurement of Stream Reaeration. I. Laboratory Studies," *J. Water Pollution Control Federation* **37**, No. 10, October, 1965, pp. 1343–1362.

4-47. Tsivoglou, E. C., *et al.*, "Tracer Measurement of Stream Reaeration. II. Field Studies," *J. Water Pollution Control Federation* **40**, No. 2, Part 1, February, 1968, pp. 285–305.

4-48. Tsivoglou, E. C. and Neal, L. A., "Tracer Measurement of Stream Reaeration. III. Predicting the Reaeration Capacity of Inland Streams," *J. Water Pollution Control Federation* **48**, No. 12, December, 1976, pp. 2669–2689.

4-49. Tsivoglou, E. C. and Wallace, J. R., "Characterization of Stream Reaeration Capacity," *Ecological Research Series Report No. EPA-R3-72-012,* U.S. Environmental Protection Agency, October, 1972.

4-50. U.S. Geological Survey, "Water Resources Data for Kentucky Water Year 1975," *Water Data Report KY-75-2*, Louisville, Kentucky, 1976.

4-51. U.S. Environmental Protection Agency, "A Guide to the Selection of Cost-Effective Wastewater Treatment Systems," *Technical Report EPA-43019-75-002*, Washington, D.C., July, 1975.

4-52. U.S. Environmental Protection Agency and U.S. Department of Agriculture, "Control of Water Pollution from Cropland: Volume I—A Manual for Guideline Development," *Report Nos. EPA-600/2-75-026a* and *ARS-H-5-1*, Washington, D.C., November, 1975.

4-53. Vollenweider, R. A., "Input-Output Models," *Schweiz. Z. Hydrol.* **37**, 1975, pp. 53–84.

4-54. Vollenweider, R. A. and Dillon, P. J., "The Application of the Phosphorus Loading Concept to Eutrophication Research," paper prepared for the Associate Committee on Scientific Criteria for Environmental Quality, Burlington, Ontario, Canada, June, 1974.

4-55. Ware, R. W. and Foree, E. G., "Reaeration Characteristics of Low Level In-Channel Dams and Waterfalls in Kentucky Streams," *Technical Report UKY TR97-76-CE25*, University of Kentucky College of Engineering, Lexington, Kentucky, March, 1976.

4-56. Water Resources Engineers, "Ecologic Simulation for Aquatic Environments," U.S. Department of the Interior, Office of Water Resources Research, December, 1972.

4-57. Water Resources Engineers, "Computer Program Documentation for the Stream Quality Model QUAL-II," U.S. Environmental Protection Agency, Washington, D.C., October, 1973.

4-58. Water Resources Engineers, "Computer Program Documentation for the Stream Quality Model DOSAG-III," U.S. Environmental Protection Agency, Washington, D.C., October, 1973.

4-59. Water Resources Engineers, "Computer Program Documentation for the Reservoir Water Quality Model EPARES," U.S. Environmental Protection Agency, Washington, D.C., January, 1974.

4-60. Water Resources Engineers, "Computer Program Documentation for the Reservoir Ecologic Model EPAECO," U.S. Environmental Protection Agency, Washington, D.C., February, 1975.

APPENDIX 4-1 BIBLIOGRAPHY

American Public Works Association, "Water Pollution Aspects of Urban Runoff," *Report WP-20-15*, Federal Water Pollution Control Administration, Washington, D.C., 1969.

Anderson, D. R., *et al.*, "Water Quality Modeling of Deep Reservoirs," *Journal of the Water Pollution Control Federation*, Vol. 48, No. 1, January, 1976, pp. 134–146.

Banks, R. B., "Distributions of BOD and DO in Rivers and Lakes," *Journal of the Environmental Engineering Division, American Society of Civil Engineers*, Vol. 102, No. EE2, April, 1976, pp. 265–280.

Bansal, M. K., "Nitrification in Natural Streams," *Journal of the Water Pollution Control Federation*, Vol. 48, No. 10, October, 1976, pp. 2380–2393.

Bedient, P. B. and Characklis, W. G., "Prediction of Water Quality Inputs from Agricultural and Urban Drainage," paper presented at the National Conference on Environmental Engineering, American Society of Civil Engineers, Seattle, Washington, July 12–14, 1976.

Betson, R. D. and McMaster, W. M., "Nonpoint Source Mineral Water Quality Model," *Journal of the Water Pollution Control Federation*, Vol. 47, No. 10, October, 1975, pp. 2461–2473.

Bradford, W. L. and Maiero, D. J., "Simplified Process Models Applied to Reservoir Management—The Case of Tocks Island Lake," paper presented at the National Conference on Environmental Engineering, American Society of Civil Engineers, Seattle, Washington, July 12–14, 1976.

Brandstetter, A., Engel, R. L., and Cearlock, D. B., "A Mathematical Model for Optimum Design and Control of Metropolitan Wastewater Management Systems," *Water Resources Bulletin*, Vol. 9, No. 6, December, 1973.

Brandstetter, A., "Comparative Analysis of Urban Stormwater Models," paper presented at the short course on Applications of Stormwater Management Models, University of Massachusetts, Amherst, Massachusetts, August 19-23, 1974.

Brown, L. C., "Statistical Evaluation of Reaeration Prediction Equations," *Journal of the Environmental Engineering Division, American Society of Civil Engineers*, Vol. 100, No. EE5, October, 1974, pp. 1051–1068.

Busch, A. W., "A Five-Minute Solution for Stream Assimilation Capacity," *Journal of the Water Pollution Control Federation*, Vol. 44, No. 7, July, 1972, pp. 1453–1456.

Canale, R. P., "Model of Coliform Bacteria in Grand Traverse Bay," *Journal of the Water Pollution Control Federation*, Vol. 45, No. 11, November, 1973, pp. 2358–2371.

Canale, R. P., *Modeling Biochemical Processes in Aquatic Ecosystems*,. Ann Arbor Science, Ann Arbor, Michigan, 1976.

Cleasby, J. L., Stangl, E. W., and Rice, G. A., "Atmospheric Reaeration in a River Using Productivity Analysis," *Journal of the Environmental Engineering Division, American Society of Civil Engineers*, Vol. 101, No. EE5, October, 1975, pp. 729–739.

Curtis, D. C., "A Deterministic Urban Storm Water and Sediment Discharge Model," *Proceedings of the National Symposium on Urban Hydrology, Hydraulics, and Sediment, Control*, University of Kentucky, Lexington, Kentucky, July 26–29, 1976.

Deininger, R. A., *Models for Environmental Pollution Control*, Ann Arbor Science, Ann Arbor, Michigan, 1974.

Donigian, A. S. and Crawford, N. H., "Modeling Nonpoint Pollution from the Land Surface," *Report No. EPA-600/3-76-083*, U.S. Environmental Protection Agency, Washington, D.C., February, 1976.

Eagleson, P. S., "Unit Hydrograph Characteristics for Sewered Areas," *Journal of the Hydraulics Division, American Society of Civil Engineers*, Vol. 88, No. HY2, March, 1962.

Foree, E. G., "Determination of Permissible BOD Loading to Streams," paper presented at the National Conference on Environmental Engineering, American Society of Civil Engineers, Seattle, Washington, July 12–14, 1976.

Foster, G. R. and Meyer, L. D., "Mathematical Simulation of Upland Erosion Using Fundamental Erosion Mechanics," in *Present and Prospective Technology for Predicting Sediment Yields and Sources*, *Report No. ARS-S-40*, U.S. Department of Agriculture, Washington, D.C., June, 1975.

Frere, M. H., *et al.*, "ACTMO, An Agricultural Chemical Transport Model," *Report No. ARS-H-3*, U.S. Department of Agriculture, Washington, D.C., June, 1975.

Fuhs, G. W. and Allen, S. P., "Limiting Nutrient Study of the Delaware River at Montague in Relation to the Proposed Tocks Island Reservoir," *Environmental Health Report No. 3*, New York State Department of Health, Albany, New York, October, 1975.

Grigg, N. S., *et al.*, "Metropolitan Water Intelligence Systems," *Completion Report, Phase II, Water Resource Systems Program,* Department of Civil Engineering, Colorado State University, June, 1973.

Herricks, E. E. and Cairns, J., Jr., "Rehabilitation of Streams Receiving Acid Mine Drainage," *Bulletin No. 66*, Virginia Water Resources Research Center, Blacksburg, Virginia, 1975.

Huck, P. M. and Farquhar, G. J., "Water Quality Models Using the Box-Jenkins Method," *Journal of the Environmental Engineering Division, American Society of Civil Engineers*, Vol. 100, No. EE3, June, 1974, pp. 733–752.

Hyden, J. W., Chisman, J. A., and Macaulay, H. H., "Simulation of Water Quality Management Policies," *Journal of the Environmental Engineering Division, American Society of Civil Engineers*, Vol. 101, No. EE4, August, 1975, pp. 623–641.

Hydrocomp International, Inc., "Hydrocomp Simulation Programming Operators Manual, Second Edition," Palo Alto, California, July, 1969.

Imboden, D. M., "Phosphorus Model of Lake Eutrophication," *Limnology and Oceanography*, Vol. 19, No. 2, March, 1974, pp. 297–304.

Jain, S. C., "Longitudinal Dispersion Coefficients for Streams," *Journal of the Environmental Engineering Division, American Society of Civil Engineers*, Vol. 102, No. EE2, April, 1976, pp. 465–474.

Jennings, M. E. and Sauer, V. B., "Flow Routing Models for Stream System Studies," *Water Resources Bulletin, American Water Resources Association*, Vol. 8, No. 5, October, 1972, pp. 948–956.

Jones, J. R. and Bachmann, R. W., "Prediction of Phosphorus and Chlorophyll Levels in Lakes," *Journal of the Water Pollution Control Federation*, Vol. 48, No. 9, September, 1976, pp. 2176–2182.

Keshavan, K., Sornberger, G. C., and Hirsberg, C. I., "Oxygen Sag Curve with Thermal Overload," *Journal of the Environmental Engineering Division, American Society of Civil Engineers*, Vol. 99, No. EE5, October, 1973, pp. 569–575.

Kirshen, P. H. and Marks, D. H., "Screening Model for Storm Water Control," *Journal of the Environmental Engineering Division, American Society of Civil Engineers*, Vol. 100, No. EE4, August, 1974, pp. 807–820.

Klym, H., Koniger, W., and Vogel, G., "Urban Hydrological Processes, Computer Simulation," Dorsch Consultants, Munich, Toronto, 1972.

Kramer, G. R., "Predicting Reaeration Coefficients for Polluted Estuary," *Journal of the Environmental Engineering Division, American Society of Civil Engineers*, Vol. 100, No. EE1, February, 1974, pp. 77–92.

Krenkel, P. A., Thackston, E. L. and Parker, F. L., "Impoundment and Temperature Effect on Water Assimilation," *Journal of the Sanitary Engineering Division, American Society of Civil Engineers*, Vol. 95, No. SA1, February, 1969, pp. 37–64.

Li, W., "Effects of Dispersion of DO-Sag in Uniform Flow," *Journal of the Sanitary Engineering Division, American Society of Civil Engineers*, Vol. 98, No. SA1, February, 1972, pp. 169–182.

Li, W. and Kozlowski, M. E., "DO-Sag in Oscillating Flow," *Journal of the Environmental Engineering Division, American Society of Civil Engineers*, Vol. 100, No. EE4, August, 1974, pp. 837–854.

Litwin, Y. J. and Donigian, A. S., Jr., "Development of a Nonpoint Source Pollutant Loading Model," paper presented at the National Conference on Environmental Engineering, American Society of Civil Engineers, Seattle, Washington, July 12–14, 1976.

Loehr, R. C., "Characteristics and Comparative Magnitude of Non-point Sources," *Journal of the Water Pollution Control Federation*, Vol. 46, No. 8, August, 1974, pp. 1849–1872.

Lung, W. and Canale, R. P. "Phosphorus Model for Eutrophic Lakes: Application to White Lake," paper presented at the National Conference on Environmental Engineering, American Society of Civil Engineers, Seattle, Washington, July 12–14, 1976.

McPherson, M. B., "Urban Hydrological Modeling and Catchment Research in the U.S.A.," *Technical Memorandum No. IHP-1*, American Society of Civil Engineers, November, 1975.

McQuivey, R. S. and Keefer, T. N., "Simple Method for Predicting Dispersion in Streams," *Journal of the Environmental Engineering Division, American Society of Civil Engineers*, Vol. 100, No. EE4, August, 1974, pp. 997–1011.

Meta Systems, Inc., "Land Use—Water Quality Relationships" *Water Quality Management Guidance Report No. WPD-3-76-02*, U.S. Environmental Protection Agency, Water Planning Division, Washington, D.C., March, 1976.

Meyer, L. D., "Soil Erosion by Water on Upland Areas," in *River Mechanics* (H. W. Shen, Editor and Publisher), Fort Collins, Colorado, 1971.

Middlebrooks, E. J., *et al.*, *Modeling the Eutrophication Process*, Ann Arbor Science, Ann Arbor, Michigan, 1974.

Middlebrooks, E. J., *et al.*, *Biostimulation and Nutrient Assessment*, Ann Arbor Science, Ann Arbor, Michigan, 1976.

Nejedly, A., "The Effect of Longitudinal Dispersion of the Deoxygenation Rate in Polluted Streams," *Journal of the Water Pollution Control Federation*, Vol. 45, No. 7, July, 1973, pp. 1601–1605.

Newbold, J. D., and Liggett, J. A., "Oxygen Depletion Model for Cayuga Lake," *Journal of the Environmental Engineering Division, American Society of Civil Engineers*, Vol. 100, No. EE1, February, 1974, pp. 41–59.

Novotny, V. and Krenkel, P. A., "Simplified Mathematical Model of Temperature Changes in Rivers," *Journal of the Water Pollution Control Federation*, Vol. 45, No. 2, February, 1973, pp. 240–248.

Novotny, V., *et al.*, "Land Use Effect on Water Quality: An Overland Nonpoint Continuous Model," paper presented at the National Conference on Environmental Engineering, American Society of Civil Engineers, Seattle, Washington, July 12–14, 1976.

Ohio-Kentucky-Indiana Regional Council of Governments, "A Method for Assessing Rural Non-Point Sources and Its Application in Water Quality Management," *Interim Report V*, Cincinnati, Ohio, September, 1975.

Olufeagba, B. J. and Flake, R. H., "Shooting Models for Salinity and Coliform," *Journal of the Environmental Engineering Division, American Society of Civil Engineers*, Vol. 102, No. EE1, February, 1976, pp. 95–109.

Omernik, J. M.. "The Influence of Land Use on Stream Nutrient Levels," *Report No. EPA-60013-76-014,* U.S. Environmental Protection Agency, January, 1976.

Papadakis, C. N. and Preut, H. C., "University of Cincinnati Urban Runoff Model," *Journal of the Hydraulics Division, American Society of Civil Engineers*, Vol. 98, No. HY10, October, 1972.

Penumalli, B. R., Flake, R. H. and Fruh, E. G., "Large-Scale Estuarine Water Quality Matrix Model," *Journal of the Environmental Engineering Division, American Society of Civil Engineers*, Vol. 102, No. EE1, February, 1976, pp. 191–209.

Pingry, D. E. and Whinston, A. B., "Application of Multigoal Water Quality Planning Model," *Journal of the Environmental Engineering Division, American Society of Civil Engineers*, Vol. 100, No. EE1, February, 1974, pp. 61–75.

Pisano, M. A., "Nonpoint Sources of Pollution: A Federal Perspective," *Journal of the Environmental Division, American Society of Civil Engineers*, Vol. 102, No. EE3, June, 1976, pp. 555–565.

Plass, W. T., *et al.*, "Water Quality Models for a Contour Mined Watershed," *Third Symposium Surface Mining and Reclamation Reprints*, National Coal Association Bituminous Coal Research, Inc., Vol. 1, Louisville, Kentucky, 1975.

Quinlan, A. V. and Paynter, H. M., "Some Simple Nonlinear Dynamic Models of Interacting Element Cycles in Aquatic Ecosystems," *Journal of Dynamic Sys-*

tems, Measurement, and Control, Transactions of the American Society of Mechanical Engineers, March, 1976, pp. 6–19.

Quirk, T. P. and Eder, L. J., "Evaluation of Alternative Solutions for Achievement of River Standards," *Journal of the Water Pollution Control Federation*, Vol. 42, No. 2, Part 1, February, 1970, pp. 272–290.

Roesner, L. A., *et al.*, "A Model for Evaluating Runoff Quality in Metropolitan Master Planning," *Technical Memorandum No. 23*, Urban Water Resources Research Program, American Society of Civil Engineers, April, 1974.

Rood, O. E., Jr. and Holley, E. R., "Critical Oxygen Deficit for Bank Outfall," *Journal of the Environmental Engineering Division, American Society of Civil Engineers*, Vol. 100, No. EE3, June, 1974, pp. 661–678.

Rutherford, J. C. and O'Sullivan, M. J., "Simulation of Water Quality in Tarawera River," *Journal of the Environmental Engineering Division, American Society of Civil Engineers*, Vol. 100, No. EE2, April, 1974, pp. 369–390.

Sartor, J. D., *et al.*, "Water Pollution Aspects of Street Surface Contaminants," *Journal of the Water Pollution Control Federation*, Vol. 46, No. 3, March, 1974, pp. 458–467.

Sauer, V. B., "The Unit Hydrograph—A Technique for Routing Reservoir Releases," *U.S. Geological Survey Prof. Paper 750-B*, 1970, pp. B259–B264.

Schofield, W. R. and Krutchkoff, R. G., "Stochastic Model of BOD and OD in Segmented Estuary," *Journal of the Sanitary Engineering Division, American Society of Civil Engineers*, Vol. 98, No. SA5, October, 1972, pp. 745–762.

Schofield, W. R. and Krutchkoff, R. G., "Stochastic Model of Dynamic Eutrophic Estuary," *Journal of the Environmental Engineering Division, American Society of Civil Engineers*, Vol. 100, No. EE3, June, 1974, pp. 613–628.

Schofield, W. R. and Krutchkoff, R. G., "Deterministic Model of Dynamic Eutrophic Estuary," *Journal of the Environmental Engineering Division, American Society of Civil Engineers*, Vol. 100, No. EE4, August, 1974, pp. 979–996.

Schroepfer, G. J., Robins, M. L., and Susag, R. H., "A Reappraisal of Deoxygenation Rates of Raw Sewage, Effluents, and Receiving Waters," *Journal of the Water Pollution Control Federation*, Vol. 32, No. 11, November, 1960, pp. 1212–1231.

Shannon, E. E. and Brezonik, P. L., "Eutrophication Analysis: A Multivariate Approach," *Journal of the Sanitary Engineering Division, American Society of Civil Engineers*, Vol. 98, No. SA1, February, 1972, pp. 37–57.

Shastry, J. S., Fan, L. T., and Erickson, L. E., "Nonlinear Parameter Estimation in Water Quality Modeling," *Journal of the Environmental Engineering Division, American Society of Civil Engineers*, Vol. 99, No. EE3, June, 1973, pp. 315–331.

Shea, T. G., Grayman, W. M., and Drucker, J. M., "Estimation of NPS Generation in Areawide Waste Treatment Management Planning," paper presented at the National Conference on Environmental Engineering, American Society of Civil Engineers, Seattle, Washington, July 12–14, 1976.

Shubinski, R. P. and Roessner, L. A., "Linked Process Routing Models," paper presented at the Spring Meeting of the American Geophysical Union, Washington, D.C., April, 1973.

Singh, K. D. and Stall, J. B., "Hydrology of 7-Day, 10-year Low Flows," *Journal of the Hydraulics Division, American Society of Civil Engineers*, Vol. 100, No. HY12, December, 1974, pp. 1753–1771.

Slawson, G. C., Jr. and Everett, L. G., "Segmented Population Model of Primary Productivity," *Journal of the Environmental Engineering Division, American Society of Civil Engineers*, Vol. 102, No. EE1, February, 1976, pp. 127–138.

Smith, S. A. and Bella, D. A., "Dissolved Oxygen and Temperature in a Stratified Lake," *Journal of the Water Pollution Control Federation*, Vol. 45, No. 1, January, 1973, pp. 119–133.

Sornberger, G. C. and Keshavan, K., "Simulation of Dissolved Oxygen Profile," *Journal of the Environmental Engineering Division, American Society of Civil Engineers*, Vol. 99, No. EE4, August, 1973, pp. 479–488.

Stall, J. B. and Terstriep, M. L., "Storm Sewer Design—An Evaluation of the RRL Method," *Report No. EPA-R2-72-068*, U.S. Environmental Protection Agency, Washington, D.C., October, 1972.

Systems Control, Inc., "Evaluation of Water Quality Models: A Management Guide for Planners," *Report No. EPA-600/5-76-004*, U.S. Environmental Protection Agency, Washington, D. C., July, 1976.

Thackston, E. L., *et al.*, "Effect of Eliminating Point Source Discharges from a Eutrophic Reservoir," paper presented at the National Conference on Environmental Engineering, American Society of Civil Engineers, Seattle, Washington, July 12–14, 1976.

Tholin, A. L. and Kiefer, C. J., "The Hydrology of Urban Runoff," *Journal of the Sanitary Engineering Division, American Society of Civil Engineers*, Vol. 85, No. SA2, March, 1959.

Thomann, R. V., DiToro, D. M., and O'Connor, D. J., "Preliminary Model of Potomac Estuary Phytoplankton," *Journal of the Sanitary Engineering Division, American Society of Civil Engineers*, Vol. 100, No. EE3, June, 1974, pp. 699–715.

U.S. Army Corps of Engineers, "Generalized Computer Program: Water Quality for River-Reservoir Systems," *Hydrologic Engineering Center Report No. 401-100, 401-100A*, Davis, California, September, 1975.

Wallace, D. A. and Dague, R. R., "Modeling of Land Runoff Effects on Dissolved Oxygen" *Journal of the Water Pollution Control Federation*, Vol. 45, No. 8, August, 1973, pp. 1795–1809.

Wanielista, M. P. and Bauer, C. S., "Centralization of Waste Treatment Facilities," *Journal of the Water Pollution Control Federation*, Vol. 44, No. 12, December, 1972, pp. 2229–2238.

Williams, B. J. and Hinwood, J. B., "Two-Dimensional Mathematical Water Quality Model," *Journal of the Environmental Engineering Division, American Society of Civil Engineers*, Vol. 102, No. EE1, February, 1976, pp. 149–163.

Willis, R., Anderson, D. R., and Dracup, J. A., "Steady-State Water Quality Modeling in Streams," *Journal of the Environmental Engineering Division, American Society of Civil Engineers*, Vol. 101, No. EE2, April, 1975, pp. 245-258.

Wischmeier, W. H., *et al.*, "A Soil Erodability Nomograph for Farmland and Construction Sites," *Journal of Soil and Water Conservation*, Vol. 26, No. 5, 1971.

Wischmeier, W. H., "Use and Misuse of the Universal Soil Loss Equation," *Journal of Soil and Water Conservation*, Vol. 31, No. 1, 1976.

Woolhiser, D. A., "The Watershed Approach to Understanding our Environment," *Journal of Environmental Quality*, Vol. 4, No. 1, 1975.

Yen, B. C., "Methodologies for Flow Predictions in Urban Storm Drainage Systems," *Report No. UILC-WRC-0072*, University of Illinois, Water Resources Center, September, 1973.

5

Management-institutional aspects of water quality management planning

Eldridge Lovelace and
T. Scott Cantine

Harland Bartholomew Associates
St. Louis, Missouri

5-1. INTRODUCTION

Management of water quality from a national perspective is a recent phenomenon. Beginning with the ineffective Water Pollution Control Act of 1948, the federal government has gradually moved toward more systematic and effective management of the quality of the nation's water. The Water Pollution Control Amendments of 1972, through the "208" Program, confront one of the basic problems which has limited effective water quality management in the past—the limited political and geographical jurisdiction of the typical wastewater management agency. The areawide approach supplements the federal and state role by providing for an areawide administration to exercise water quality management over an entire areawide problem—or at least as much of the problem as is reasonable (Ref. 5-1).

An effective areawide waste treatment management system would result only when management, legal, and financial elements of the system are developed concurrently with facility and regulatory programs. Many public planning programs in the United States are ineffective because the physical planning portions are put together first as a separate entity; and then, later on, the management, legal, and financial measures required to carry them out are added as separate packages (or ignored completely). Consequently, proposals of the physical plan may be found to be impracticable when implementation is attempted and the process becomes one of frustration and futility. This pitfall may be avoided, possibly, by integrating the management, institutional and financial aspects with the other phases of a planning program. While not entirely novel, except for its mandatory features, the requirement that this be done is progressive. While it may not insure success, it should help in avoiding complete failure, and experience should improve the performance.

The management system, with its many parts, is an important key to the success of the overall Wastewater Management Program. Development of the management plan is the most politically sensitive and difficult element for most regional areas, with their many independent jurisdictions, to agree upon.

5-1.1 Management Aspects of the Statutes

Section 208(c) (2) of the Federal Water Pollution Control Act Amendments of 1972 requires that a regulatory program be established to implement waste treatment management requirements, to direct the location of land uses that produce wastewater discharges, and to assure that industrial or commercial wastes discharged into any treatment works meet applicable pre-treatment requirements.

Agencies that would build, operate and maintain facilities and undertake the regulatory measures are to be identified. Finally, when the plan is completed, the governor of the state is to designate one or more waste treatment management agencies which will be charged with carrying out the plan. The Administrator, subsequently, is not to approve the designation of these agencies unless he finds that the management system has adequate authority:

"(A) to carry out appropriate portions of an areawide waste treatment management plan developed under subsection (b) of this section;

"(B) to manage effectively waste treatment works and related facilities serving such area in conformance with any plan required by subsection (b) of this section;

"(C) directly or by contract, to design and construct new works, and to operate and maintain new and existing works as required by any plan developed pursuant to subsection (b) of this section;

"(D) to accept and utilize grants, or other funds from any source, for waste treatment management purposes;

"(E) to raise revenues, including the assessment of waste treatment charges;

"(F) to incur short- and long-term indebtedness;

"(G) to assure in implementation of an areawide waste treatment management plan that each participating community pay its proportionate share of treatment costs;

"(H) to refuse to receive any wastes from any municipality or subdivision thereof, which does not comply with any provisions of an approved plan under this section applicable to such area; and

"(I) to accept for treatment industrial wastes."

5-1.2 Management Guidelines

The Environmental Protection Agency has prepared a series of guidelines (Ref. 5-2) for operations under the 208 programs. These include sections on Management Responsibilities, Institutional Arrangements, and Financing Arrangements.

The guidelines state that in some areas there will be management agencies that will be able to undertake all of the activities required under the Act as cited above. However, in most wastewater management areas, new management and institutional arrangements will be needed.

Highlights of the guidelines include:

1. That the institutional arrangements assure that the overall program of waste management be coordinated; the plan and program implemented; and its performance monitored. The planning program is to be continuous and not just a "one-shot" program, so that the continuing planning responsibility must be placed within the proposed institutional framework. A major responsibility of 208 management will be the obtaining and budgeting of the financial resources for all of the many types of implementation required under the plan and the program. The wastewater management planning agency is to have a regulatory program which includes identification of pollution sources, assignment of regulatory responsibilities, and identification of types of regulation–land use permits, licenses, standards, etc., that will be utilized.
2. That the responsibilities may be divided among a number of agencies. Where special sewage districts are established, for example, some non-point source regulation might need to be assigned to other state agencies such as Soil Conservation Districts or Fish and Game Departments, with land use regulations assigned to cities or counties. Such divisions of responsibility could be effected by interagency agreements. Controls and limitations of the powers of local government are usually established by state legislation or by city or county charters. A management program may require changes in these.

Objectives of the program include providing for waste treatment management on an areawide basis to the extent practical; regulation of the location and construction of any facilities which result in any type of wastewater discharge; and assurance that the industrial or commercial wastes meet pre-treatment standards.

Regulatory controls include those over land use, the issuance of permits and licenses, establishment of water quality standards, and establishment of fiscal policies such as taxation and user charges that may be required to complement the regulatory requirements. Enactment of the federal legislation was in response to a public demand that the waters of the United States be cleaned up. Unless public support for these goals is applied with equal enthusiasm to measures required for their achievement, it will be difficult to obtain substantial accomplishment.

5-2. THE MANAGEMENT PROBLEM

For the most part, domestic sewage, some kitchen wastes, and industrial wastes in western society are carried off by water. These wastes are diluted to a considerable extent, put in a pipe, and carried to one or more places where they may be treated or dumped untreated into a receiving stream, lake, or ocean. While it may be debatable as to whether or not this is the most efficient possible way to get rid of these wastes, at this point in time we are pretty well removed from considering any other greatly different option. Our multitude of urban complexes have been developed in reliance upon this type of disposition of these wastes. This involves a rather complex system, as the wastes are first diluted and then taken to a plant where they are removed.

Rainfall comes to the land, part disappears through evaporation, part is transpired through plant life, and the remainder runs off. Part of the runoff water soaks into the ground and goes into the groundwater, part flows directly into the streams, reservoirs, lakes, or ocean, and the remainder may be collected as "urban runoff" in a storm drainage or storm sewer system. Sometimes this is combined with the sanitary sewage system, but more frequently, except in the larger, older metropolitan areas, it is a separate system, with storm water in one system of pipes and sanitary sewage in another. As the water runs off the urban area, it picks up dust and dirt from roofs, from pavement, and various other materials, and frequently reaches the receiving waters in a polluted condition.

As the water runs off rural areas, it picks up soil from erosion, fertilizers, pesticides, herbicides, and animal waste.

5-3. THE 208 MANAGEMENT SYSTEM

5-3.1 The Water Use Cycle and Water Quality Management

As is clear from the statutes and guidelines, the areawide management system should be comprehensive and provide maximum control, in order to achieve water quality goals. To effectively plan the management system, some understanding of the complex water use and treatment cycle, and the points in the cycle at which the management system can direct activities which affect water quality is needed.

5-3.2 Water Use and Treatment Cycle

The water use and treatment cycle is complex, and most waste treatment areas include many variations (see Figure 5-1). Basically, the cycle consists of four functional areas: 1) the natural system from rainfall to recharge; 2) water supply, including quality treatment and transmission; 3) distribution and use; and 4) disposal. These four functional areas encompass many individual elements. The chart indicates where point and non-point source discharges take place in the cycle.

5-3.3 Management Application to Improve Water Quality

Areawide management may be applied at 18 major points in the water use and treatment cycle. Proper regulation, financing, and construction of physical improvements or decision making as to options at each of these points affects water quality. These points, in the process, may be grouped as follows:

1. Stormwater Runoff (Non-Point Source). Rainfall or snowmelt runoff over land is an aspect of the cycle that is difficult to manage (Step 4). Possible management steps involve providing the opportunity for water to soak into the ground or reduce transport of sediment or other pollutants through crop management programs, retention reservoirs, design considerations optimizing recharge, rural drainage programs, and other similar approaches.

2. Collection of System Flows (Point Source). At the time of collection of sanitary, combined or storm water, management action can influence in a number of diverse ways the quality of water that emerges (Steps 19, 21 and 24). Decisions as to combined or separate systems, acceptance of wastewater (from wastes generated through use of foundation drains to required quality of industrial wastes) are involved. Available also is the option of diverting and treating the "first flush" of storm water.

3. Treatment of System Flows (Point Source). An important management control point exists at the time of treatment of wastes (Steps 22 and 25). A key decision here involves the treatment of wastewater from combined sewers; financing alternatives, selection of regionalization options, and establishment of improvement priorities.

4. Discharge of System Flows (Point Source). Management can make decisions relative to discharge standards, monitoring programs and subplan restructuring, and stream water quality (Steps 27, 28, 29, 30 and 31). Clearly, most of these actions involve state and federal management; however, areawide management can play a role also.

5. Treatment of Individual Wastewater Flows (Point Source). Treatment of these individual point sources involves management of treatment options such as septic tanks, financing improvements, treatment standards, etc. (Steps 42, 43,

44). Although treatment of non-profit runoff from cropland is not really managed at present, it is an area of concern.

6. Discharge of Individual Wastewater Flows (Point Source). Discharge from these often scattered and diverse sources is also managed through discharge standards and permits, monitoring, etc. (Steps 46, 47, 48 and 49). Again, these are normally the prerogative of federal and state management, but also involve areawide management.

Five of the six groupings above are point source categories. They are the less complex elements of the cycle and are easier to control. However, the first group, the non-point sources, although much more complex, becomes a more important part of the cycle as pollution from point sources is reduced.

For an areawide management agency to function effectively, it must be able to manage each of the groups of elements effectively and in a coordinated manner. This affects, and is affected by, the type of institutional arrangements, the legal authority, and the financing and regulatory options the management system is able to utilize.

5-4. OVERVIEW OF ISSUES

Essentially, two different systems are involved. There is a *closed system*, in which the water is put into a pipe, piped to the user, used (and polluted), put back in another pipe, taken to a central place, treated, and discharged. (The closed system or systems are generally referred to as the "point sources"). The second is the *open system* (now referred to as the "non-point system"), in which the rain comes down on the urban or rural area and picks up a number of man-induced pollutants. The closed system is simple in concept and susceptible to straightforward measures. The open system is made up of tens of thousands of parts and is not susceptible to simple management measures.

The public systems in the usual urban area take the water that has been used and send it through a sanitary sewer collection system, or a combined sanitary and storm sewer system, to a treatment facility. The treatment facility then discharges the treated wastewater to the surface water, groundwater, or to land through irrigation. It is then evaporated or reused. Water picked up by the storm runoff collection system of an urban area is usually not treated and is discharged into surface water or groundwater. The storm runoff water could be treated; the problem is that there may be so much of it. Treatment of either the sanitary sewage or the combined runoff may require overflows during periods of heavy rain, when it is necessary to bypass great quantities of untreated water around a treatment plant, and then to discharge it directly.

There are incredible complexities and difficulties in this simple process. The sanitary sewers frequently leak and pick up groundwater. Roof downspouts and

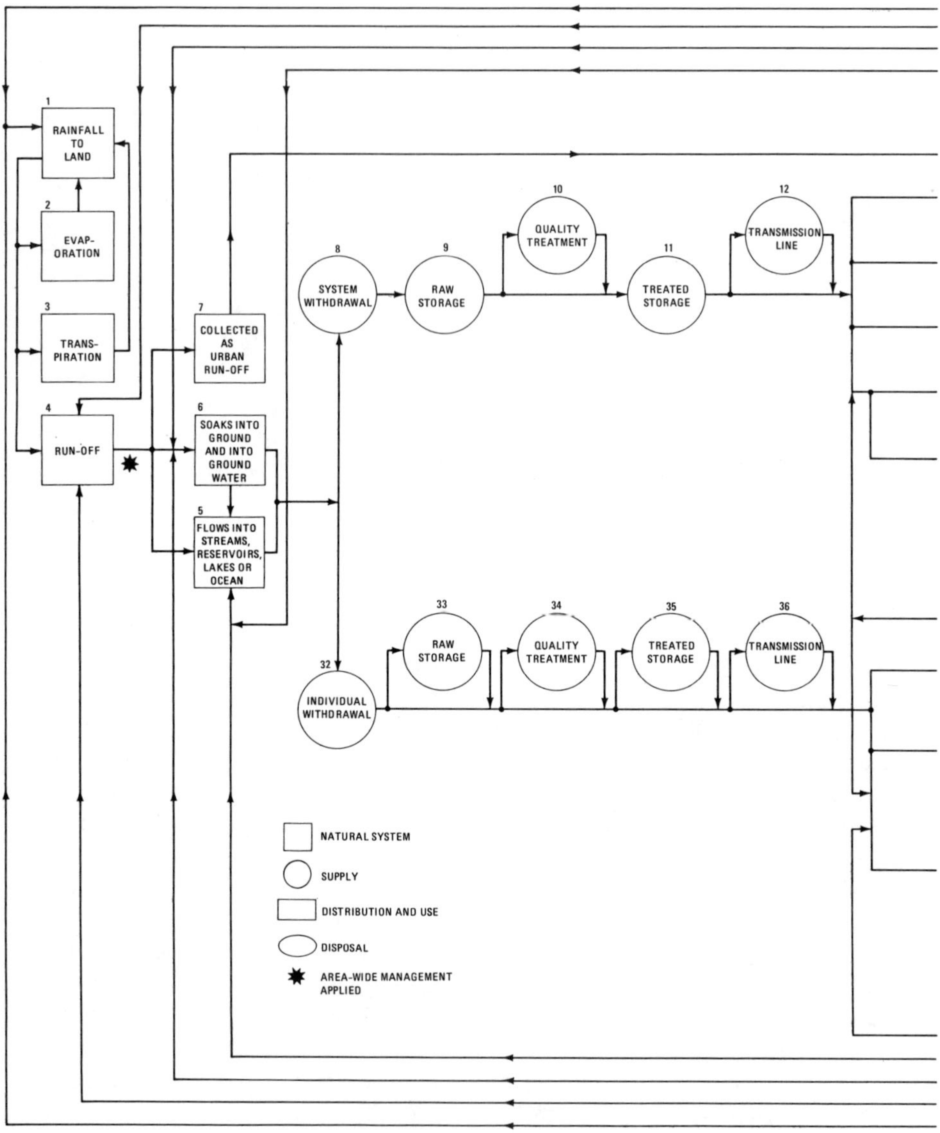

Fig. 5-1. Diagrammatic water use and treatment cycles.

foundation drains may be connected to them illegally. Collection systems may have been poorly designed. Topography may require pumping. Every system is different. Most have had a gradual evolution, and some are quite old.

Individual treatment facilities may be utilized, particularly in rural or in sub-

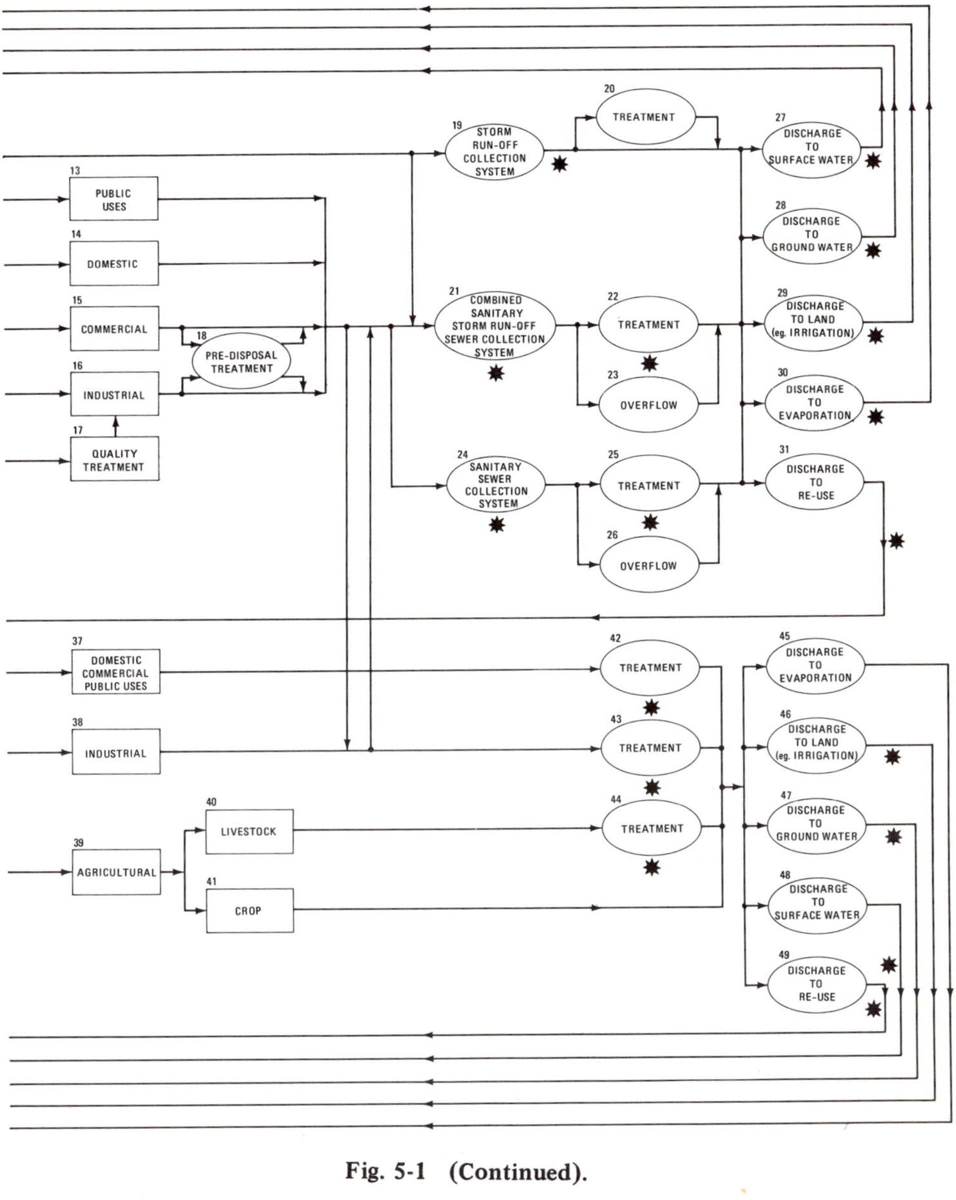

Fig. 5-1 (Continued).

urban areas, for the treatment of domestic, commercial, or industrial wastes. For some agricultural uses such as feedlots, special treatment may be needed. However, irrespective of whether the water is treated or untreated, or comes from a public water system or an individual water system, the general discharges are to evaporation, land through irrigation, groundwater, surface water, or for reuse purposes.

Complex and unrelated public actions are being applied to the water use and treatment cycles at various points. These are not a planned attack on a complex problem, but a legacy of many decades of "band-aid" measures. The current "system," as it is generally found, lacks coordination and adherence to a specific set of well defined objectives. It is precisely to bring some order out of this enormous clutter of unrelated independent public actions that the Regional Wastewater Management Program was developed.

Section 208(c) (2) sets forth specific requirements for the management system. Clearly, few existing wastewater management agencies are areawide in scope, or have a combination of municipalities, agencies, and special districts involved in managing point sources, each having differing authority, differing financial investments, and generally a local perspective with little or no coordination. These organizations have evolved over many years and are extremely jealous and protective of their authority. Very few areas do any significant management of non-point sources.

The three key management issues are: 1) who is to control wastewater management?; 2) who is to pay for wastewater management?; and 3) who and what will be regulated?

5.5. INSTITUTIONAL–LEGAL CONSIDERATIONS

5-5.1 Relation to System of Local Government

Political scientists agree that the typical system of local government in our major urban areas is obsolete. The system was designed to function in an agrarian society 150 to 200 years ago. It is not well suited to the demands of an urban society in the last part of the 20th century. The typical metropolitan area is a patchwork of cities and counties bearing very little relationship to geographic features or to public needs. While only two states, Virginia and Hawaii, do not have overlapping cities and counties, the other 48 states do. This duplication is an unnecessary complication. The illogical, geographic fragmentation of urban governments makes it extremely difficult to deal with any program on a unified basis, as is necessary if water quality is to be improved. Quite a number of our metropolitan areas are in more than one state, which further complicates the problem.

In addition to the geographic fragmentation, many of our local governments are fragmented functionally as well. The fragmented system of cities and counties has in many instances been "locked in" by legislation passed making it very difficult for municipalities to merge or for annexation to take place.

To solve regional problems, it has been our recent practice to create "metropolitan districts." A separate governmental entity or "district" is created and part or all of a local governmental function is transferred to this district. Sewage districts have been created to provide wastewater collection and disposal and to overcome this fragmentation of the local government. At the present time, there are approximately 50 large metropolitan sewer districts* in the United States, with a total population of approximately 60,000,000 persons.

Sewers are not the only function that have gone to special districts. We have special districts for parks, for junior colleges, and occasionally such as in Minneapolis and St. Paul, for a group of related functions all in the same district. The St.Louis Metropolitan Area, for example, has six of these special regional districts at the present time; to solve all regional problems in the same way would require the establishment of a total of 19 such districts.

A difficulty with the special district is that it usually only has one function. It is charged with carrying out this single function and does not relate its activities to anything else.

The geographic and functional fragmentation of our typical metropolitan area is both expensive and inefficient. The water quality management system is to utilize this fragmentation and to make the necessary changes required to bring about an areawide program. This is extremely difficult to do.

In most instances, the wastewater quality management programs have been turned over to the regional planning agencies or "councils of government." These are agencies created by local cities and counties, under the urging of the federal government, in order to deal with regional problems. None of these agencies has much more than advisory power, except in unusual cases, such as that of Minneapolis-St. Paul. In a few areas, such as Jacksonville, Florida, Nashville, Tennessee, or Baton Rouge, Louisiana, there has been a complete consolidation of local government, which greatly simplifies wastewater management.

In the usual metropolitan area, the only *regional* planning agency is a voluntary "council of government" (COG) which has been created in order to qualify the region for federal grant money and which is not really supported either by the people of the metropolitan area or by the individual local governments. If federal financial support were withdrawn, most of these agencies would dissappear.

5-5.2 Land Use Relationships

The character and intensity of land use affect both the open and the closed system (see Chapter 2). In the urban area, the closed system is customarily designed

*Population of 500,000 or more.

in relation to the existing and the anticipated land use pattern, with the pipes being of the dimensions, placed at the depth, and with the pressure as required to serve the complex of land uses and their day and nighttime occupancy. This is an established, recognized engineering system that has been developed through experience, and is reasonably efficient and straightforward.

However, in the urban area, the open system that takes the storm drainage away is usually not nearly as efficient. Frequently, storm drainage is merely directed toward the closest channel and turned loose. The paving of parking areas and streets and the impervious surface of the roofs of buildings increase the speed with which the water drains off. This places a heavy burden on both the storm sewers and upon the open channels. Where storms are of unusual severity, flooding results. The intensity of development determines the percentage of the land that is placed under impervious surface and this, in turn, together with the type of soil, slope, and similar factors, determines the volume of water that must be handled and the amount of pullutants that go into the streams.

In the agricultural areas, a rather similar problem occurs in that the methods used by the farmer in plowing his land, utilizing his land, and in spreading fertilizers, herbicides, and pesticides (including the time of the year that he does all of these things), affect the quantity of storm drainage and its pollutant load. Farming practices may be very carefully done in order to minimize soil erosion. On the other hand, careless farm practices may result in excessive erosion and this, in turn, in highly turbid and muddy water flowing through the streams and rivers.

Because of the intimate relationship between water use, treatment cycle, and land use pattern, one postulate is that in order to obtain clear water it is necessary to control land use, both urban and rural.

Land use controls are essential components of a water quality management system. These are imposed by the individual units of local government. This makes it difficult for an areawide wastewater program to work. Land use control measures watered down enough to receive the support of 25 to 250 individual units of government are not likely to be truly effective.

5-6. MEASURES REQUIRED TO PRODUCE CLEAN WATER

To produce clean water in the closed system, the basic problem is that of any public utility. To each home, business, industry, and institution in the community, we need to deliver an adequate quantity of potable water, have the customers use the water, dirty it up, return it to another set of pipes, carry it to a place or places, clean it up, and return it to the river, lake, or ocean. The techniques for doing this are known. The only element required is an adequate amount of money to build, operate, and maintain it.

Even so, there are problems with the closed system. One stems from the general practice of separating water supply from sewage collection and disposal. It would seem most logical to have both of these systems operated by the same management; the user to pay one bill for his water—for its coming and for its going. Arrangements this simple and this sensible, however, are seldom found in our metropolitan areas. Water is essential to life. The ability to "cut off the water" if a bill is not paid makes financing easier. Some sewer districts, such as that of Hampton Roads, have this power, although they exercise it indirectly through the water supply agency.

The second problem with the closed system comes in irregularities and inequities in the system. Part of a metropolitan area will be doing a good job at collection and treatment of its wastes; another part will not. One watershed will bc in satisfactory condition—another will not. People in an area with a good system dislike helping an area with a bad system pay for its needed improvements.

While there are problems in the closed (point) system, we are used to dealing with these; the techniques are known. The biggest need is financing and an equitable financing system.

In contrast, the open system (non-point) is extremely complex. We don't know as much about it as we should, and doing anything with it may require consistent applications of controversial and unpopular regulations. Some studies, for example, show that running stormwater over planted areas such as forests or lawns takes pollutants out of the water. Some improvement in water quality may result from retention basins. Controls over *urban* fertilizer use might be appropriate. Erosion control ordinances for urban areas have been effective. There are some implications to these open system problems, an important one being that a low density scattered development pattern (urban sprawl) may produce less non-point pollution.

Detailed land use controls placed over *new* land uses (retention basins, porous pavements, spreading of water from downspouts over lawn areas) will greatly reduce non-point pollution loadings from *new* development. Yet, the growth of our larger metropolitan areas is slowing down and some are losing population. What will we do about non-point pollution from the existing urban areas which will comprise 75 to 90 percent of the "year 2000" urban area?

5-7. REQUIREMENTS OF THE MANAGEMENT SYSTEM AND GUIDELINES FOR DEVELOPMENT

The management system needs to have authority to implement the plan, to manage the system, to design, construct, operate and maintain treatment works, to accept grants, raise revenue wastes, and to accept industrial wastes. The guidelines identify three basic options for the management system:

1. A single planning and management agency;
2. A single planning agency and a single management agency;
3. A single planning agency and plural management agencies.

Possible institutional arrangements include utilization of existing units of local government, modification of these by establishment of inter-agency agreements, establishment of a complete new management agency, and finally, a combination of several of these approaches.

In addition, the guidelines have been supplemented by a management agency's handbook (Ref. 5-3). The handbook cites and discusses a relationship of planning to management; outlines the requirements the management agencies must meet; lists questions that are to be answered when a management structure is designed; establishes criteria for effective water quality management; describes methods of making changes in governmental institutions charged with wastewater management; and then gives four examples of the type of management charges that would be required for four typical wastewater treatment management areas.

Clearly, one of the important aspects of the management system planning is that careful scrutiny be given to determine the authority of existing and potential agencies and their ability to meet the requirements.

Over the years, a number of innovative management agencies have been created in metropolitan areas to deal with water quality and/or other similar areawide problems. Probably the most widely known of these is the Minneapolis-St. Paul Metropolitan Council and its Metropolitan Waste Control Commission. Created in 1969, the Metropolitan Waste Control System is the operation area of the Council. The Council has wide authority in planning and management, and has been used as an example for those metropolitan areas contemplating a centralized wastewater management system. While the Metropolitan Council structure does not have the authority to implement a plan, it does not qualify as a 208 Planning Agency (Ref. 5-4). However, it does have the basic structure, experience in formation, and some seven years' experience in operation to provide valuable input to management planning programs. Other metropolitan areas such as Seattle, Toronto, Houston–Galveston, and Dade County, Florida, have developed wastewater management systems which provide insights to the needed institutional arrangements system.

5-8. FINANCIAL CONSIDERATIONS

5-8.1 The Cost of Water Quality

With unlimited money, most water quality problems can be solved. Yet, the financing of the programs and improvements may be the most serious impediment to reaching water quality goals.

A study of eight midwestern states indicated that local taxes ranged from 9.8 percent of personal income in Ohio to 12.3 percent in Michigan. Dollar-per-capita local taxes ranged from $713 in Kentucky to $1,030 in Michigan (Ref. 5-5).

Sewage expenses per capita per year ranged from $10.25 in Missouri to $30.09 in Michigan. For the 13 largest Standard Metropolitan Statistical Areas (SMSA's), sewage expenses per capita per year ranged from $6.23 for Miami to $9.81 in St. Louis, to $44.21 in the District of Columbia (Ref. 5-6).

A study of eight smaller midwestern SMSA's (Des Moines, Omaha, Peoria, Madison, Appleton-Oshkosh, Duluth-Superior, Evansville, and Ft. Wayne) indicated an average total sewage expenditure of 0.43 percent of personal income, of which 0.30 percent was for capital outlay and the remainder–0.13 percent–was for operation and maintenance (Ref. 5-7).

There are limits to the amount we can spend for wastewater treatment. The limits may vary from one area of the U.S. to another; however, they are a product of the individual's ability to pay and his perception of need. The amount we pay for improved water quality in the form of increased prices for consumer goods is somewhat hidden. The amount we pay in taxes or user charges is not. No one knows exactly where the limit is; however, it is probably in the vicinity of one-half of one percent of family income per year. Thus, correction of old, combined sewers in an inner city low income area would probably never be possible without other assistance. In developing suburban areas, modest improvement programs might not seriously affect resident spending habits.

It should be kept in mind, also, that wastewater treatment is only one of several pressing urban problems and probably not the most important, at that, with crime and education, for example, being given a higher priority in the public mind.

5-8.2 Financial Concerns in the Management Program

Finances are one of the most important parts of the management system. Certainly, a unified system that can draw on all of the resources of the urban area will be better than a fragmented system. A management system that provides the maximum options for financing and the maximum flexibility will be best. Provisions for assessing costs by watersheds will be of help also.

A large and efficient system will be able to establish a record of prudent management and be able to finance bond issues at a lower interest rate. This could well have the greatest single influence on the cost of a needed program. This results from the characteristics of the management system to be established and the abilities of those that manage.

Without an ability to enforce collection of user fees, revenue bonds may be difficult to sell. It is possible to issue general obligation bonds (which can sell at a lower interest rate) and retire these from user fees. Institutional arrangements must be approved by recognized financial counsel; the agency must be in sound financial condition, and the income ample to "cover" the bond amortization before the bonds can be sold at reasonable interest rates.

Major financial difficulties in the wastewater program are found in three areas: 1) financial interests of existing institutions; 2) methods of raising the required monies; and 3) methods of allocating costs.

1. *Financial Interests.* Most wastewater management areas, and particularly the larger metropolitan areas, have a large number of agencies with vested interest in physical improvements. In the Des Moines, Iowa 208 Area, as of January 1976, the value of gravity sewers, lift stations, force mains, and treatment facilities (not including federal or state grants) totaled over 40 million dollars, divided among 17 different municipalities. At the same time, municipalities and sanitary districts have 12 million dollars in debt outstanding, with approximately five million of these in general obligation bonds. Thus, almost all the jurisdictions have an ownership interest in physical facilities, most of which may be a part of the wastewater management plan (Ref. 5-8). How will they be compensated for this investment?

2. *Raising Money.* Methods of raising money are fairly standard, although they tend to vary slightly, depending on the particular state legislation. Expenditures for capital projects will be from state and federal grants, with the local part from issuance of bonds, although the option of pay-as-you-go does exist for some communities. Bonds are normally paid back by user charges levied against the benefited portion of the system. Operation and maintenance on a year-to-year basis is covered by user fees or by advalorem taxes or both.

The management system itself will have to be a financially self-sustaining operation, and thus the system will have to have the power to raise revenues for administration and the continuing planning process, unless the system is to be at the mercy of federal or state funding. Decisions will have to be made as to whether this is in the form of an areawide advalorem tax, a user fee, or some other method of raising revenue.

Developing urban areas will most likely be able to raise sufficient revenue to pay for improvements. Small rural areas may not be able to. The cost of installing a collection and treatment system in a small rural community may be beyond the ability of the town to pay. Does the remainder of the metropolitan area subsidize it in the interest of water quality? Similarly, an inner city area may need to be assisted to bring about needed improvements.

Financing of point source improvements is a much less complicated task than financing of non-point source projects. Nonstructural measures are not necessarily cost-free (Ref. 5-9). How are these to be financed and who eventually pays for them? Can they be justified in terms of improved water quality? With the amount of research now going on in the non-point field, a body of knowledge will be available to attack non-point problems; however, the financing problem will still have to be addressed and solved.

3. *Distributing Costs.* The manner in which costs are distributed will be a key

factor in the success of the management system. This affects everyone—the resident who is there now, the future industry, the neighboring and competing metropolitan area—all have a stake in the decisions.

The management agency must be able to assure that each participating community pays its proportionate share of the treatment costs. Two factors are involved: 1) the character of the waste; i.e., the volume, strength or time flow; and 2) the transportation difference; i.e., the distance from entering the wastewater system to the treatment plant. Arriving at an acceptable formula for distributing costs of treatment will be a difficult task and will be an important part of the negotiation/compromise process leading to approval of the wastewater management plan.

On a broader scale, distribution of costs for elimination of non-point source pollution, dealing with prior investments in physical facilities, as well as treatment costs, will be perplexing.

5-9. REGULATORY CONSIDERATIONS

The management system is to include a regulatory program to:

1. Implement the waste treatment management requirements. (To the extent practicable, waste treatment management shall be on an areawide basis and provide control or treatment of all point and non-point sources of pollution, including in-place or accumulated pollution sources.)
2. Regulate the location, modification, and construction of any facilities within such area which may result in any discharge in such area, and
3. Assure that any industrial or commercial wastes discharged into any treatment works in such area meet applicable pre-treatment requirements.

The regulatory program would include controls over land use, issuance of permits and licenses, establishment of water quality standards, and establishment of fiscal policies such as taxation and user charges that may be required to complement the regulatory requirements.

One of the major points of emphasis of the wastewater management program as it has developed has been the encouragement of innovative and effective controls and regulations as a means of improving water quality. Considerable attention has been given to such topics as growth management, soil loss limitations, and other such measures.

To see program options in their broad perspective, it is important that the program be viewed in relation to the previously discussed water use and treatment cycle (see Table 5-1). The water use and treatment cycle includes 49 separate stages or steps. An inventory was prepared to determine the public actions that are or could be applied, at each step of the cycle, to improve water quality. Some type of public action is or could be undertaken on 43 of the 49 steps (see Table 5-1). An extraordinarily wide variety of federal, state, and local political

TABLE 5-1 Diagrammatic Water Use and Treatment Cycles

Summary of Public Actions

Stage in Flow Chart No.	Description	Type of Public Actions Taken	Public Agency Taking Action
1.	Rainfall to land	A. Cloud seeding	None
2.	Evaporation	A. Design standards for reservoirs	None
		B. Covers for reservoirs	
3.	Transpiration	A. Landscape planting requirements	
		B. Open space requirements	
		C. Buffer area requirements	Federal-state-local
		D. Agricultural zones	
		E. Inducements for agriculture or forests	
		F. Public parks and forests	
		G. Wetlands protective laws	
		H. Purchase of development rights	
		I. Scenic easements	
4.	Runoff	A. Urban growth management	State-local
		B. Requirements for amount of impervious surface	Local
		C. Requirements for porous pavements	Local
		D. Requirements for detention reservoirs and other devices to slow flow	Local
		E. Development controls	Local
		F. Crop management	Soil Conservation District
		G. Watershed programs	Soil Conservation District
		H. Soil loss limits	State
		I. Soil bank programs	Federal
		J. Drainage programs	Drainage Districts-Federal
5.	Flows into streams, reservoirs, lakes, oceans	A. Water rights legislation	State
		B. Water withdrawal legislation	State
		C. Flood control	Federal-state
		D. Reservoirs	Federal-state
		E. Navigation improvements	Federal
		F. Land development controls	Local
6.	Soaks into ground	A. See Item 4	See Item 4
		B. Groundwater supplements	State
7.	Urban runoff	A. Development control	Local
		B. Grading-construction controls	"
		C. Street paving and cleaning	"
		D. Snow and ice control	"
		E. Foundation and roof gutter systems	"

TABLE 5-1 *(Continued)*

Stage in Flow Chart		Summary of Public Actions	
No.	Description	Type of Public Actions Taken	Public Agency Taking Action
		F. Mowing and trimming	"
		G. Litter removal	"
		H. Hillside development regulation	"
		I. Floodplain regulations	"
		J. Urban growth managements	State-local
8.	System withdrawal	A. Well regulations	State
		B. Withdrawal regulations	State
		C. Water rights	State
9.	Raw storage	A. Design and financing of facility	State-local
10.	Quality treatment	A. Removal of substances	Federal-state
		B. Addition of substances	Federal-state
		C. Monitoring	State
		D. Design-financing	Local
11.	Treated storage	A. See Item 10	See Item 10
12.	Transmission line	A. See Item 10	See Item 10
		B. Rate structure	State-local
13.	Public uses	A. Growth management	State-local
14.	Domestic uses	B. Rate structure	State-local
15.	Commercial uses	C. Peak period restrictions	Local
16.	Industrial uses	D. Lawn watering restrictions	Local
		E. Garbage disposal restrictions	Local
17.	Quality treatment for industrial use	A. Control of quality	Federal (food and drug processing regulations)
18.	Predisposal treatment	A. Control of quality	Federal-state-local
19.	Storm runoff collection system	A. Design-finance-operate	Federal-state-local
		B. Road culverts-bridges	State-local
		C. Development control	Local
		D. Detention facilities	Local
20.	Storm runoff treatment	A. Design-finance-operate (1) Retention basins for first flush (2) Underground storage for first flush (3) Manhole diversions for first flush (4) Screen separators	Federal-state-local
21.	Combined sanitary-storm system	A. Separation of system	Federal-state-local
22.	Treatment of combined system	A. See Items 20 and 25	See Items 20 and 25
23.	Overflow of combined system	A. Pull first flush and treat	Federal-state-local

TABLE 5-1 *(Continued)*

Summary of Public Actions			
Stage in Flow Chart			
No.	**Description**	**Type of Public Actions Taken**	**Public Agency Taking Action**
24.	Sanitary sewer collection system	A. Design-finance-operate	Federal-state-local
		B. Standards for service	Local
		C. Requirements for connections	Local
		D. Disposal restrictions	Local
		E. Sewer use regulations	Local
25.	Treatment of sanitary sewage	A. Design-finance-operate	Federal-state-local
		B. Requirements for degree of treatment	Federal-state
		C. NPDES permits	Federal
26.	Overflow of sanitary sewage treatment	A. Eliminate	Federal-state-local
27.	Discharge to surface	A. Required effluent standards	Federal-state
		B. Water quality standards	Federal-state
28.	Discharge to groundwater	A. Prohibit, or	State
		B. Require standards, or	State
		C. Require the discharge	State
29.	Discharge to land	A. Require standards	State
		B. Monitor	State
30.	Discharge to evaporation	A. Review and approve	State
		B. Supervise	Local
		C. Monitor	State
31.	Discharge to reuse	A. Required effluent standards	State
		B. Control of type of reuse	Federal-state
32.	Individual withdrawal	A. Regulation and prohibition of wells	State-local
		B. Monitoring of wells	State-local
		C. Water rights	State
33.	Individual raw storage	A. No action	None
34.	Individual quality treatment	A. Quality standards in some cases	State-local
35.	Individual storage	A. Standards in some cases	State-local
36.	Individual transmission line	A. Standards in some cases	State-local
37.	Domestic, commercial, public use of individual supply system	A. Standards required in some cases	State-local
38.	Industrial use of individual supply systems	A. Standards required in some cases	Federal-state-local
39.	Agricultural use of individual supply system	No action required	None
40.	Agricultural use of individual supply system for livestock	No action required	None

TABLE 5-1 *(Continued)*

Summary of Public Actions			
Stage in Flow Chart			
No.	Description	Type of Public Actions Taken	Public Agency Taking Action
41.	Agricultural use of individual supply system for crops	No action required	None
42.	Treatment of individual domestic systems	A. Standards required B. Location control	State-local Local
43.	Treatment of individual industrial systems	A. Standards required B. Location control	State-local (Sometimes federal)(NPDES permit might be required)
44.	Treatment of animal wastes (feedlots)	A. Standards required if feedlot large enough B. Location control	State Local
45.	Discharge of individual systems to evaporation	No action required	None
46.	Discharge of individual system to land	No action required, except major uses would be monitored	State
47.	Discharge of individual system to groundwater	A. Standards required	State
48.	Discharge of individual system to surface water	A. Required effluent standards B. Water quality standards	State State
49.	Discharge of individual system to reuse	A. Required effluent standards B. Control of type of reuse	State Federal-state

units have authority to undertake various parts of the program. Over 100 different categories of public action that are or could be taken in connection with the water use and treatment cycles were found. On four percent federal action alone is exercised, on 23 percent state action alone, and on 29 percent local action alone. Forty-five percent involve actions by more than one level of government.

The types of action that can be taken vary greatly between point sources of pollution and non-point sources of pollution. Dealing with point sources through such methods as permits, treatment requirements, and land use controls is relatively straightforward. However, dealing with the myriad of parts which make up the non-point system is a far more complex task.

Public actions contained in the table can generally be broken down into five categories as follows:

1. *Permits.* Permits, together with standards, are the basic framework of the regulatory system today. The NPDES permit procedures allow EPA to im-

plement water quality standards and effluent requirements for both point and non-point sources. Permits for hook-ups provide control over industrial waste discharge and private system discharge.

2. *Water Quality Standards.* Water quality standards are the criteria and implementation plans for accomplishing water quality goals and consist of stream, effluent and pretreatment standards. Standards can be applied to almost any type of pollution and no doubt will as we come to grips with non-point problems.
3. *Land Use Controls and Growth Management Techniques.* The broad category of land use controls includes many procedures for influencing the location and quality of development which will in turn influence water quality. These include zoning, subdivision regulations, floodplain regulations, hillside development regulations, erosion and sediment controls, etc. Growth management practices and policies, including a variety of techniques being used in various parts of the country following the Ramapo decision, are available as well.
4. *Project Design Standards.* Design standards involving open space, drainage design and material (porous vs. nonporous pavements), etc. can be used to reduce non-point flows and improve non-point quality. These are generally enforced at the local government level and have a project level impact.
5. *Others.* Many techniques have not been grouped into one of the above categories, yet can have a significant effect on pollution. These include urban programs such as streetsweeping, snow and ice removal, litter removal, policies regarding foundation and roof gutter discharges, and policies regarding agricultural practices.

5-9.1 General Principles for Regulatory Programs

In order to obtain the requisite benefit from a regulatory water quality program, the control must be exercised consistently over the entire water quality management area. Furthermore, because of our competitive economy wherein one community is competing with another to obtain business and industry, it would not be fair to require one community to exercise a type of control that would result in a higher building cost than that which would be required of a competing community.

Federal and state controls and standards and their enforcement are a primary element in the regulatory system and no matter what type of a management option is chosen, its successful operation will depend upon the elements imposed "from above" by the federal and state governments.

5-9.2 Commentary on Regulatory Programs from a Land Use Perspective

Generally, local control of land use in the United States has been controversial and ineffective. Only in a few places, where community objectives were straightforward and quite simple, has it been possible to apply the public support required for long-term consistent enforcement of land use regulations. There is a considerable amount of question as to whether the public might support land use controls based on bringing about improved water quality. Planners are generally quite nervous when more land use control is proposed, and it will be necessary to demonstrate a compelling public purpose in order to bring about the requisite degree of land use control required to improve water quality.

In the urban area, there are two types of land use control that are essential in order to improve water quality. The first is a coordination of the land use pattern with the facilities for removal and treatment of the wastewater emanating from the land use pattern. For example, a high density development in a lower part of a watershed is much more efficiently located because the wastewater travels a shorter distance to the waste treatment plant than would be the case if this same high density development were placed at the far upper limits of the watershed. To develop a watershed with the higher densities in the lower parts and the lower densities in the upper parts naturally brings about efficiencies in the sizes of the pipes required to carry the waste away.

As a second general land use control, development could be kept completely out of an individual watershed and take place on a watershed by watershed basis, so that the use of pumping stations could be minimized and a gravity system be as efficient as possible. In only a few instances in American communities has there been an endeavor to control the land use pattern so that the community develops progressively, watershed by watershed. Particularly damaging is a situation where development is allowed in only the upper part of the watershed and then has to be pumped over to another watershed or carried a long distance through undeveloped property to reach a treatment facility.

Where sewers have to be extended through large areas of vacant land in order to reach isolated pockets of development, the presence of the sewers themselves (and the capacity built into them to handle ultimate development of the watershed) may in turn induce changes in the urban pattern. The urban area will follow the sewer instead of the sewer and the urban area following a land use plan. Thus, a coordination of the sewer program with the urban land use pattern will greatly assist in bringing about a planned urban land use pattern. Both must be planned and programmed at the same time.

A second component of the land use control system in relation to wastewater is directed to the detail of land development. There are many measures that may be imposed in order to slow down the flow of stormwater. These can take the form of baffles on roofs, of retention basins, and even of storage tanks for portions of the stormwater. Instead of creating such a volume of storm drainage by impervious paving, it is possible to provide new types of surfacing that allow the water to go through the pavement to the groundwater either continuously over a porous pavement or by picking up the storm drainage at frequent intervals and discharging it into the groundwater. Through controls of such details of land use development, it is possible to greatly improve the quality and to bring about a lesser peak volume of stormwater runoff. Most of these measures involve higher costs on the part of the land developer, however.

Finally, special land use controls may be placed over special uses that cause considerable difficulty, such as feedlots, construction projects, mining, sand and gravel extraction, etc. The special uses may be controlled insofar as location is concerned, and then be required to be developed in accordance with practices that would minimize their damage to water quality.

The land use controls described above represent a large number of public actions which can be directed at assisting in the achievement of water quality goals. However, they will be ineffective unless implemented as a part of an overall management and regulatory program.

5-10. MANAGEMENT SYSTEM OPTIONS

In looking over the options available in institutional arrangements, financing, regulatory programs, and other essential parts of the 208 management system, some broad comments would seem to be in order.

The basic question appears to be: How are we to modify an obsolete and inefficient system of local government to enable it to handle a complex water quality improvement program? It is difficult to conceive of providing a sufficient inducement (federal money or state moratoria) to induce each local agency to play its exact and proper part—similar to a symphony orchestra with 15 to 200 players. A few discordant notes could wreck the performance. To orchestrate the existing obsolete system is the most simple answer, but possibly the least effective.

On the other hand, it is not practical to expect the 208 Program to be used as the instrument to completely remodel our obsolete system of local government. We will have to fall back upon a less sweeping approach. We should do this in such a manner as to not further fragment the local government system, and in a manner that would foster the eventual establishment of a local government system, such as that of Jacksonville, Nashville, Minneapolis-St. Paul, or Dade County, which is more suited to the needs and demands of our own times and to solving our complex problems (such as water quality).

Guidelines suggest the consideration of options that separate *planning* and *management.* This is probably to be avoided. The agency doing the planning should be intimately aware of management problems, and management people should be heavily involved in planning. Experience has indicated that this is difficult to accomplish where the two functions are separated. However, *planning* cannot be single-purpose planning, as wastewater programs cannot be separated from land use, transportation, or open space planning. Thus, there are real dangers in both of these options.

Effective management in the end will have to fall back on state agencies to provide (or lend) some powers and to undertake some activities, such as control of agricultural pollution. Each management system will have to be tailored to the particular circumstances of the area. Each plan and program will require public support for any real accomplishment. Thus, sweeping changes and idealistic (textbook) solutions are not likely. The management system cannot depart too much from present arrangements. This constraint, and that imposed by financing, may appear to make the first steps in the program weak and halting. The important consideration is that they be in the right direction.

REFERENCES

5-1 Indiana University, School of Public and Environmental Affairs, *Problems and Approaches to Areawide Water Quality Management,* prepared for the U.S. Environmental Protection Agency, October, 1973, p. 16–17.

5-2 U.S. Environmental Protection Agency, *Draft Guidelines for State and Areawide Water Quality Management Program Development,* Washington, D.C., February, 1976.

5-3 U.S. Environmental Protection Agency, *Management Agencies Handbook for Section 208 Areawide Waste Treatment Management,* Washington, D.C., September, 1975.

5-4 Einsweiller, Robert C., *What Is Needed to Implement the Management Plan,* April, 1976.

5-5 U.S. Bureau of Census, *Governmental Finances in 1973–74,* Tables 17 and 26.

5-6 U.S. Bureau of Census, *Census of Governments*—Local Government in Metropolitan Areas 5, Table 12.

5-7 *Ibid.* 5, Table 12.

5-8 Central Iowa Regional Association of Local Governments, *Des Moines 208 Areawide Waste Treatment Management Plan, Phase I Interim Report,* Chapters VII and IX, January, 1976.

5-9 Einsweiller, Robert C., *Op. cit.,* p. 3.

6

Public participation in water quality management planning

Charles C. Schimpeler and Marva Gay

Schimpeler—Corradino Associates
Louisville, Kentucky

Adelbert L. Roark

A. L. Roark and Associates
Lexington, Kentucky

6-1. INTRODUCTION

Urban planning or engineering must today consider public participation in selecting the programs to be implemented. This emphasis reflects basic trends in American society over the last twenty years.

Much of the emphasis on public participation has been generated by federal and state requirements; planners and engineers have also come to believe that plans for the urban environment stand a better chance of being implemented if the general public is involved in their development. This latter conviction is supported by experience, such as recent lawsuits that have effectively halted major public works programs. Many also feel that only through citizen participation can the best possible plan be developed.

In order to understand public participation, we must develop a definition of the process. For the purpose of this chapter, public participation is defined as an open process in which the rights of the community to be informed and to influence and receive a response from government are exercised through the interaction of a cross section of affected citizens, with appointed and elected officials on policies relative to projects under consideration (Ref. 6-1). The principal purposes of such a program are to:

1. Achieve a better plan in keeping with the goals and objectives of the general community;
2. Increase the probability of implementing the plan; and
3. Increase the likelihood that the benefits to be derived from the program are actually captured by various segments of the population (Ref. 6-2).

Secondary goals are achieved in terms of education of the public, resolution of conflict among various segments of the population, and increased credibility of the planning agency.

The concept of participation is a cornerstone of the American form of government; the town meeting was an early example of direct citizen involvement in governmental decision-making. One of the early projects in urban planning that was influenced by citizen involvement was the development of the country's first zoning ordinances in New York City in 1916.

The current emphasis on citizen participation grew out of the general unrest of the 1960's. The late President John F. Kennedy showed an awareness of this by his "those who make peaceful revolution impossible, make violent revolution inevitable" (Ref. 6-3). The Department of Housing and Urban Development was one of the first agencies to react to these pressures by requiring that citizen participation be part of its programs.

The move toward more active citizen involvement continued in the new environmental and consumer activism which grew out of the increased American awareness of "quality of life." While the decisions on many environmental issues must be governed by technical considerations, the general public is also concerned about the outcome of various planning and engineering programs, particularly in urban areas where it is obvious that a single subsystem can greatly affect all other parts of the total system.

In this chapter, citizen participation programs will be discussed as they relate to water quality management planning. Beginning with an analysis of the requirements of programs in this area, we will move toward definition of the basic structure of the citizen participation program; later sections will deal with specific technical aspects of each component. Of particular importance are the analytical procedures used to determine the relative utilities of various criteria applied to water quality planning and how the utilities will be used in the total cost effectiveness evaluation. The chapter will conclude by discussing two specific water quality management planning studies in which these procedures have been applied—a 201 facilities planning study and a 303(e) river basin planning study.

6-2. REQUIREMENT FOR PUBLIC PARTICIPATION IN THE WATER QUALITY MANAGEMENT PLANNING PROCESS

The urban environment is influenced directly by its wastewater management system, because it affects both development patterns and public health issues. The importance of this relationship is recognized in the requirement of public participation in water quality management planning.

6-2.1 Desirability of Public Participation

The necessity for resolving conflicting needs and desires within a community of varied social and economic interests has recently been met effectively through public participation, revealing its potential for promoting the best interests of the total community. The public participation process occurs in many different forms, but generally desires to:

1. Insure that relevant, social, economic, and environmental factors are adequately considered;
2. Maximize the satisfaction of all interested individuals and groups with respect to thc proposed plans;
3. Provide a medium for identification of community goals and objectives that allows citizens to share in directing proposed programs and developments;
4. Encourage plan acceptance and thereby foster rapid implementation; and
5. Promote the planning process and associated principles by educating citizens in sound community development practices.

6-2.2 Legal Requirements for Public Participation

The Federal Water Pollution Control Act, as amended in 1972 (Public Law 92-500), contains one of the strongest requirements for participatory democracy in the entire federal statute book. In the words of Section 101(e) of this law:

> "Public participation in the development, revision, and enforcement of any regulation, standard, effluent limitation, plan, or program established by the administrator (of EPA) or any state under this Act shall be provided for, encouraged, and assisted by the Administrator and the State. The Administrator, in cooperation with the State, shall develop and publish regulations specifying minimum guidelines for public participation in such processes (Ref. 6-4)."

In response to this part of the federal law, the Environmental Protection Agency (EPA) issued Regulation 40 CFR 105, in August, 1973. This regulation, which applies to all EPA water pollution control programs, requires that state agencies do the following:

1. Provide technical information at the earliest possible time at places easily accessible to interested or affected persons or organizations, and assist the public in understanding and responding to water programs;
2. Establish permanent arrangements for consultation and exchange of views with interested or affected persons and organizations prior to reaching any decision on a course of action;
3. Develop and maintain a current list of interested persons and organizations to be notified concerning any action to be taken under Public Law 92-500;
4. Provide at an appropriate location one or more central public collections or depositories of water quality reports and data pertinent to the geographical areas concerned;
5. Provide for full and open information on any legal action taken under the Act, consistent with requirements of the court aimed at avoiding disclosures that might prejudice the outcome of litigation; and
6. Provide opportunities for the public (e.g., public hearings) to express their opinions on the actions being anticipated by the state.

The regulations also outline procedures for organizing and conducting public hearings.

The state and each EPA unit or office are required to report at the end of each operating year on the public participation process and the activities undertaken. The regional administrator or other approving official must review and evaluate this report. A reviewing official may disapprove or suspend activities undertaken under Public Law 92-500 if he determines that there has been inadequate opportunity for the public to become involved in the formulation and/or adoption of the program.

The enforcement sections of these regulations require that states encourage the public to report violations of water quality laws, and develop procedures for considering evidence submitted in such matters.

6-3. FRAMEWORK FOR CITIZENS' PARTICIPATION PROGRAM

Increased education and mass communication have now made people more aware of local and national issues than ever before. Citizens are more likely not only to know they can influence community decision-making, but also to be aware of the procedures by which they can make themselves heard. There is a heightened "social consciousness," manifested partially by an increased concern for the less fortunate. It is more and more recognized that:

> "A compartmentalized approach is obsolete . . . The provision of facilities and services also needs to be responsive to the physical and social targets established by national and regional development agencies" (Ref. 6-5).

Therefore, the personnel conducting the citizen participation program should assume the position of "advocate planners" for various citizen groups. In this role, the planner acts as the proponent of public views to the review bodies functioning at the management and decision-making levels. In water quality management planning, the major public input is evaluation and generation of alternatives. One theory on the evaluation of systems that the individual engineer/planner should evaluate the alternatives from the perspective of *each* special interest group. Citizen participation requires a communications program organized so as to provide the public with adequate information concerning plan development, allow for significant public response in terms of both time and substance, and to increase the decision-making bodies' acceptance of public attitudes and comments. This communications program is founded on the basic concepts of transmission and feedback.

One of the first considerations in establishing the citizen participation program is to determine what portion of the population actually constitutes the "public." In a study performed for the state of California relative to water quality management decisions, two conclusions were reached.

> "While there are many publics, only those which have organized interest groups with defined power bases participate effectively in the decision-making process; the "mass public" and individuals speaking on their own behalf seldom are either involved or influential.
>
> There is a significant "communications gap" between the engineers/planners responsible for water quality management, and those interest groups which do not deal in "technical" terms towards which the engineers/planners are traditionally oriented; as a consequence, those groups which do have the technical communications capacity are more capable of influencing the decision process. Additionally, it should be noted that this communications gap is both threatening and friction-producing to the parties involved (Ref. 6-5)."

The major objective in the participation process is to insure that each segment of the public is provided with an opportunity to be informed on the water quality planning process and to express itself on all issues involved. This view, together with the California study, emphasized a program encouraging individuals to act through organizations and to deal directly with various organized publics.

6-3.1 Transmission

The public involvement program should be structured to encourage maximum public participation in the technical and management planning process. Initial efforts should center on implementing this public involvement structure as soon as possible and developing a smooth, efficient format for accommodating citizen input. The central effort must focus on educating the community in all facets of relevant water resource plans and coordinative procedures from the National Pollution Discharge Elimination System (NPDES), and on stimulating widespread interest in the planning program. General activities involved in this program design are described below:

- Effective use of the media;
- Provision for special exhibits in commonly used community facilities, with extensive circulation of publications describing activities to be undertaken and activities completed;
- Distribution of a brochure concisely describing the purpose and scope of the planning program, with a direct invitation to community residents to participate in one or all of the following interactive processes:
  - Correspondence
  - Membership on citizen committees
  - Public hearings
  - Workshops
  - Direct contact with the planning/engineering staff.

Specific activities throughout the study should follow a similar pattern of media use and direct/indirect contact, but must also respond to differences in specific project tasks and to the progress toward completion of the total plan.

6-3.2 Feedback

There must be mechanisms that assure feedback from the public so that the public's response may be interpreted and evaluated. The information derived from the public must be prepared by the planners for presentation to the committees established for a review and discussion of the progress, direction, and results of the planning effort. These presentations should be designed to reveal, clarify, and promote the views and attitudes of the public so that the results of the study reflect the needs and desires of the community. Elements of advocacy presentations might include:

- Results from surveys;
- Summaries of public opinion expressed in the news media;
- Results from interviews;
- Documented public comment and/or criticism;
- Analysis of local and regional issues as they relate to the study program; and
- Anticipated reactions of the public to alternative decisions.

The presentations should involve both written and graphic material and should be comprehensive, so that the committees are able to understand the complex social, economic, and political elements of the study.

6-4. PUBLIC EDUCATION IN THE PUBLIC PARTICIPATION PROCESS

Effective public participation in water quality management planning cannot be achieved unless there is an aware and active public. The public must be aware of two things—the existence of a water quality management planning project, and the public's ability to influence decisions concerning that project. The public must also be motivated to maintain its awareness and to participate in decision-making activities.

To achieve these goals of awareness and participation, the public must be informed. All segments of the public must be reached. The segments can be broken down into two main groups:

1. Those persons already interested in the water quality management planning project; and
2. Those persons not interested in the project.

This second group must also be reached because they will eventually be affected by the project and, upon the realization that the project does affect them personally, they may raise objections jeopardizing the implementation of the project.

There are three steps in creating awareness and participation:

1. Use of the news media: This step is of great importance because information distributed through the news media reaches the mass public and can be geared to both the interested and uninterested segments of the public. The media is the primary means of drawing the uninterested into the participation process.
2. Use of public workshops: The workshops are primarily directed at further educating those persons who have become interested in the project and in obtaining their participation in the decision-making process.
3. Use of public hearings: The hearings are required by law and are primarily mechanisms for obtaining citizen opinions for use by government officials in the following decision-making process.

The strategy for the most effective use of the news media, public workshops, and public hearings is as follows:

1. News media: All forms of the news media should be used to reach the largest possible number of people. The forms of media to be used include newspapers and broadcast stations (radio and television). These can be reached by the following:
   a. press conferences;
   b. meetings with press members;
   c. press releases—feature and straight news;

d. interviews;
e. news making events;
f. paid advertisements; and
g. public service broadcast announcements.

Cooperation should also be solicited from other news gathering and disseminating organizations to publicize the water quality management planning project. Information distributed for the press and for public consumption should emphasize dramatic statements about the effects of the project. Dramatic statements must be used to gain the attention of the media and the public. Plan alternatives and their effects should be presented as if they were an existing reality, to enable individuals to identify how that reality affects them personally. Very generalized statements about effects should be avoided.

2. Public workshops: After an interested public has been created by the news media, public workshops can be used to further educate the public and to obtain opinions from the public sector to be used in decision-making. Public workshops should be designed to offer the persons attending a picture of the project's background, goals, status, and technical aspects. The workshops should also educate the public in the techniques used to formulate decisions and in the methods by which the public might modify decisions. The public workshops can also help to educate the public, enabling citizens to present more valuable testimony at public hearings. By use of the news media and letters to civic and government organizations, as well as to individuals who have expressed an interest in the project, attendance at the workshops can be solicited.
3. Public hearings: These constitute the final step in obtaining citizen opinions before a decision is reached by government officials. Attendance and participation in the hearings should be encouraged through extensive use of the news media and public workshops. At the hearings, opinions will be received from the public. There is little direct feedback from the consultant or from government decision-makers during the hearings. However, feedback should be assured by obtaining news coverage of the hearings and of the ultimate decisions made about the water quality management planning project. Changes made in the project as a result of citizen participation should be reported in the news media.

6-5. ANALYTICAL TECHNIQUES FOR USE IN PUBLIC PARTICIPATION PROGRAMS

Much emphasis has been placed on determining and quantifying criteria and goals for urban planning.

These analytical approaches have been valuable in such diverse fields as social service delivery systems, education, transportation, and site selection. They are now being applied to water resources and wastewater systems planning. Many of these approaches quantify attributes usually assumed to be qualitative.

This section outlines a procedure that has been used in several studies for evaluating water quality management plans. It has played an important role in these studies by allowing feedback of a very precise nature to be incorporated into both the overall planning process and the cost effectiveness analysis.

6-5.1 Evaluation Methodology

The first step is establishment of an alternatives evaluation committee, with membership formulated as discussed in Section 6-3. In many studies, the evaluation committee is essentially a citizen's advisory committee.

The activities of this committee include the following major steps:

1. Review and modification of the criteria deemed important in water quality planning;
2. Ranking and rating of the criteria to establish relative importance through assignment of a numerical weight to each criterion;
3. Critical review of each of the alternative systems; and
4. Evaluation of each alternative system, regarding the extent to which a particular criterion might be satisfied by that alternative. This evaluation is performed for all alternatives, one criterion at a time.

The ranking technique is essentially a classification into quantitative categories. Complete discrimination in rank is asked of each judge, in that he or she is instructed to place only one criterion in each category. The ranking technique assumes equal intervals of importance among criteria. The most important criterion receives a rank of 1; the least important is given a rank of n, where n is the number of criteria being ranked. These ranks are then converted by subtracting the raw ranks from n. Thus, the most important criterion has a converted rank of $n - 1$ and the least important a rank of zero. The conversion is necessary to insure that the most important criterion has the highest weight or utility value.

A composite rank, R_j, is determined for each criterion by summing the converted ranks of the judges:

$$R_j = \sum_{i=1}^{m} R_{ij}, j = 1, 2, \ldots, n;$$

R_j = composite rank of criterion j;
R_{ij} = converted rank of criterion j established by judge i;
n = number of criteria;
m = number of judges.

The composite ranks thus determined are normalized in the following manner:

$$u_j = \frac{R_j}{\sum_{j=1}^{n} R_j},$$

where u_j is the composite weight or utility value associated with the jth criterion. Of course, the summation of the u_j from $j = 1$ to n must equal 1.

The rating scale technique is probably the most popular of the procedures used for collecting the judgments of individuals. A numerical rating scale is used, but the descriptors normally associated with integer values on the scale are not.

A list of criteria to be weighted is placed in columnar fashion down the right side of the page. A rating scale marked in units continuously from 0 to 100 (bottom to top) is placed at the left side. An example of the rating form is shown in Figure 6-1. A rating of 0 indicates there is no value associated with a given criterion; a rating of 100 is the highest value that may be assigned. Any value in the unbroken continuum may be assigned to any criterion. The judge assigns a rating by simply drawing a line from the criterion to an appropriate point on the rating scale.

These data are then treated in much the same manner as the ranking data; however, there is no need to convert the rating data, since the highest-valued criterion receives the highest rating. A composite rating, V_j, is determined for each criterion by summing the ratings of the judges:

$$V_j = \sum_{i=1}^{m} V_{ij}, j = 1, 2, \ldots, n;$$

where each term is defined just as those for the ranking technique. The composite utility values associated with each criterion are determined by normalizing:

$$u_j = \frac{V_j}{\sum_{j=1}^{n} V_j}.$$

The rating and ranking techniques just described result in generation of two utility vectors. To combine the best features of both techniques, the two vectors are averaged to produce yet a third utility vector, referred to as the composite or average utility vector. The average utility vector is the only one used in subsequent calculations of total alternative utility.

After the criteria have been weighted, the committee members now consider one criterion at a time. Each committee member is provided with information on each alternative. Each member assigns an effectiveness value to each of the alternatives based on how well he or she thinks each alternative would fare relative

Please assign a value (rating) to each of the goals and objectives.

A rate of zero (0) indicates that there is no value associated with a given goal or objective and a rate of 100 is the highest that may be assigned. Any value on the scale may be assigned to any goal or objective. The value should reflect the *importance to you* of the goal or objective. Two or more goals or objectives may be assigned the same value. (This gives you more flexibility than the ranking technique). Accomplish the desired rating by drawing a line from a particular goal or objective to the selected position on the rating scale as shown in the example below.

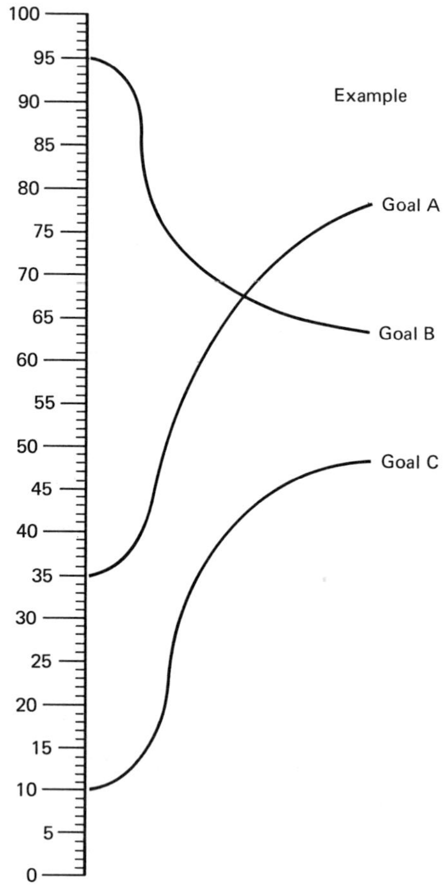

In the example shown above, the following rates were assigned:

GOAL A: 35
GOAL B: 95
GOAL C: 10

Fig. 6-1. Rating form.

Community Goal Areas

Please rank the most important goal 1, the second most important 2, etc. Assign to the least important goal a rank of 11.

Place the rank you assign in the box at the right of each goal statement.

Goal	Raw Rank
A. Public Safety Program Development	
B. Public Utility and Transportation Development	
C. Economic Development Programs	
D. Cultural Development	
E. Health Program Development	
F. Land Use Development	
G. Education Program Development	
H. Human Resource Program Development	
I. Recreation Program Development	
J. Political Framework	
K. Housing Development	
L.	
M.	

Fig. 6-1. (Continued)

to that criterion. A value of 1.0 implies virtual certainty that all aspects of that criterion could be adequately and efficiently provided for; a value of 0.0 implies such provision is almost impossible; a value of 0.5 implies that the alternative holds no particular advantage or disadvantage relative to the criterion.

When each judge has finished with the effectiveness values for the criteria, the forms are collected and analyzed, while the judges go on to the next criterion under consideration. The average effectiveness value is represented by:

$$e_{ij}, \quad \begin{matrix} i = 1, 2, \ldots, s \\ j = 1, 2, \ldots, n \end{matrix},$$

where e_{ij} is the average effectiveness value assigned to the ith alternative for the jth criterion. The individual effectiveness values are summed for all judges, and then divided by the number of judges to produce the average effectiveness values. The average effectiveness values are arrayed in an effectiveness matrix having the following form:

$$
\begin{array}{c|c|ccccc}
 & & \multicolumn{5}{c}{\text{Proposed Criteria}} \\
 & & 1 & 2 & \cdot\cdot j & \cdot\cdot n \\
\text{Alternative Systems} & 1 & e_{11} & e_{12} & \cdot\cdot e_{ij} & \cdot\cdot e_{1n} \\
 & 2 & e_{21} & e_{22} & \cdot\cdot e_{2j} & \cdot\cdot e_{2n} \\
 & \vdots & \vdots & \vdots & \vdots & \vdots \\
 & i & e_{i1} & e_{i2} & \cdot\cdot e_{ij} & \cdot\cdot e_{in} \\
 & \vdots & \vdots & \vdots & \vdots & \vdots \\
 & s & e_{s1} & e_{s2} & e_{sj} & e_{sn}
\end{array}
$$

In generalized vector notation, the effectiveness matrix may be represented by $[E]$. Each e_{ij} in the above effectiveness matrix represents the average effectiveness value as determined by all judges.

The total effectiveness of an alternative is determined by multiplying the effectiveness matrix and the criterion utility vector. The product is an alternative utility vector in which each entry represents the utility of an individual alternative or, as stated above, the total effectiveness of an alternative.

The product mentioned above is actually a summation of the products formed by multiplying the average effectiveness with which an alternative satisfies a criterion by the utility value of that criterion. This may be stated as:

$$U_i = \sum_{j=1}^{n} e_{ij}u_j, \; i = 1, 2, \ldots, s;$$

where

U_i = total utility of alternative i;

e_{ij} = average effectiveness of alternative i in satisfying criterion j;

u_j = utility value of criterion j.

Since the summation of u_j over j must be 1 and the maximum value for e_{ij} is 1, then the maximum value for any U_j is also 1. Thus, a perfect alternative would have a value of 1, and a totally worthless alternative would have a value of 0. Obviously, most alternatives will have utility values higher than zero and less than 1.0–probably in a range from 0.2 to 0.7.

The above decision model may be stated in generalized vector notation as follows:

$$[E]\;[u] = [U],$$

where $[u] = (u_1, u_2, \ldots, u_i, \ldots, u_n)^T$ is a column vector whose components represent the utility values associated with each of the n criteria, and $[U] = (U_1, U_2, \ldots, U_i, \ldots, U_s)^T$ is a column vector whose components represent the total utility associated with each of the s alternatives.

The matrix $[E]$ is the s-by-n effectiveness matrix defined earlier. In matrix multiplication, components of the product vector, U_i, are formed by the summation of the products of respective elements of the ith row of the effectiveness matrix and the criteria utility column vector.

6-5.2 Statistical Tests

Basically, the statistical tests to be performed on the data previously collected are nonparametric in character. Nonparametric techniques are often called "ranking tests" or "order tests" (Ref. 6-6). They are also called "distribution free," because no assumptions relative to the nature of the population need be made.

In the case of the ranking of criteria, only nonparametric methods can be used. This is because the ranking simply specifies one criterion is more important than another, saying nothing about how *much* more important. Use of parametric methods necessitates unrealistic assumptions.

In the analyses of data on rankings and ratings, the information desired is whether the sets of rankings are related or unrelated. Hopefully, the inference can be drawn that the rankings of the judges are all related. If they are related, then the composite ranking of all the judges can be used as the best estimate of the true ranking of the criteria. Two nonparametric measures of correlation that were used in this analysis were Spearman's rank correlation coefficient and Kendall's coefficient of concordance.

The ratings, however, allow the judges not only to rank the criteria in order of importance, but to place relative measures of importance on each criterion. If one assumes the ratings and rankings of the criteria constitute a bivariate case, then the Spearman rank correlation coefficient may be used to test whether the two methods are related for an individual judge. However, there is no available nonparametric technique to test the relatedness of the utility vectors determined by rankings and ratings, because of the number of judges used in arriving at these utility vectors.

Neither is there a way to test whether the committee's rankings are related to those of the experts who might be considered. However, the product-moment correlation coefficient, r, can be used to test whether there is any degree of relatedness between the aforementioned utility vectors.

The effectiveness values, as determined by the judges, are not actually rankings, but estimates of how well an alternative might satisfy a given criterion. For each criterion, the variances of the effectiveness values assigned to each alternative are tested for equality using Bartlett's test for homogeneity of variance. This indicates consistency of the judges in evaluating a given alternative. Then the mean effectiveness values for each alternative are tested for equality. If the means are not equal, the process indicates that the judges were able to discriminate between alternatives. Brief descriptions of these statistical tests follow.

Spearman: Rank Correlation Coefficient. The situation, simply put, is this: n items (criteria) are ranked from 1 to n according to some specified characteristic by m judges; it is desired to know whether the m rankings are substantially in agreement. Spearman's coefficient is calculated to resolve this point when there are only two rankings; i.e., $m = 2$. Whenever m is equal to 3 or more, Kendall's coefficient of concordance must be calculated to resolve the issue.

Spearman's coefficient is determined by:

$$r_s = 1 - \left(6 \sum_{i=1}^{m} d_i^2/(n^2 - n)\right),$$

where d_i = the difference in the two ranks assigned to the ith criterion. A derivation of r_s may be found in either Kendall (Ref. 6-7) or Siegel (Ref. 6-6).

The value of r_s may range from -1 to $+1$, where -1 indicates perfect disagreement, 0 no agreement whatever, and $+1$ perfect agreement. Suppose three criteria are being ranked by two judges. For a given ranking by judge A, there are $n!$, or six possible rankings by judge B. There are also six possible rankings by judge A for a total of $(n!)^m$ or 36 possible sets. The reader may verify that the only possible values of r_s for these 36 sets are -1, $-\frac{1}{2}$, $+\frac{1}{2}$, and $+1$, which occur with respective probabilities of $\frac{1}{6}$, $\frac{1}{3}$, $\frac{1}{3}$, and $\frac{1}{6}$.

In a random ranking of a set of criteria, any given ranking is equally likely. Each combination of two rankings gives rise to values of r_s that occur with calculable probability, as shown above. In an actual ranking supplied by two judges, the null hypothesis is that there is no agreement between them. The hypothesis is rejected if r_s is close to 1.

The calculation of appropriate significance levels is a laborious process as n becomes large. Kendall gives tables for n up to 10. Siegel gives tables of .05 and .01 significance levels for values of n from 4 to 30.

Kendall: Coefficient of Concordance. Kendall's coefficient of concordance is used to test agreement among judges when there are three or more rankings, say m, of n criteria. The method considers the sum of ranks given to each criterion by the m judges. The sum of ranks given by one judge to n criteria is equal to $\frac{1}{2}$ $(n\ [n + 1])$, and the sum of ranks given by m judges is then $\frac{1}{2}$ $(mn\ [n + 1])$. The mean sum associated with the n criteria equals $\frac{1}{2}$ $(m\ [n + 1])$. If all the rankings are identical, the sum of squares of deviations about the mean are maximum. For any set of m rankings, the sum of squares of deviations may be calculated and termed S. The maximum sum of squares of deviations (when all rankings are identical) may be calculated as $(m^2 \ [n^3 - n])/12$. The coefficient of concordance, W, is then calculated as the ratio of the observed sum of squares, S, to the maximum sum of squares:

$$W = (12S)/(m^2 \, [n^3 - n]).$$

If there were no agreement at all, the sum of ranks would be equal for each criterion and S would be zero. Hence, W would also be zero. W increases from zero to 1 as the deviations from the mean become greater. Consequently, the relative agreement between rankings is also greater.

It is, of course, desirable to test the significance of W; i.e., what value of W is significantly different from chance at some significance level such as 0.05. When n is greater than 7, the quantity

$$\chi_r^2 = m(n - 1)W$$

is distributed as Chi Square, with $n - 1$ degrees of freedom. Thus, a table of the Chi Square distribution, as found in many references, may be used to test hypotheses concerning the significance of W, provided $n > 7$. Kendall gives some tables for $n = 7$ or less.

Pearson: Product-Moment Correlation Coefficient. Any measurable quality provides what, in the theory of statistics, is termed a variate. The ratings of the n criteria on a continuous scale of 0 to 100 thus constitute variates. The criteria may be ranked according to their relative position on the scale, such that variate values are replaced by ranks. A ranking is considered less accurate than a rating (on a scale) due to the loss of information on how close criteria may fall on the scale. But the ranking is more general, in that it is invariant no matter what the scale or how the scale may change.

Suppose a sample is drawn from a univariate population of X-values and ranked in ascending order. According to Kendall, the product-moment correlation coefficient between the rating values and the ranks is sometimes quite high. In a practical sense, this means that for a high product-moment correlation coefficient, the same conclusions may be drawn whether variate values (ratings) or rank values (rankings) were used in weighting any set of criteria. The best weighting of the criteria set may then be determined, perhaps, as being the average weighting of the two techniques.

The Pearson product-moment correlation coefficient, r, may be used to test the degree of relatedness between the aforementioned weightings, or what were earlier referred to as criteria utility vectors. Sampling from a bivariate normal population for the rankings and ratings may be assumed. In this instance, r is an estimate of a population parameter, and hypotheses about its true value may be tested; particularly the null hypothesis that the true coefficient equals zero. In most cases, we prefer to be able to reject this hypothesis.

Bartlett: Homogeneity of Variance. Bartlett's method is used to test the homogeneity of the variances of several normal populations. The mechanics of Bartlett's method are found in many standard statistical texts (Ostle for one). Bartlett's method involves calculation of sample variance for each of the populations. If k is the number of populations under consideration, the calculated value of Chi Square is distributed with $k - 1$ degrees of freedom. Therefore, if

$$\chi^2 \geqslant \chi^2_{(1-\alpha)(k-1)},$$

the null hypothesis of equal variances would be rejected.

***F*-Test: Equality of Means.** The equality of means of several normal populations may be tested by calculating the F-statistic and comparing it to the F-distribution for appropriate degrees of freedom and significance levels. Detailed procedures for performing the test may be found in Ostle and Brownlee (Ref. 6-8), among others.

In the alternative evaluation process, each judge assigns an effectiveness value to each of the alternatives on the basis of a subjective evaluation of the alternative with respect to a particular criterion. The average effectiveness values of the alternative are tested for equality. Rejection of this hypothesis implies that the judges were able to discriminate between alternatives, relative to a given criterion. The test assumes equality of variances within groups or within alternatives. However, the analysis of variance is insensitive to unequal variances when the sample sizes of each group are equal or very nearly so.

6-6. APPLICATION OF METHODOLOGIES TO SPECIFIC WATER QUALITY MANAGEMENT PLANNING STUDIES

In this section, use of the techniques previously discussed will be demonstrated by summarizing two specific water quality management planning studies. The first is a 201 facility planning study conducted for Louisville, Kentucky and its surrounding area. This community of approximately 713,300 people has a variety of water quality problems, as reflected in the large number of package treatment plants and the numerous rivers and creeks receiving their discharges.

The second example is a 303(e) river basin planning study conducted for the state of Kentucky. Unlike many states, Kentucky chose to perform all its river basin planning at one time. The citizen participation portion was coordinated, even though the various river basins were being studied by separate consulting organizations. The unity achieved yielded results that can be used in other studies.

6-6.1 201 Facilities Planning—Louisville, Kentucky and Surrounding Areas

In this section, we will discuss the citizen participation program conducted as part of Louisville's 201 facilities planning study, including public information and value scaling techniques for cost effectiveness evaluation.

Before involving the general public, the study program was coordinated with governmental agencies through a Technical Advisory Committee. This committee formed the nucleus of an expanded Water Quality Management Alternatives Evaluation Committee. Identification of a group of citizens representing various public and private interests throughout the study area was a continuing process that included development of a list of concerned agencies and persons and a preliminary public hearing. After the basic data analyses had been completed and the preliminary hearing held, the Water Quality Management Alternatives Evaluation Committee was formed.

A. PRELIMINARY PUBLIC HEARING

On April 8, 1974, following completion of the data collection and field investigation stages of the Water Quality Management Study, the Kentuckiana Regional Planning and Development Agency (KIPDA) conducted a preliminary public hearing. There were two purposes: first, to inform the general public of the work already accomplished and to outline the scope of work proposed; second, to give both individuals and organizations an opportunity to respond formally to any of the study findings or any portion of the Water Quality Management Study yet to be completed. It was hoped that holding a preliminary hearing after the data collection and inventory stages but before generation and evaluation of alternatives would encourage increased citizen involvement and input during the critical final stages of the study.

B. ALTERNATIVES EVALUATION COMMITTEE

Membership on the Alternatives Evaluation Committee included: 1) those public officials directly affected by or interested in the development of water quality collection and treatment facilities; 2) special interest groups, such as environmental organizations, Chambers of Commerce, etc., and 3) interested individuals.

Selection of members for the Alternatives Evaluation Committee (other than members of the Technical Advisory Committee) proceeded with identification of persons in the study area who had expressed interest in the 201 facilities plan during the primary data collection and analysis phases. Formation of the Committee was announced at the preliminary public hearing. A letter of invitation to participate in active evaluation of alternative water quality plans was sent to each potential member. The first meeting of the committee was held on May 1, 1974.

Approximately forty persons participated in the Committee at one time or another during the four and one-half months of full, intensive Committee activity. The organizations represented on the Committee are listed below.

- Acting Director, Kentuckiana Regional Planning and Development Agency
- Clark County Health Department
- Clark County Regional Planning Board
- Concerned Citizens of Fairdale
- Environmental Protection Agency Region 4, Atlanta
- Environmental Protection Agency Region 5, Chicago
- Floyd County Commissioner's Office
- Floyd County Planning Commission
- Indiana State Board of Health
- Jefferson County Environmental Coordinator
- Jefferson County Government Conference
- Jefferson County Public Works Department
- Kentuckians for Environmental Planning
- Kentucky Department for Natural Resources and Environmental Protection
- League of Women Voters
- Louisville Area Chamber of Commerce
- Louisville Audubon Society
- Louisville Board of Aldermen
- Louisville-Jefferson County Air Pollution Control Commission
- Louisville-Jefferson County Health Department
- Louisville-Jefferson County Planning and Zoning Commission
- Metropolitan Sewer District
- New Albany and Jeffersonville Wastewater Treatment Plants
- Sierra Club
- U.S. Army Corps of Engineers
- U.S. Department of Housing and Urban Development

Initially, the Alternatives Evaluation Committee was assisted by KIPDA officials, staff, and consultants. The first meetings involved formal establishment of the committee membership, statement of its purpose, and general orientation to water quality management planning. Thus, the committee became familiar with the overall planning process, its role in that process, and the specific activities carried out up to the time of the committee's involvement. This early orientation was critical, for it provided members of the committee with background data that would enable them to play an effective role in the planning process and, ultimately, to evaluate alternative water quality plans.

C. ALTERNATIVES EVALUATION

The primary goal and responsibility of the committee was to formulate a set of water quality management recommendations for review by appropriate federal, state, and local agencies. The sequence of activities included the following major steps:

1. Review and modification of a set of criteria deemed important in evaluation of alternative sewage treatment systems;

2. Rankings and ratings of the criteria to establish their relative importance through assignment of numerical weights to each criterion;
3. A critical review of each alternative treatment system with particular emphasis on the criteria; and
4. Evaluation of each alternative treatment system in terms of the extent to which a particular criterion might be satisfied by that alternative. (Ideally, the evaluation should be performed for all alternatives, one criterion at a time.)

The committee's first task was to determine important factors for selection of a feasible, cost-effective treatment facilities plan. An initial listing of fifteen criteria was taken from Guidance for Facilities Planning (Ref. 6-9), a publication of the Environmental Protection Agency, which provides advisory data on procedures. The committee had the option of revising the set of criteria any way they wished; however, the committee elected to retain the original set.

The fifteen criteria and types of information provided by the committee relative to each are:

- Water Quality
- Land Use
- Environmental Effects
- Financing Capabilities
- Reliability
- Government, Management, and Enforcement Considerations
- Project Timing
- Elimination of Existing Wastewater Sources
- Service Area Coverage
- Phasing Out or Upgrading Existing Facilities
- Project Cost
- Energy Availability and Conservation
- Availability of System Resources
- Resource Recovery and Reuse
- Water Supply Compatibility

When the committee members were satisfied with the set of criteria and their understanding of each, they began the process of determining the relative weight of each criterion. The criteria were ranked in order of importance, and then rated on a continuous scale from 0 to 100, where the most important criterion could receive a rating as high as 100. The ranking and rating utility values were averaged to yield the average utility vector. The average vector was the only vector used in subsequent calculations of overall utility for each alternative plan.

The committee members then reviewed the results, and decided a second iteration would be needed to reflect more accurately their thinking on the criterion weights. The calculated utility vectors for the second iteration are shown in Table 6-1.

TABLE 6-1 Criteria Utility Vectors.

Evaluation Criteria	Ranking[a]	Rating[a]	Composite
1. Water quality	0.1314 (1)	0.0950 (1)	0.1132
2. Land use	0.1006 (2)	0.0798 (2)	0.0902
3. Environmental effects	0.0857 (3)	0.0771 (4)	0.0814
4. Financing capabilities	0.0853 (4)	0.0745 (5)	0.0799
5. Reliability, flexibility, and operability	0.0811 (5)	0.0776 (3)	0.0794
6. Government management and enforce-	0.0750 (6)	0.0745 (5)	
ment considerations	0.0750 (6)	0.0745 (5)	0.0748
7. Project timing	0.0724 (8)	0.0738 (7)	0.0731
8. Elimination of existing wastewater sources	0.0735 (7)	0.0697 (8)	0.0716
9. Service area coverage	0.0541 (11)	0.0679 (9)	0.0612
10. Phasing out or upgrading existing facilities	0.0598 (9)	0.0610 (11)	0.0604
11. Project cost	0.0545 (10)	0.0617 (10)	0.0581
12. Energy availability and conservation	0.0411 (12)	0.0495 (13)	0.0453
13. Availability of system resources	0.0339 (13)	0.0521 (12)	0.0430
14. Resource recovery and reuse	0.0328 (14)	0.0481 (14)	0.0404
15. Water supply compatibility	0.0137 (15)	0.0376 (15)	0.0281

[a]Relative order of ranking and rating scores in parentheses.
Source: Schimpeler-Corradino Associates

At the meeting following the ranking and rating exercises, the committee requested that the consultant give a detailed explanation of the criteria. The committee wanted to know the type of information to be used in evaluating each criterion, the source of the information, and something about the technical expertise required to develop the information. The consultant explained these points for each of the fifteen criteria involved.

Before the next meeting, the consultant generated a set of alternative treatment plans for the geographic area under consideration; the data on the criteria were distributed to each judge. Members of the consultant's staff briefly explained each criterion, and the committee members asked many questions and participated in sometimes lively discussions.

Each alternative was then evaluated relative to a given criterion. Each judge assigned an effectiveness value to each alternative on the basis of how well he or she thought the alternative would meet the criterion under consideration. An effectiveness value of 1.0 implied virtual certainty; a value of 0.0 implied virtual impossibility; a value of 0.5 implied no special advantage or disadvantage.

The average was calculated for every alternative of each criterion; the judges then went on to the next criterion and evaluated the alternatives relative to it (Table 6-2).

The alternative plan utility vector was calculated by multiplying the effectiveness matrix by the average criteria utility vector. Each element of the alternative plan utility vector represented the overall effectiveness with which a particular

TABLE 6-2 Criteria Evaluation—Louisville/Jefferson County, Kentucky.

Alternative	Effectiveness Matrix														
	Availability of System Resources	Elimination of Existing Wastewater Sources	Energy Availability and Conservation	Environmental Effects	Financing Capabilities	Land Use	Government Management and Enforcement Considerations	Project Cost	Project Timing	Reliability, Flexibility, and Operability	Resource Recovery and Reuse	Service Area Coverage	Phasing Out or Upgrading Existing Facilities	Water Supply Compatibility	Water Quality
LJC-1	0.5350	0.5582	0.5812	0.6924	0.8288	0.6500	0.5537	0.8535	0.6156	0.5788	0.5662	0.7744	0.6315	0.4797	0.6747
LJC-2	0.7169	0.7588	0.6912	0.6206	0.6900	0.5444	0.7931	0.6406	0.4894	0.6912	0.6015	0.8200	0.6635	0.6187	0.7812
LJC-3	0.8081	0.8018	0.6350	0.6011	0.5565	0.5344	0.8462	0.5388	0.3619	0.7553	0.5915	0.8112	0.4832	0.6687	0.7794
LJC-4	0.5837	0.6076	0.6775	0.5882	0.8129	0.6087	0.6569	0.8018	0.5906	0.6288	0.6185	0.7700	0.6135	0.4625	0.7071
LJC-5	0.4887	0.4812	0.5187	0.5076	0.5982	0.5750	0.4475	0.7071	0.5739	0.3794	0.5685	0.7469	0.6265	0.3313	0.4365
LJC-6	0.5469	0.5718	0.5712	0.5888	0.3424	0.4675	0.5562	0.2894	0.5800	0.7576	0.5685	0.7606	0.5771	0.4906	0.8024
LJC-7	0.5362	0.5582	0.5712	0.5159	0.4759	0.3750	0.5137	0.4729	0.5703	0.5835	0.5669	0.7687	0.5671	0.4812	0.7894
LJC-8	0.7204	0.7479	0.7306	0.7653	0.8529	0.7259	0.7591	0.8329	0.6759	0.7024	0.6162	0.8219	0.6582	0.5831	0.7906

Source: Schimpeler·Corradino Ass.

alternative satisfied the fifteen criteria. The magnitude of the plan effectiveness value ranged from 0.0 for a worthless plan to 1.0 for a perfect plan; most plan utilities varied from about 0.3 to 0.7.

An important aspect of water resource planning is the cost effectiveness ratio. Estimated project cost was divided by overall plan utility and a cost effectiveness ratio determined for each alternative plan. The cost effectiveness ratio is theoretically equivalent to the cost of producing an alternative with a total effectiveness of 1.0. This is, of course, impossible to achieve even with unlimited resources.

D. STATISTICAL ANALYSIS

The committee members generated large quantities of data during the ranking and rating of criteria and determination of the effectiveness values associated with each alternative plan. Much of the data was subjected to statistical analysis in an effort to determine the consistency with which the judges applied the several techniques employed in this management study. Each of the statistical tests discussed previously was used in this analysis.

When a group of individuals ranks or rates a set of criteria, it is often of interest to test whether the rankings and ratings of the several judges are related or unrelated.

The committee decided to use two iterations in the ranking and rating of the criteria; only the results of the second iteration are presented here. The magnitude of W for the rankings was 0.4704, and that of Chi Square was 164.6. This was overwhelmingly significant, inasmuch as tabular Chi Square with 14 degrees of freedom was 29.14 at the 0.01 significance level. Thus, it is concluded that substantial agreement among the judges existed when using the ranking technique.

The ratings were converted to ranks and the magnitude of W was calculated as 0.4693. The value of Chi Square was 164.26. This also is overwhelmingly significant; it was concluded that substantial agreement existed among the judges when using the rating technique as well.

The highly significant values of W for both the ranking and rating technique allows calculation of the utility vectors as the best estimate of the relative importance or weight of each of the fifteen criteria. It may also be concluded that there is substantial agreement between the two techniques.

On this basis, the best estimate of the criteria utility values was considered to be the average value of the ranking and rating utility values. This vector was the only one used in subsequent calculations of overall alternative plan utility. The average utility vector represents a consensual ordering of the opinions of 25 judges utilizing two different weighting techniques.

The within-alternative variances of the effectiveness values assigned to an alternative were computed for each alternative plan. Bartlett's test for homogeneity of variance was performed for each of the fifteen criteria. The results are shown in Table 6-2 for all of Jefferson County.

The F-test was used to test the hypothesis that the mean effectiveness values associated with each alternative were equal. This result is shown in Table 6-2 as well.

In general, a desirable result is the acceptance (nonsignificance) of the hypothesis of equality of variance and rejection (significance) of the hypothesis of equality of means. This situation implies the judges had a clear understanding of the relationship of the criteria to an alternative plan, and were able to discriminate among plans in assigning effectiveness values. Table 6-3 shows that in the majority of instances, the within-alternative variances were equal, indicating stability of preference among the judges, and the among-plan mean effectiveness values were different, indicating the ability of the judges to discriminate among plans.

TABLE 6-3 Louisville/Jefferson County Equality of Mean and Variance Alternative Plan Effectiveness Values.

Criterion	F-Statistic (Analysis of Variance)	Chi Square (Bartlett)
1. Water quality	11.74[a]	2.94
2. Land use	7.85[a]	10.83
3. Environmental effects	2.63[a]	7.90
4. Financing capabilities	3.56[a]	8.68
5. Reliability, flexibility, and operability	17.26[a]	9.95
6. Government management and enforcement considerations	5.41[a]	2.08
7. Project timing	12.27[a]	27.44[a]
8. Elimination of existing wastewater sources	23.68[a]	7.50
9. Service area coverage	2.35[a]	2.73
10. Phasing out or upgrading existing facilities	8.29[a]	2.09
11. Project cost	0.16	0.76
12. Energy availability and conservation	0.37	1.50
13. Availability of system resources	1.10	18.32[a]
14. Resource recovery and reuse	4.07[a]	1.68
15. Water supply compatibility	9.82[a]	9.75

[a]Significant at X-level of 0.05.
TAB Chi Square (0.95) (s-1) = 14.1
TAB F (0.95) (s-1, $s[m$-1]) = 2.05
(s = number of alternative plans; m = number of judges).
Source: Schimpeler·Corradino Associates.

TABLE 6-4 Alternative Selection—Louisville/Jefferson County, Kentucky.

Alternative	Total Plan Effectiveness	Rank	Project Cost	Project Cost ÷ Total Utility = Cost Eff. Ratio	Cost Eff. Ratio Rank
LJC-1	.6507	5	382,549,000	587,904,000	3
LJC-2	.6787	2	428,269,000	631,014,000	4
LJC-3	.6532	4	521,708,000	798,696,000	6
LJC-4	.6580	3	385,170,000	585,365,000	2
LJC-5	.5317	8	391,172,000	735,701,000	5
LJC-6	.5776	6	829,440,000	1,436,011,000	8
LJC-7	.5636	7	549,449,000	974,892,000	7
LJC-8	.7446	1	383,323,000	514,804,000	1

Source: Schimpeler•Corradino Associates.

Significant values in Bartlett's test are not overly important in the conduct of the evaluation methodology. However, such values did indicate broad disagreement among the judges, relative to the effectiveness values assigned to a few of the alternatives. This disagreement could have resulted from a lack of understanding of the criterion and its relationship to the plan. These same arguments could hold for nonsignificant values in testing the equality of mean effectiveness values. However, a more logical explanation is that, in certain instances, the alternative plans were actually, as a group, indifferent toward the criteria. For example, only two criteria, 6 and 9, exhibited homogeneity of variance in all five evaluations. Based on this analysis, it appears that two criteria, Energy Availability and Conservation and Resource Recovery and Reuse, were of little value in the evaluation process.

The mean effectiveness values were arrayed in a matrix of s rows (s = number of alternative plans) by n columns (n = number of criteria). The product of the effectiveness matrix and the average criteria utility vector of Table 6-1 yields the overall or total alternative plan utility, as shown in Table 6-4. Table 6-4 includes the cost effectiveness of each plan, which was calculated by dividing total project cost by overall plan utility.

In summary, analysis of the data supports the general conclusion that the committee members were able to perform adequately the many tasks demanded by the evaluation methodology.

6-6.2 303(e) River Basin Planning—State of Kentucky

The Public Participation Subcommittee (PPS) was established at the beginning of Kentucky's river basin management planning effort. Its purpose, membership, and activities during preparation of Kentucky's river basin management plan are an example of implementation of provisions of Section 101 (e) of the Federal Water Pollution Control Act Amendments of 1972 and applicable rules and regulations promulgated by the Administrator of the Environmental Protection Agency.

One of the most important objectives of the PPS was coordination of the activities of the eight consultants involved in the statewide river basin program to assure compliance with legal requirements. Coordination was also obtained with appropriate public information and supporting planning agencies. In this capacity, the Subcommittee was charged with outlining the public participation efforts of the eight consultants throughout the planning study and reviewing the execution of public information actions. Activities by the consultants included development of news releases for state and regional press agencies, scheduling and conduct of official meetings between consultants and government agencies, and workshops for general public participation. In addition to reviewing the activities of the individual consultants, the Subcommittee assisted the Department for Natural Resources and Environmental Protection in setting up the statewide public hearing.

Early in the course of the study, the PPS defined and transmitted to all participating consultants a public participation program consisting of three levels of interaction among the Kentucky Department for Natural Resources and Environmental Protection, the consultants conducting the individual basin studies, and the state's citizens and public officials. Level One was designed as a public information function, providing the public with pertinent facts concerning the river basin management planning effort. Dissemination of information was considered an educational process.

Level Two established the need for direct contact between consultants and public officials responsible for planning and development activities in the various river basins. Level Two also included a regional workshop concept, through which the general public and interested officials of local government and quasi-governmental agencies could personally interact with planning consultants in an informal atmosphere of discussion and learning.

Level Three was the public hearing phase of the planning effort. At this level of participation, the public was offered the opportunity to comment formally on the planning effort as a matter of public record. The public hearing phase was conducted at the end of the initial planning period.

During the planning period, the PPS attempted to provide guidance to all consultants as they undertook the recommended public participation program. To assist the consultants in presenting material to the Area Development District Boards, a general outline providing for presentation of uniform yet flexible information was developed and disseminated. Sample articles for basin-specific news releases were also distributed, to assure uniform exposure of basic facts concerning each river basin plan and the state's total planning effort. Finally, due to the desire for uniformity in publication of all river basin plans, a standard format was developed for the chapter on public participation to be included in final documents on each of the ten river basin plans.

Level One: Public Information and Education. The Subcommittee's recommendations for public information and education envisioned a reciprocal process between the river basin management and the statewide and regional news media. Press releases were prepared explaining the general purpose and nature of the statewide planning effort. "Clipping services" and other similar sources were used to monitor the growth of an informed public.

In cooperation with the State Public Information Department, the Subcommittee prepared two news releases dealing with KDNREP's overall planning efforts. In addition, an article of general information appeared in the publication *Focus* (prepared and distributed by the Office of Local Government). These articles provided the basic background information on the river basin planning program, characterized the efforts of the various consultants and the Division of Water Quality of KDNREP, indicated how the program was funded, and listed the expected results of the study. Figure 6-2 reproduces the first in this series of articles.

The Subcommittee also advised participating consultants to prepare summary articles for each river basin, to be reviewed by KDNREP. The Subcommittee coordinated dissemination of these articles through the Kentucky Department of Public Information to regional news agencies in specific basin areas. These articles presented general information similar to the statewide news releases described above, in addition to specific facts and preliminary conclusions relating to individual basin areas. A published sample article was forwarded to all consul-

10 September, 1974

PLANNING FOR WATER QUALITY IMPROVEMENT IN KENTUCKY

FRANKFORT, KENTUCKY. Representatives of the U.S. Environmental Protection Agency, Kentucky Department for Natural Resources and Environmental Protection (DNREP), and major national and local consulting firms met at the Capitol Plaza Tower today to discuss preparation of River Basin Plans for Kentucky. The meeting is part of a $930,000 effort of the State and Federal Government to prepare plans for the improvement of water quality in Kentucky's ten river basins.

River Basin Plans are being prepared by consulting engineering firms for the Big Sandy, Green, Kentucky, Licking, Lower Cumberland, Mississippi, Ohio, Salt, Tennessee and Upper Cumberland River Basins. The effort is being directed by Mr. Walter C. Martin, Chief Sanitary Engineer for the Division of Water. The University of Kentucky is also participating in the preparation of material which will be included in the Plans.

Fig. 6-2. News release, fall, 1974

River Basin planning is an activity required under the Federal Water Pollution Control Act Amendments of 1972. The Act provides substantial Federal Funds for the improvement of municipal wastewater treatment plants and Kentucky's water quality management policy. Since 1972, Kentucky has received almost 65 million dollars to distribute to towns and cities in Kentucky needing improved wastewater treatment. Another 65 million dollars has been authorized for distribution in 1975. The River Basin Plans prepared this year will establish the basis for the distribution of State and Federal Funds for municipal wastewater treatment over the next five years.

A major objective of the river basin planning process is the preparation of each plan in such a way that the information may be combined in a meaningful fashion with similar information from other plans. This objective addresses the need of the Division of Water to standardize and coordinate its assistance to local communities throughout Kentucky.

Funds for water quality improvement are limited. Some worthy improvement of municipal wastewater treatment plants must now wait several years for State and Federal money. Funding the most worthy projects first is the objective of the Division of Water. Each River Basin Plan will identify the standards for maintaining water quality, sources of water pollution, and a schedule of actions which must be conducted to prevent continued water pollution. This information will be used by the Division of Water to establish priorities for water pollution control actions in Kentucky.

Achieving similarity in the description and evaluation of pollution problems and water pollution control needs among the River Basin Plans will assist the Division of Water to allocate funds to worthy water pollution control projects throughout Kentucky. This will bring the greatest return in water quality improvement. Furthermore, this compatibility among the River Basin Plans will reduce the time and effort expended by the Division of Water to prepare its annual revision and update of its priorities. The River Basin Plans are expected to be formally updated every five years.

Compatibility among the River Basin Plans is being achieved by standardizing planning methods and specialized techniques for establishing water pollution control requirements. Planning methods are being standardized by committees comprised of the River Basin Plan consultants and representatives of Kentucky DNREP and the Environmental Protection Agency. As at today's meeting, these men discuss the goals, methods, specialized techniques and progress of the River Basin Plans. They meet monthly to discuss the efforts of subcommittees and finalize the format of specific elements of each plan.

One of the most unique elements of the planning process to date has been the development of a specialized technique for the determination of water pollution control actions required in Kentucky's polluted streams. This work has been conducted at the University of Kentucky under the direction of Dr. Edward G. Foree. Using measures of water pollution and water quality in actual Kentucky streams, a technique was developed which would allow the Division of Water to predict the level of improvement in municipal wastewater treatment needed to correct water pollution from sewage. Special attention was given to the development of this technique for use on the many Kentucky streams which cease to flow in the summer. This work will standardize the evaluation of municipal wastewater treatment needs throughout the State.

Forthcoming meetings on River Basin Planning will be focused at the development of environmental inventories and means for public participation in the planning process.

Fig. 6-2. (Continued)

tants to help them prepare uniform descriptions of their river basin management planning efforts. Main elements of the article included:

Identification of the consultant(s) working in the individual basins;

Specific analysis of basin water quality problems;

A description of expected results;

An outline of the general work program performed in the basin area; and

A brief definition of the planning effort's purpose and ultimate impact.

The sample article used in assisting consultants in this phase of the public information and education activity is reproduced here as Figure 6-3.

contact Marva Gay (502) 564-3350
Kentucky Department for Natural Resources and Environmental Protection

(*note to editor: for distribution in western, south central and north central regions*)(*Release by April 25*)

Approved by: Ed Findley & George Nobel of
Weston, Inc.
Walter Martin

WATER NEEDS INDICATED BY RIVER BASIN STUDY

By Marva Gay

Frankfort, Ky.–Madisonville, Elizabethtown and a few smaller communities in the Green River Basin are reaching the limits of their local water resources according to preliminary findings of an environmental consulting firm.

The Wilmette, Illinois firm of Roy F. Weston, Inc. is studying the Green River Basin to determine its existing water quality and the effect each significant discharge into a stream has on its water quality. The firm will analyze that information to determine how much the waste water discharge must be reduced to meet state and federal water pollution standards.

According to project engineer Edward Findley, when the water quality management planning is finished, probably in July, a management strategy for water pollution control and a ranking of municipalities according to their need for federal funds for construction of sewage treatment facilities will be the major feature of the plan.

The Green River Basin plan will then be combined with similar plans being done by other consulting firms in each of Kentucky's ten river basins.

The $830,000 statewide effort is being funded from the governor's contingency fund and the division of water quality in the Kentucky Department for Natural Resources and Environmental Protection. A part of the planning effort is being supplemented through a grant of $80,000 from the U.S. Environmental Protection Agency.

A statewide water quality management plan will result from the information contained in the studies of each river basin. This management plan is required by the Federal Water Pollution Control Act of 1972 and will be used by the division of water quality to administer the construction grant program.

Fig. 6-3. News release, April 14, 1975

Municipalities throughout the state will be ranked according to their need for federal grants to construct waste water control facilities. The top of the list will have first priority to receive the federal grants.

A public hearing will be held on that ranking when it is complete, sometime this summer.

According to Findley, indications are several towns in the Green River Basin will need larger water supply and greatly improved treatment and disposal of sewage if they are to combine economic growth with a wholesome urban environment.

Findley added that the early findings suggest that pollution of surface water by acid mine drainage in the western portion of the basin adds to the problem of water quality improvement in the basin. Findley said the acid drainage is caused primarily by abandoned mines and, to some extent, by other mining operations.

Weston's study of the Green River Basin began in 1974. Initially, the firm's representatives began a campaign to inform federal, state and local agencies in the basin about the goals and objectives of the study and to obtain their advice about major economic, water resources and environmental issues in the area.

Those contacts included the state office of the U.S. Department of Agriculture Soil Conservation Service, the Louisville district of the U.S. Army Corps of Engineers, various divisions of the state environmental protection agency and the state commerce department plus Area Development Districts having jurisdiction in the basin. Because the Green River Basin takes in portions of 27 counties, the Barren River, Green River, Lake Cumberland, Lincoln Trail and Pennyrile Area Development Districts were contacted.

Next, an engineer and a naturalist photographer from Weston, Inc., plus a terrestrial and aquatic biologist from the state division of water quality surveyed the Green River and its tributaries which included the three major hydro-geographic regions of the basin. Those three regions are the headwater impoundment at Green River Reservoir, navigable channels through Mammouth Cave National Park and from Paradise to Calhoun, and the Barrens limestone region drained by the Barren River.

A photographic record and log was kept of the industrial, recreational and municipal uses of the waterways. That record was explained at the first 1975 meetings of the Green River and the Barren River Area Development Districts.

According to Findley, as the firm approaches the end of the planning process it will make greater effort to obtain support from public officials for the development of the management strategy.

Fig. 6-3. (Continued)

Level Two: Public Involvement. The second level of public participation encompassed three phases of consultant contact with residents and public officials. Phase One consisted of presentations to Area Development District (ADD) Boards. Guidelines for these presentations were prepared and distributed by the PPS to encourage uniformity in advising the Boards of the central purposes of the basin planning program. Figure 6-4 reproduces a sample of these guidelines. The Sub-committee coordinated these meetings between ADD Boards and consultants through the Office for Local Government. An introductory memoran-

I. Background
   A. Review of PL 92-500, Federal Water Pollution Control Act Amendments of 1972.
   B. Discussion of the Scope and Objectives of the Kentucky River Basin Management Planning Study.
   C. Relationships Between and Among Water Quality-Related Planning Programs, such 303e, 402, 201, and 208.
   D. Funding Levels (State and Federal)

II. Specific Basin Planning Process
   A. State Role
   B. Consultant Role (Scope of Services)
   C. University Role

III. Conclusions
   A. Outputs from the Basin Planning Program
   B. Establishment of Priorities for Capital Funding

Fig. 6-4. Suggested outline for ADD board presentations

dum, shown in Figure 6-5, was transmitted from the Executive Director of the Office for Local Government to all ADD Directors. This was followed with direct contact by the consultant.

The Subcommittee considered it essential that each presentation stress the "theoretical" interrelationships of the various water quality programs (303e, 402, 201, 208), although the outline was only a recommendation. The use of graphic illustrations was encouraged. The progress of these Board/consultant meetings was monitored; consultants were encouraged to hold the meetings early in plan development.

CAPITOL ANNEX–3rd Floor

MEMORANDUM

TO: Area Development District
Executive Directors
FROM: Allan D. Spader, Executive Director
Office for Local Government
DATE: December 17, 1974
SUBJECT: River Basin Management Plans

As many of you are aware, the Federal Water Pollution Control Act Amendments of 1972 (PL 92-500) require each state to complete river basin plans for each river basin in the state. The Department for Natural Resources and Environmental Protection has the responsibility for developing these plans in Kentucky. The Department has contracted with eight consulting firms (see attached list) to perform the ten basin plans.

Fig. 6.5. Introductory memorandum

As I indicated at your meeting on Friday, part of the public involvement and participation process requires that the consultants make contact with local officials in each basin. Therefore, as a means to acquaint these officials with the purpose of the plans, their status, and to solicit initial questions and concerns, the consultants are requesting an opportunity to appear before each ADD Board of Directors to make a brief presentation. Some of you have already been contacted, others have not. Each of you should be contacted in the very near future. (These presentations will be followed by other individual contacts, public hearings, and other efforts.)

The Department for Natural Resources and Environmental Protection and our Office would appreciate your assistance in scheduling consultants to appear before your boards. Should you have questions or concerns about the river basin effort, please feel free to relay them to me or Elvis Simmons. They can be addressed at the monthly meetings of the River Basin Management Team which meets in Frankfort.

ADS:ELS:es

Fig. 6-5. (Continued)

Phase Two of public involvement included seven public workshops conducted under the formal sponsorship of the Kentucky Environmental Quality Commission, which is advisory to the Governor and KDNREP. The workshops had two purposes: first, to provide the public, governmental, and quasi-governmental agencies information concerning the river basin management process; and, second, to provide an opportunity for informal discussion between citizens and consultants on issues of concern to the general public.

Organization. To foster participation in the workshops, the Environmental Quality Commission developed a basic mailing list for specific invitations. The list was divided into two general categories: groups and individuals throughout the state to be notified of the entire schedule; and groups and individuals to be notified of one or more specific workshops in their immediate locale. The first category included Kentucky citizens involved in the Ohio River Valley Water Sanitation Commission (ORSANCO) and Ohio River Basin Planning Programs, as well as representatives of the Audubon Society, Sierra Club, sewer treatment plant operators, U.S. Army Corps of Engineers, U.S. Geological Survey, U.S. Soil Conservation Service, Kentucky Municipal League, and state universities. The second category was composed of elected officials, Area Development District Boards, appropriate environmental committees and officials, local Chambers of Commerce, local Leagues of Women Voters, and local sportsmen's clubs.

In some cases, more than one consultant conducted a river basin study within the regional area encompassed by the workshop; therefore, a lead consultant was designated by the Subcommittee to initiate the presentation and discussion. The presentation developed around the following program elements recommended by the subcommittee:

1. Introductory remarks by the Environmental Quality Commission member chairing the workshop;

2. A general discussion including legal and administrative requirements by the lead consultant;
3. Discussion of the relevant critical issues involved in each basin by the respective consultant.

To provide citizen participants with substantive material for discussion, the Subcommittee recommended that specific information regarding the basin study be reproduced for distribution at the workshop sessions. A ten page handout containing a summary of information was distributed. The Subcommittee advised the consultants to be prepared to follow through on significant comments or questions raised at the workshop sessions. Consultants were specifically instructed to forward significant comments or questions to the state for review and to respond directly if necessary.

At the close of the session, the Environmental Quality Commission member chairing the workshop explored with those attending the idea of forming a permanent citizens advisory group for basin-specific projects. It was hoped this would lay the groundwork for a set of permanent citizens advisory groups, each representing the public interest in individual basin areas. Finally, the participants in the workshops were informed of the public hearing to be covered by KDNREP, its location, and date.

Conduct. The seven workshops were conducted during the month of June, after an announcement in regional newspapers and a mailing of approximately 850 special letters of invitation. Figure 6-6 reproduces the letter of invitation sent under the dual sponsorship of the Kentucky Environmental Quality Commission and the Kentucky Department for Natural Resources and Environmental Protection. These letters were sent to organizations and individuals identified by the Environmental Quality Commission. The general mailing list for the letters of invitation is presented below.

Area Development District Boards
Audubon Society
Chambers of Commerce
County Judges
Environmental Quality Commission Mailing List
Kentucky Members of Ohio River Basin Planning Program
Kentucky Members of ORSANCO
Leagues of Kentucky Sportsmen
Leagues of Women Voters
Local and Regional Health Departments
Municipal Government Officials
Sierra Club
U.S. Soil Conservation Service

In addition, during the study process, consultants held informal review sessions with various local officials throughout the state.

RIVER BASIN MANAGEMENT PLANNING WORKSHOP

A one million dollar study of water quality in Kentucky, begun months ago, is almost completed. The ten river basins in Kentucky have been closely monitored by ten consulting firms, testing sewage treatment, water recycling and water usage. These results are ready for presentation to administrators, environmental groups, legislators, and interested citizens.

The workshop will offer information to prepare you for the public hearing on the river basin management process, the choices available for implementation of federal regulations, and a suggested ranking, in order of need, of communities that will be eligible for federal grants to improve local water supplies and treatment. The meeting is an open forum–the consulting firm needs your ideas and suggestions in order to consolidate requests and suggestions to make to the government. In addition, the Environmental Quality Commission, sponsor of the workshop, hopes to establish a permanent citizen advisory group for the basin project to participate in future basin management programs.

The River Basin Management Planning Workshop will be on June 3, from 10 a.m. until 2 p.m., at Eastern Kentucky University in Richmond, KY. Participants will meet in the Adams Room, Wallace Building. Luncheon will be available from the college cafeteria in the Powell Building. Kindly advise, as soon as possible, on the enclosed postcard if you will attend.

John Hoffman, Secretary
Department for Natural Resources
and Environmental Protection

Jackie Swigart
Chairperson

Fig. 6-6. Workshops–letter of invitation

Level Three: Public Hearing. The Public Participation Subcommittee assisted KDNREP in reviewing the need for a public hearing process to meet the requirements of the federal law. This process endorsed the concept of one formal public hearing. Figure 6-7 reproduces the notice issued by KDNREP announcing the hearing.

NOTICE

Notice is hereby given that the Kentucky Department for Natural Resources and Environmental Protection will hold a public hearing on June 27, 1975, in the Auditorium of the Capital Plaza Tower, Frankfort, Kentucky at 10:00 AM, EDT, to receive public comment on the fiscal year 1976 ANNUAL PROGRAM PLAN for the Division of Water. The plan is intended as a management tool which will establish goals, allocate resources and assess progress. It will contain a strategy to accomplish water quality improvement. The plan will include a priority ranking for applicants for construction grants for public treatment works under Title II of the Federal Water Pollution Control Act, 1972 of 33 USCA 1251, et. seq. The Department solicits oral or written comment regarding the plan at the hearing. In preparing comment, consideration should be given to Title I, II and III, 33 USC 1251, et. seq., (P. L. 92-500) 40 CFR 35.400 to 35.560-2, 40 CFR 35.900 to 35.960 and 40 CFR 130 and 131. The hearing record will be held open for 3 days following the hearing to receive written comment. Written comment should be addressed to Mr. Harold T. Snodgrass, Director, Division of Water, Department for Natural Resources and Environmental Protection, 6th Floor,

Fig. 6-7. Public Hearing Notices

Capital Plaza Tower, Frankfort, Kentucky 40601. Copies of the plan may be reviewed at the Division of Water, Department for Natural Resources and Environmental Protection, 6th Floor, Capital Plaza Tower, Frankfort, Kentucky 40601.

PUBLIC NOTICE

Notice is hereby given that the Kentucky Department for Natural Resources and Environmental Protection will hold a public hearing on June 27, 1975, in the Auditorium of the Capital Plaza Building, Frankfort, Kentucky at 1:30 P.M., EDT, to receive public comment on the River Basin Water Quality Management Plans for the Division of Water. The plans are intended as a management tool which will establish goals, allocate resources and assess progress. They will contain data from which a strategy to accomplish water quality management can be developed. The plans will include a priority ranking for construction grants for public treatment works under Title II of the Federal Water Pollution Control Act, 1972 of 33 USCA 1251, et. seq. The Department solicits oral or written comment regarding the plan at the hearing. In preparing comment, consideration should be given to Title III, 33 USC 1251, et. seq., (P. L. 92-500) 40 CFR 130 and 131. The hearing record will be held open for 3 days following the hearing to receive written comment. Written comment should be addressed to Mr. Harold T. Snodgrass, Director, Division of Water, Department for Natural Resources and Environmental Protection, 6th Floor, Capital Plaza Tower, Frankfort, Kentucky 40601. Copies of the plans may be reviewed at the Division of Water, Department for Natural Resources and Environmental Protection, 6th Floor, Capital Plaza Tower, Frankfort, Kentucky 40601.

Fig. 6-7. (Continued)

Conclusion: Continuing Program. The requirements for public participation extend beyond the period of the river basin management plan's preparation and adoption. The continuing planning program of the state, which must be annually reviewed by the U.S. Environmental Protection Agency, envisions a "continuing program for public participation" that is both required by law and necessary for adoption and effective implementation of water pollution control and overall water quality planning in the Commonwealth, including the river basin management plan. Legal requirements for public participation in water quality and pollution control planning for the continuing planning program are identical to those outlined earlier.

The Management Subcommittee submitted four recommendations to the Department for Natural Resources and Environmental Protection to follow in expediting an effective, continuing river basin management program. The recommendations are:

1. Formation of 303(e) Technical Advisory Committee;
2. Establishment of decentralized Division of Water Quality regional basin offices responsible for one or more river basins;
3. Formation of both regional and statewide River Basin Citizen Advisory Committees to provide public participation in continuous updating of the state's river basin plans; and

4. Use of a computerized data management system for data distribution to 201 and 208 planners and for updating of management information.

In elaborating on the establishment of a citizens advisory committee, the Management Subcommittee developed an approach for continuing citizens participation. The Management Subcommittee stated:

> "Since the continuous updating of river basin planning efforts should include viable citizen input and participation, it is proposed that two distinct levels of citizen input be provided to insure continuing river basin planning activities through the formation of regional and statewide citizen advisory groups.
>
> Regional River Basin Citizen Advisory Committees would be composed of representatives from various public sectors in the region including: interested citizens, public officials, industries, environmental groups, individuals possessing technical expertise, etc. A Regional River Basin Citizen Advisory Committee would provide external advice to the regional basin office through the respective basin engineer with regard to the review and updating of basinwide plans and priorities.
>
> One representative from each regional River Basin Citizen Advisory Committee would be selected to serve on the statewide River Basin Citizen Advisory Committee. The function of this statewide committee would be to provide external advice to the Division of Water (Quality) in development and review of its Annual 106 Program Plan.
>
> The formation of these two citizen advisory groups should provide viable public input to the river basin planning process from both a regional or river basin basis and an overall statewide basis."

REFERENCES

6-1. Highway Research Board, *Citizen Participation in Transportation Planning*, Highway Research Board Special Report 142, Washington, D.C., 1973, p. 4.

6-2. Willeke, Gene E., "Citizen Participation: Here to Stay," *Civil Eng., ASCE*, January, 1974, pp. 78–82.

6-3. Kennedy, J. F., "Address on the First Anniversary of the Alliance for Progress," presented before the Diplomatic Corps, Washington, D.C., March 13, 1962.

6-4. Public Law 92-500, Federal Water Pollution Control Act, Amendments 1972.

6-5. Males, R. M., "Design of Mass Media Programs." Submitted to the Office of Water Resources Research, Department of Interior, Engineering Sciences, Inc., and W. E. Gates and Associates, February 1974.

6-6. Siegel, S., *Nonparametric Statistics*, McGraw-Hill Book Company, Inc., New York, 1956.

6-7. Kendall, M. G., *Rank Correlation Methods*, Griffin, London, 1948.

6-8. Brownlee, Kenneth A., *Statistical Theory and Methodology in Science and Engineering*, John Wiley and Sons: New York, 1960, in cooperation with the University of Louisville Institute for Industrial Research.

6-9. Schimpeler-Corradino Associates, *Plan Development and Implementation, Water Quality Management Plan*, Vol. II, Kentuckiana Regional Planning and Development Agency, Louisville, Ky., 1974.

7

Environment assessments in water quality management planning

John E. Heer, Jr.
and D. Joseph Hagerty

Civil Engineering Department
University of Louisville
Louisville, Kentucky

7-1. INTRODUCTION

Comprehensive information and widespread publicity on various environmental problems have made the American public aware of some of the difficulties of life in the 20th Century. Today's citizen is well informed concerning the problems of population explosion, shortages in energy supplies, depletion of mineral resources, urban sprawl, and a host of other problems, many of which were created by today's technological society. The industrial revolution brought with it a revolution in environment, with a dramatic increase in the number and quantities of pollutants. This text is devoted to the management of water resources. Thus, it addresses problems of resource exploitation as well as difficulties with pollution.

As Americans have become aware of the problems of environmental degradation and depletion of resources, they have become increasingly aware of the need for comprehensive environmental planning. There are a number of different philosophies concerning either oil, uranium, wilderness, or water.

Even with this wide diversity in opinion, however, almost all Americans now agree that it is necessary to "look before we leap"—to investigate the consequences of any proposed action before we undertake that action. This concept has been expressed in law in the National Environmental Policy Act of 1969. This law contains a requirement for the preparation of comprehensive Environmental Impact Statements on any actions significantly affecting the human environment. Environmental Impact Statements and associated environmental assessments can be used to identify the detrimental consequences of particular plans for water resources utilization, and can be used to develop efficient means for the prevention and abatement of water pollution. Because the environmental assessment is such a powerful tool in environmental planning and water resources planning, this tool should be familiar to all individuals engaged in water resources planning.

7-1.1 History of the National Environmental Policy Act

In 1969, the National Environmental Policy Act, the culmination of almost 15 years of attempts by the Congress to define a national policy which would guarantee conservation and improvement of environmental quality, was passed. The most significant single provision in the National Environmental Policy Act is a requirement for an impact statement addressing the predicted impacts of a proposed action. This impact statement requirement actually was a last-minute addition to the bill introduced by Senator Henry Jackson. When this requirement was added to the basic bill, it had been given little consideration as a measure which could have far-reaching consequences of great magnitude. It is ironic that the most significant portion of this law was passed as a last-minute measure, since the passage of the law had been preceded by a long history of formulation, analysis, and compromise.

In 1959, in the 86th Congress, Senate Bill 2549, entitled the "Resources and Conservation Act," had been proposed as a measure to coordinate the many federal programs which were supposed to guarantee environmental conservation. A number of federal programs for environmental conservation and improvement were in existence at that time, with their inevitable overlap and conflict. Senate Bill 2549 failed to pass the 86th Congress, and several years elapsed before another bill was introduced. The "Ecological Resources and Surveys Bill" was proposed in 1965 as Senate Bill 2282 of the 89th Congress. This bill contained provisions for the coordination of programs within various federal agencies for the use and exchange of information on environmental quality. Unfortunately, this bill also failed to pass the Congress. Finally, Senator Jackson introduced Senate Bill 1075 in 1969. When he introduced this bill, Senator Jackson stated that the proposed law was intended to accomplish three goals: 1) to formulate a National

Environmental Policy; 2) to authorize research on the conservation and preservation of natural resources; and 3) to establish a council of environmental advisers to the President.

In its original form, the draft of the National Environmental Policy Act contained no operational procedures to ensure implementation of the environmental policy which was formulated in the statement and philosophy of the law. The bill simply required ". . . a national strategy for management of the human environment" (Ref. 7-25). Although a national strategy for management of environmental quality was required, no specific procedures for controlling continuing federal agency activities in management of the environment were included in the bill. In no instance was a requirement stated for the review and coordination of federal activities affecting the environment. After the bill was introduced by Senator Jackson, it was referred to the Senate Interior and Insular Affairs Committee. During committee hearings, the bill was modified to include a provision which would require certain operational procedures. This provision was refined to become the now famous Section 102(2) C, the action-forcing section of the law. This section requires all federal agencies to prepare Environmental Impact Statements for any proposed actions which are considered to have significant effects on the human environment.

Further modification of Senator Jackson's bill occurred during committee hearings as a result of the introduction of a similar measure in the House of Representatives (HR Bill 6750) by representative Dingall of Michigan. After the introduction of the similar measures in the House and in the Senate, a combined House and Senate conference committee synthesized the two proposals into the text, which appeared as the National Environmental Policy Act of 1969. After passage through the Congress, the bill was signed into law by President Nixon on January 1, 1970.

The passage of NEPA marked a major revision in federal policy concerning environmental conservation and protection. Before 1969, federal policy had been directed toward dealing with environmental crises individually as they occurred. The new legislation generally was directed toward anticipating and preventing environmental degradation. The National Environmental Policy Act was proposed by its initiators as a measure to guarantee an interdisciplinary approach to decision-making by all agencies of the federal government concerned with management of the nation's environment. Because of its great significance to environmental management and because of its wide application to federal activities, NEPA has great significance for water resources planning.

7-1.2 NEPA in a Nutshell

The more important features of the National Environmental Policy Act are described in the following paragraphs.

Section 101. According to statements by the initiators of this law, a major goal of the legislation was a coherent statement of National Environmental Policy. This statement of National Environmental Policy is contained in Section 101 of NEPA. At first glance, this section of the law seems to be a rather trite reiteration of basic national beliefs. Such a restatement of well-known truths was not the intention of the individuals who formulated this law, according to comments by Senator Jackson and other members of the House and Senate committees involved in the development of NEPA. According to statements by these lawmakers, Section 101 was intended to establish *definite priorities* in environmental protection. This section was intended to give expression to national goals and ideals with respect to environmental preservation. Additionally, it was intended to provide a *legal foundation* upon which federal administrators could base any decisions in regard to federal actions significantly affecting environmental quality. In other words, this statement of national policy was intended to provide guidance to these administrators in their efforts to evaluate environmental quality and the means to preserve environmental quality, especially when environmental values appear to be in conflict with other societal values. In essence, the policy statement can be viewed as a national commitment to avoid or prevent any actions which would endanger the health or the continued existence of mankind. Section 101 of NEPA was intended to be a statement that Americans no longer would intentionally initiate any action which would do irreparable harm to the land, air, and water of their homeland.

Section 102. The second section of Title I of NEPA contains the "Action-forcing" procedures of the Act, which, as mentioned previously, were added almost as an afterthought. This section of the law contains a requirement that all federal agencies proposing significant actions(significant with respect to environmental quality) first consult with all other appropriate agencies having jurisdiction over, or expertise in environmental matters, concerning the contemplated action. In other words, the administrators of any federal agency must examine all proposed actions and, for those actions considered to have possible significant environmental effects, these administrators should consult their counterparts in other federal agencies concerning the possible environmental effects of the proposed action. Also, this section of the law requires federal administrators to prepare a comprehensive statement on the predicted environmental effects of the proposed action. Detailed provisions for the preparation of this impact statement are contained in the law. These detailed provisions are described in a later paragraph.

The last three sections of Title I of NEPA, Section 103, 104 and 105, contain provisions requiring the administrators of all federal agencies to review agency procedures to ensure that these agency procedures comply with the requirements of the National Environmental Policy Act. Also, in this portion of Title I, the developers of NEPA included a statement to the effect that the statutory obligations previously existing for any federal agency were not to be supplanted by the provisions of NEPA; that is, the requirements in the National Environmental Policy Act were to be considered to be supplementary to existing obligations for federal agency administrators.

Sections 201-207. In Title II of NEPA, a Council on Environmental Quality (CEQ) was established as an advisory body in the Office of the President. The three members of this council are to provide the President with advice on the significance of various federal activities, with respect to their possible effects and impacts on the human environment. Since the purpose of this advisory council included a requirement to maintain continued supervision of federal activities (for possible environmental consequences), the council was required to establish a system for monitoring environmental quality. This system of environmental indicators was required specifically by provisions of the law. Furthermore, in this section of the law, a requirement was stated for the council to maintain comprehensive records on the state of the environment and to ensure that a comprehensive data base is maintained, so that decisions on environmental problems can be made rationally and logically on the basis of sufficient data. The three members of CEQ are appointed by the President to serve at his pleasure, by and with the advice and consent of the Senate. The Council has the duty, in addition to analyzing environmental information, to predict trends in environmental quality, and to assist the President in formulating national policy to improve the quality of the environment.

If the text of the National Environmental Policy Act is read closely, the wording of the law seems to indicate that the members of Congress wished to establish a group of environmental experts (the Council on Environmental Quality), who would examine the national environmental situation in a broad study and who would then establish specific environmental goals and objectives. To date, the Council on Environmental Quality has not functioned in this manner. Since the council is located within the Office of the President, its position as an independent group of advisors has been seriously compromised. In the period since the passage of NEPA, the members of the council often have appeared to speak *for* the President, as his spokesmen, rather than *to* the President, as his advisors.

The intentions of the developers of NEPA seem to have been thwarted with respect to the role of the CEQ; more importantly, the intentions of the federal legislators with respect to impact statements also appear to have been misinterpreted, if not circumvented. According to comments made by the developers of the law during committee hearings and at subsequent times, the provision for a comprehensive statement of expected impact was inserted into NEPA with the intention of developing a concise summary statement (two or three pages) on the predicted environmental effects of proposed federal actions. At the present time, the Environmental Impact Statements which are being developed by federal agencies and submitted for review usually contain hundreds of pages of information and interpretations. These voluminous documents are difficult to read and certainly do not furnish concise information to the public on the possible consequences of proposed federal actions. Because of the wealth of information in many of these statements, it is certainly difficult for a lay reader to ". . . see the forest for the trees."

7-1.3 Consequences of NEPA

The most important consequence of the National Environmental Policy Act, to date, has been the preparation of numerous detailed assessments of the possible effects of actions proposed by federal agencies. Perhaps contrary to the expectations of Senator Jackson and his colleagues, the environmental impact statement (EIS) has emerged as the most important tool in comprehensive environmental planning. Because it is so important, it merits careful examination. When, why, and how are such statements prepared? When is it necessary, particularly with respect to water resources projects, to prepare an Environmental Impact Statement?

In Section 102(2) C of NEPA, the requirement is stated that a detailed report must be prepared on " . . . proposals for legislation or other major federal actions significantly affecting quality of the human environment . . . " The words "major" and "significantly" are somewhat ambiguous and certainly are open to interpretation. For this reason, controversy arose in the days following the passage of NEPA concerning interpretation of these terms. Any ambiguities in the wording of NEPA concerning the requirements for impact statements have since been clarified in federal court decisions. The precedents for analysis of the significance of federal actions with respect to environmental effects were established in two major cases, *Hanly v. Mitchell* (Ref. 7-32) and *Scrap v. United States* (Ref. 7-35). In these decisions, the judiciary branch of the government assumed the role of implementing NEPA by defining the terms in the law. In these decisions, the terms "major federal action" were interpreted in the following way. This language ". . . refers to the cost of the project, the amount of planning that preceded it, and the time required to complete it, but does not refer to the impact of the project on the environment." Also, in these decisions, guidelines for determining the environmental significance of a project were given. The environmental significance of a project could be determined in the following way: ". . . in deciding whether a major federal action will 'significantly' affect the quality of the human environment, the agency in charge, although vested with broad discretion, should normally be required to review the proposed action in the light of at least two relevant factors; 1) the extent to which the action will cause adverse environmental effects in excess of those created by existing uses in the area affected by the action, and 2) the absolute quantitative environmental effects of the action itself, including the cumulative harm that results from its contribution to existing adverse conditions or uses in the affected areas" (Ref. 7-32). Also, in the *Hanly v. Mitchell* decision, the court declared that close public interaction must be maintained with agency officials throughout the evaluation of specific projects, in order to determine the importance and significance of a particular proposed action. In essence, this decision guarantees public participation in environmental planning, in agreement with the intentions of the initiators of the National Environmental Policy Act.

The actions for which federal agencies must prepare impact statements include: 1) proposals or recommendations for legislation; 2) agency reports and evaluations on legislation initiated in other agencies; and 3) projects or continuing programs undertaken by an agency, if these programs have been judged to have significant environmental effects. Up to the present time, in court decisions on this matter, actions considered to have major environmental significance have included actions whose impact is highly controversial on environmental grounds as well as those actions which are precedents for continuing activity or for more sweeping activity, and actions which involve a number of federal agencies. The last-named type of actions is considered to be significant because of the level of agency effort involved (many agencies, much effort).

According to the provisions of the National Environmental Policy Act, responsible officials of federal agencies are required to prepare a draft environmental statement concerning a proposed action and to circulate this draft for evaluation for a period of 90 days. This review period is designed to assure sufficient time for examination of the proposed actions and their possible effects. A final statement, including answers to questions and comments raised by individuals who have examined the draft statement, must be circulated to the public for final review at least 30 days before the implementation of any proposed action. Thus, a review period of at least 120 days is guaranteed for all major federal actions significantly affecting the human environment. During this review process, each federal agency is required to obtain comments and reactions from other agencies (federal, state, and local), from interested parties such as environmentalist groups, and from the general public, at the earliest possible time. In accordance with the requirements of NEPA, the Council on Environmental Quality has issued guidelines for federal agencies in order to ensure an agency review of proposed actions is complete and appropriate (Ref. 7-5). With respect to actions which involve water resources management, the CEQ guidelines specifically require consultation with the following agencies, with respect to any proposed water resources management activity:

For water quality and water pollution control—

- Department of Agriculture
  - Soil Conservation Service
  - Forest Service

Department of the Interior
Bureau of Reclamation
Bureau of Land Management
Bureau of Sports Fisheries and Wildlife
Bureau of Outdoor Recreation
Geological Survey
Office of Saline Water
Environmental Protection Agency
Water Quality Office
Department of Health, Education, and Welfare (Health Aspects)
Department of Defense
Army Corps of Engineers
Department of the Navy (Ship Pollution Control)
Department of Transportation
Coast Guard (Oil Spills, Ship Sanitation)
Department of Commerce
National Oceanic and Atmospheric Administration
Marine Pollution
Department of Commerce
National Oceanic and Atmospheric Administration
Department of Transportation
Coast Guard
Department of Defense
Army Corps of Engineers
Office of Oceanographer of the Navy
River and Canal Regulation and Stream Channelization
Department of Agriculture
Soil Conservation Service
Department of Defense
Army Corps of Engineers
Department of the Interior
Bureau of Reclamation
Geological Survey
Bureau of Sports Fisheries and Wildlife
Department of Transportation
Coast Guard

For comprehensive environmental assessments on particular projects, other agencies which must be consulted are identified by concern in areas of air quality and air pollution control, weather modification, environmental aspects of electric energy generation, toxic materials, pesticides, etc.

Under the requirements of the National Environmental Policy Act, the role of the general public in formulating environmental policy is significant. Every draft Environmental Impact Statement prepared by a federal agency, by law, must be made available to the general public at the same time that it is circulated to other federal and state agencies for review and comment. The average citizen should have ample time to prepare comments on draft impact statements during the 90-day delay between the time of circulation of the draft statement and the date of implementation of any proposed action. Additionally, a 30-day review period is included for review of the final impact assessment. In this connection, it is important to note that federal agencies are obliged to retain and make available, upon demand, all comments received from other agencies and from members of the public on draft and final environmental statements. Also, the final EIS must include an evaluation of all comments received on the draft statement and an attempt to resolve any objections to the proposed action which were raised in the comments received. If a citizen believes that his or her comments and questions made during the 90-day draft review or during the 30-day review of the final statement have not been adequately addressed, he may notify the Council on Environmental Quality concerning this oversight. This public availability of impact assessment reports and statements allows members of the public to be informed concerning the complete scope of the proposed action and the possible effects of various alternatives. Because of this, the public availability of impact statements is perhaps one of the most significant advances in water resources planning which has resulted from passage of NEPA.

All proposed actions covered by Environmental Impact Statements may be reviewed by governmental agencies, by environmental groups, and by the general public; additionally, many actions covered by Environmental Impact Statements must be reviewed by the United States Environmental Protection Agency. This review may not be specifically required by law for every action, but the EPA is charged to review all statements prepared in connection with possible effects from air pollution, water pollution, solid wastes, radiation hazards, noise pollution, pesticides, and drinking water supplies. Almost any major undertaking, in the field of public works, would have significant effects in at least one of these areas. Thus the great majority of impact statements filed with the CEQ will have received review by the Environmental Protection Agency.

After the draft Environmental Impact Statement has been reviewed by all concerned agencies and individuals, the initiating agency can prepare answers to the comments received on the draft, and can incorporate the comments and the answers in a final impact statement submitted to CEQ. The initiating agency can pursue a proposed action, even in the face of unfavorable comment and opposition from all reviewing agencies and individuals, even the Environmental Protection Agency. The reviewing agencies have no authority to stop any project proposed by another federal agency; the reviewers simply function as advisers to the CEQ, who in turn advise the President. In fact, the Council on Environmental Quality itself does not possess statutory authority to stop a project or prevent its initiation. The Council can only make recommendations to the President, who does have the appropriate authority to stop projects which he judges significantly detrimental to national environmental quality. The Council functions primarily as a coordinating body, during the preparation of impact statements; it is not a reviewing agency in the sense that its comments will be appended to impact statements. The Council makes recommendations directly to the President. The absence of review comments on draft impact statements or on final impact statements cannot be taken to indicate Council approval or disapproval of any proposed action.

The effects of the National Environmental Policy Act have been extended and intensified in a number of states wherein state laws have been passed to require environmental assessments on actions other than federally-funded projects. Environmental policy legislation requiring some form of impact assessment has been passed in California, Washington, Delaware, Montana, Wisconsin, North Carolina, and Indiana. Approximately a dozen other states have evaluated proposed environmental assessment laws. The overall effect of the passage of state and local laws to complement the federal laws will be a requirement for comprehensive land use planning at federal, state, and local levels. In other words, comprehensive impact assessment may accomplish the goals of many of the land-use planning laws which have been passed in former years or which are proposed for passage. Comprehensive land-use planning, including that resulting from environmental assessments, can have great social and political importance, as well as environmental and economic significance.

7-2. DEVELOPMENT OF THE IMPACT STATEMENT

As mentioned in the preceding paragraphs, the requirement for full disclosure of the expected impacts from proposed federal actions has proven to be the most significant result of the passage of NEPA. The required impact statement has become significant, principally through interpretation of NEPA and subsequent enforcement of the provisions of the Act by the federal courts. Judicial rulings have established the ground rules for adherence to the provisions of NEPA and, in some cases, have even extended to interpretation of the requirements for contents of an Environmental Impact Statement.

7-2.1. Enforcement of NEPA

During the committee hearings on Senate Bill 1075, several committee members and a number of witnesses testifying to the committee expressed the idea that the National Environmental Policy Act would be enforced principally through actions of the Office of Management and Budget (OMB). The Office of Management and Budget has the institutional power to regulate expenditure of funds in federal agencies. This power appeared to many individuals to give OMB the authority necessary to enforce the stipulations of NEPA (Ref. 7-6). However, the lawmakers who framed the National Environmental Policy Act did not delegate specific authority for enforcement of NEPA to any particular agency. Since the passage of the law, personnel of the Office of Management and Budget have been very reluctant to assume any responsibility for enforcement of NEPA through withholding or allocating funds on the basis of agency compliance with the requirements of the Act. Moreover, OMB personnel have been reluctant even to create guidelines for assessment of the impact of OMB policies, even though such guidelines have been required by the Council on Environmental Quality (Ref. 7-2). No executive office or federal agency has come forward to take a leading role in implementing NEPA.

NEPA was a reform measure, with respect to governmental agency procedures. Because of this characteristic, the task of enforcement of the Act devolved upon the judicial branch of government. The federal courts, through interpretation of the provisions of the Act, have accepted the responsibility for implementing this law. Apparently, many federal judges have considered NEPA to be a powerful tool to wield in creating a general reform of governmental policy. Much of the judicial review of federal agency actions in recent years has been directed toward achieving systematic decision-making on the part of federal agency personnel. In several cases, federal judges have strongly stressed the need for full evaluation of alternative courses of action and more comprehensive assessment of the public interest. Federal judges have called for greater public participation in governmental decision making. NEPA has proven to be a powerful and convenient mechanism for review of agency activities, especially with regard to public participation in decisions affecting environmental quality.

The judicial branch of government first became involved in interpreting the requirements of NEPA as a result of challenges to federal agency actions by citizen groups. Early court decisions, made soon after the passage of NEPA, were based generally on matters of procedure, and on considerations of whether or not the procedures called for in NEPA had been followed by the initiating agency. In more recent decisions, however, federal judges have issued injunctions against actions proposed by federal agencies, on the grounds that certain crucial elements have not been considered in environmental assessments, or that prejudicial factors had entered into agency planning (Ref. 7-28). Several important trends are apparent in the history of judicial interpretation of NEPA.

For example, federal judges have been zealous in establishing lower limits of applicability for NEPA. The necessity to consider projects of small or moderate size has been emphasized. These projects must be considered because of their possible long-term consequences, and because of the secondary impacts which may be created by relatively minor primary effects. Also, the initial guidelines issued by the Council on Environmental Quality (calling for the formation of specific guidelines on the methodology of environmental assessments by every federal agency) required an impact statement when the agency policy specifically required preparation of such a statement. Federal agencies are being held to the letter of the guidelines that they themselves prepared.

Another trend which is apparent in the judicial interpretation of NEPA is a move to identify controversial or unique types of projects. In identifying these special classes of projects, the courts have based their decisions on previous rulings, on agency requirements, and on the stated intentions of the lawmakers who framed the National Environmental Policy Act. All highly controversial projects are considered to require impact assessments. In this context, federal judges have interpreted the word "controversial" to refer to ". . . size, nature or effect of the major federal action rather than to the existence of opposition to a use, the effect of which is relatively undisputed" (Ref. 7-31). Nevertheless, if the guidelines of an initiating agency include a definition of "controversial" as "meeting strong public opposition," then the courts have ruled that for the agency in question, all projects meeting strong public opposition must be considered controversial and must be the subjects of environmental assessments.

One of the most important results of judicial reviews of the National Environmental Policy Act has been the establishment of limits to the judicial authority to review impact assessments and statements, and the definition of agency responsibilities in this area. The question of judicial authority with respect to interpretation of NEPA was decided principally in two landmark cases, *Environmental Defense Fund v. Hardin* (Ref. 7-29) and *Scrap v. United States* (Ref. 7-35). The essence of these court decisions was that the wording and the spirit of NEPA itself provided grounds for judicial review. Next, the delineation of agency responsibilities, with respect to implementation of NEPA, was decided in a series of very important cases. In Section 102 of NEPA, the phrase ". . . to the fullest extent possible. . . " was taken to mean exactly what it said, in the strictest and most literal interpretation. In the widely quoted case of *Calvert Cliffs' Coordinating Committee v. Atomic Energy Commission* (Ref. 7-27), the courts declared that the text of NEPA itself established very important procedural guidelines, and that federal agencies must comply with those guidelines. In his decision on the case, Judge Wright stated, "Section 102 Duties are not inherently flexible. They must be complied with to the fullest extent, unless there is a clear conflict of *statutory* authority. Considerations of administrative difficulty, delay and economic cost will not suffice to strip the section of its fundamental importance" (Ref. 7-27). This decision and several other landmark cases indicate that it is very important to know precisely what requirements are stated in NEPA for the contents of an Environmental Impact Statement.

7-2.2 Impact Statement Contents

Three basic requirements have been identified by the federal courts in the wording of Section 102(2) C of NEPA: 1) integration of the impact assessment procedure into governmental environmental decision-making; 2) a thorough review of all possible alternatives in the environmental assessment; and 3) the use of appropriate detail in preparing the actual impact statement. In the case of *Environmental Defense Fund v. Corps of Engineers* (Ref. 7-30), the court spelled out the minimum requirements for an impact statement by stating that, "At the very least NEPA is an environmental full disclosure law. . . the 'detailed statement' required by Section 102(2) C should, at a minimum, contain such information as will alert the President, the Council on Environmental Quality, the public, and, indeed, the Congress, to all known possible environmental consequences of proposed agency action." In other cases, including *Natural Resources Defense Council v. Morton* (Ref. 7-33), the courts have indicated that agencies must evaluate all plausible alternatives to a proposed action, including not only those actions within the purview of the agency, but also those actions which are beyond the present scope of the initiating agency's jurisdiction. This stipulation by the courts apparently was intended to force increased interactions among federal agencies, in evaluating environmental alternatives. The courts specifically mentioned that this evaluation of alternatives should not be limited only to those plans offering universal solutions to the particular problem under consideration, but should include alternatives offering only partial solutions. Alternatives which offer only partial solutions to a particular problem could be more advantageous in some cases than the proposed action, because of decreased environmental damage in the particular situation being examined.

In defining agency responsibilities to comply with the letter of the law, federal judges have been very strict in evaluating the overall adequacy of filed Environmental Impact Statements. In the case of *Environmental Defense Fund v. Corps of Engineers* (Ref. 7-30), the court ruled that ". . . the EIS must be written in language that is understandable to non-technical minds and yet contain enough scientific reasoning to alert specialists to particular problems within the field of their expertise." The federal courts had previously indicated that all alternatives to a proposed action, including the null or do-nothing alternative, should be described in sufficient detail that a reviewer could make an independent, responsible decision concerning the advisability of proceeding with the proposed action and the relative merits of each alternative to the proposed action. In the *Calvert Cliffs'* decision mentioned previously, the courts ruled that not only should the environmental statement enumerate all alternatives, but also that it should contain an evaluation of the unique impacts of each alternative. For the most part, federal judges have held that an evaluation and balancing of feasible alternatives must be performed at every major stage in the decision-making process. The possibilities for evaluation of alternatives are shown, in general, in Figure 7-1. This figure is a simplified version of a flowchart for preparation of Environmental Impact Statements, prepared by a district office of the Environmental Protection Agency.

Federal courts have been extremely strict in their review of agency compliance

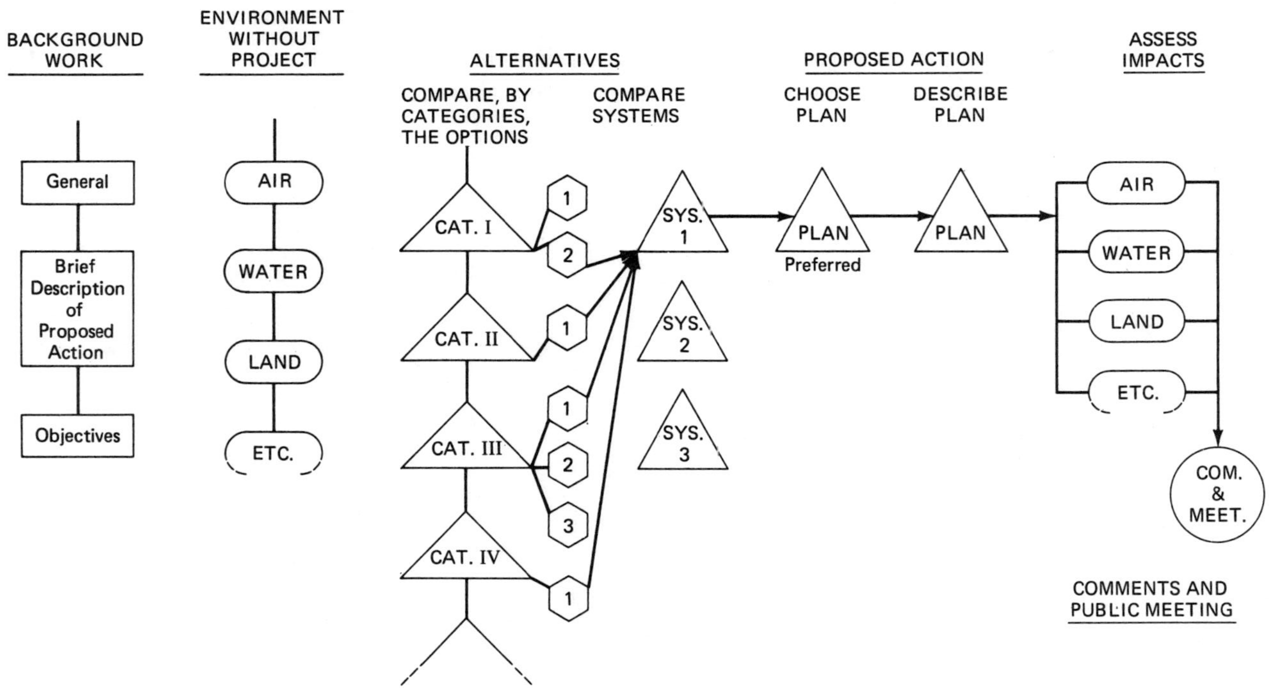

Fig. 7-1. Flow chart for EIS preparation, after EPA (Ref. 7-21).

1. PROJECT DESCRIPTION
   a. Purpose of action
   b. Description of action
      (1) Name
      (2) Summary of activities
   c. Environmental setting
      (1) Environment prior to proposed action
      (2) Other related Federal activities
2. LAND-USE RELATIONSHIPS
   a. Conformity or conflict with other land-use plans, policies and controls
      (1) Federal, state and local
      (2) Clean Air Act and Federal Water Pollution Control Act Amendments of 1972
   b. Conflicts and/or inconsistent land-use plans
      (1) Extent of reconciliation
      (2) Reasons for proceeding with action
3. PROBABLE IMPACT OF THE PROPOSED ACTION ON THE ENVIRONMENT
   a. Positive and negative effects
      (1) National and international environment
      (2) Environmental factors
      (3) Impact of proposed action
   b. Direct and indirect consequences
      (1) Primary effects
      (2) Secondary effects
4. ALTERNATIVES TO THE PROPOSED ACTION
   a. Reasonable alternative actions
      (1) Those that might enhance environmental quality
      (2) Those that might avoid some or all adverse effects
   b. Analysis of alternatives
      (1) Benefits
      (2) Costs
      (3) Risks
5. PROBABLE ADVERSE ENVIRONMENTAL EFFECTS WHICH CANNOT BE AVOIDED
   a. Adverse and unavoidable impacts
   b. How avoidable adverse impacts will be mitigated
6. RELATIONSHIP BETWEEN LOCAL SHORT-TERM USES OF MAN'S ENVIRONMENT AND THE MAINTENANCE AND ENHANCEMENT OF LONG-TERM PRODUCTIVITY
   a. Trade-off between short-term environmental gains at expense of long-term losses
   b. Trade-off between long-term environmental gains at expense of short-term losses
   c. Extent to which proposed action forecloses future options
7. IRREVERSIBLE AND IRRETRIEVABLE COMMITMENTS OF RESOURCES
   a. Unavoidable impacts irreversibly curtailing the range of potential uses of the environment
      (1) Labor
      (2) Materials
      (3) Natural
      (4) Cultural
8. OTHER INTERESTS AND CONSIDERATIONS OF FEDERAL POLICY THAT OFFSET THE ADVERSE ENVIRONMENTAL EFFECTS OF THE PROPOSED ACTION
   a. Countervailing benefits of proposed action
   b. Countervailing benefits of alternatives

Fig. 7-2. Outline for EIS content.

with the specific requirements of NEPA, especially with respect to the five requirements for impact statement content given in Section 102(2) C of the Act. These requirements are shown, together with later modifications for required contents of impact statements, in Figure 7-2. The general attitude and mood of the federal courts can be seen from the text of the decision handed down in the case of *Scrap v. United States* (Ref. 7-34): "It should go without saying that the mere filing of a document labeled 'Final Impact Statement' is insufficient to shield an agency from judicial review if the document fails to comply with standards outlined in NEPA." This seems to be a clear, and extremely significant statement of judicial attitudes toward the preparation of Environmental Impact Statements.

7-3. IMPACT ASSESSMENT PREPARATION

As mentioned previously, the first step in comprehensive environmental assessment is a decision as to whether or not it is necessary to prepare a detailed Environmental Impact Statement on a particular proposed activity. If it is decided that an Environmental Impact Statement must be prepared, certain procedures must be followed in accordance with the letter of the National Environmental Policy Act. The general sequence of preparation, and the time periods for review of draft documents are specified in the Act itself. However, the methods to be used in accomplishing environmental assessment, and the methods for organization of an assessment team have not been described in the Act.

7-3.1. General Sequence of Preparation

Figure 7-3 shows the general sequence of events in the preparation of an Environmental Impact Statement for an action proposed by a federal agency. In the first step, the initiating agency prepares an environmental assessment of the proposed action. Included in this assessment is the accumulation and interpretation of all collected data describing the existing quality of the environment in the vicinity of the project site. In step two, the personnel of the initiating agency review the proposed action at a very early stage in the overall project development, in order to decide whether or not an environmental impact statement is required. If the proposed project is considered to be highly controversial, or if agency personnel feel that the project will have a significant impact on the environment, an environmental impact statement would be prepared. If agency personnel feel that no impact statement is necessary, a brief appraisal of the predicted environmental impact is made and a document describing that appraisal is prepared. In step three in the overall process, a declaration that no EIS is required (negative declaration) is prepared and circulated to federal, state, and local agencies; interested persons; local newspapers; and other media. The appraisal

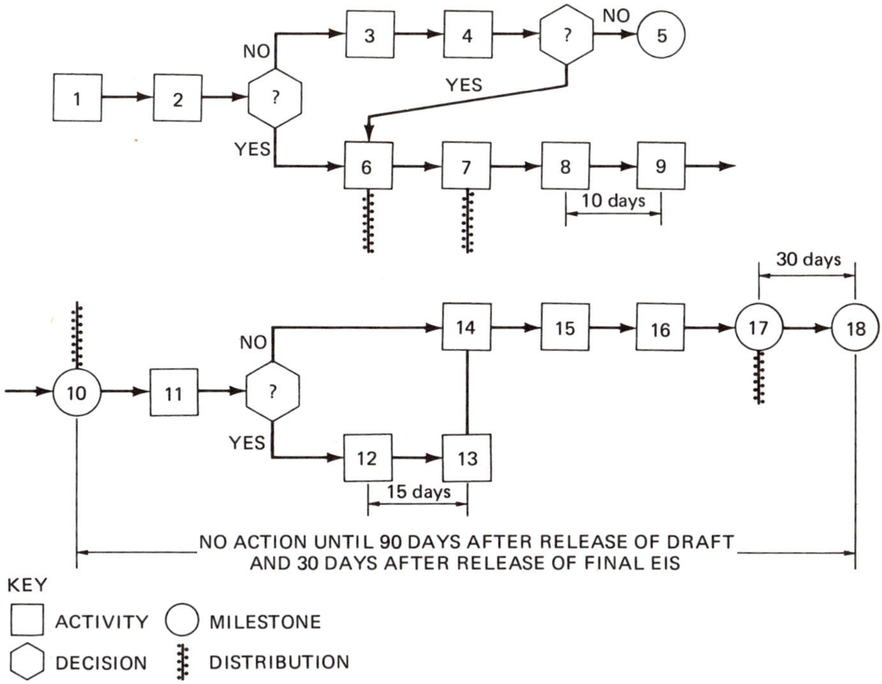

Fig. 7-3. Sequence of EIS preparation in federal agency (Ref. 7-20).

document, the negative declaration, and supporting baseline environmental information are made available for inspection by other agencies or by individuals. If a negative declaration has been made, the declaration will be circulated for comments and, in step four, the initiating agency will evaluate these comments. If, on the basis of reviewers' comments, the initiating agency personnel re-evaluate the situation and decide that an impact statement is required, a full review of the project will be made. If the agency personnel maintain that no statement will be required, they will proceed with administrative action to implement the project. If, on the other hand, they decide that an impact statement is necessary, they will proceed to step six. At this point, the agency personnel prepare a notice of intent to prepare an impact statement, and they circulate this notice to the appropriate agencies and individuals. In step seven, agency personnel prepare a preliminary draft environmental impact statement. This statement should cover all of the issues raised in Section 102(2) C of NEPA as shown in Figure 7-2. In most federal agencies, an internal review of the preliminary draft statement would be made at this point in the procedure. Also, the draft impact statement is submitted to the Office of Federal Activities for further review. These actions are represented by step eight, in Figure 7-3. In step nine, agency personnel review the comments received from the Office of Federal Activities and revise the draft environmental impact statement accordingly. The revised draft is then made available to other agencies and if requested, to the news media. In step 10, the draft environmental statement is distributed to appropriate federal, state, and local agencies, interested individuals, and the Council on Environmental Quality. At this time, agency personnel proceed to step 11; if the agency personnel feel that a public hearing on the proposed action is required, they proceed to step 12. In this phase, public notices are circulated to news media and, 15 days later, a public hearing on the proposed action is held (step 13). If a hearing is held, all suggestions, criticism, and comments which are received on the draft environmental impact statement are reviewed and evaluated. This review includes any comments generated at the public hearing, as well as those submitted in writing from persons who have received the draft environmental impact statement. The review and evaluation process is shown in Figure 7-3 as step 14. In the next step, the initiating agency personnel prepare a final environmental impact statement. A news release is submitted to appropriate media to announce the completion of a final EIS (step 16). The finalized EIS is distributed to the various agencies and individuals mentioned previously (step 17). This final environmental impact statement is reviewed for a period of 30 days. If the initiating agency personnel maintain that all comments have been answered adequately and that the proposed action should be pursued, administrative action to implement the proposed procedures may be initiated.

The sequence of events shown in Figure 7-3 corresponds to the general procedures currently used by most federal agencies in environmental impact assessment and review. Detailed procedural guidelines have been prepared by each federal agency, and specific steps in these detailed guidelines may vary somewhat from the general procedure. Also, the type of guidelines developed by the federal agencies, in response to the requirements for such guidelines by CEQ, vary considerably in the depth and breadth of their requirements and in their overall direction. In some cases, specific methodology is described. In other cases, only general procedures for environmental assessment are indicated. For this reason, it is important to examine the methods which can be used in environmental impact assessment.

7-3.2 Methods of Impact Assessment

Environmental impact assessments and the investigations associated with such assessments are quite complex. In order to facilitate such assessments, and to provide some means of comparing the results of assessments, several attempts have been made to develop standard methods for such assessments. It is too much to hope that any of these methods could be adequate for all types of assessments and for all categories of projects. Nevertheless, each of the methods described in the following paragraphs has some particular merit.

Traditional economic analysis methods (equivalent uniform annual cost, present worth, equivalent uniform annual net return, net present value, benefit-cost ratio, rate of return, etc.) have been found inadequate for environmental assessments, because of the difficulty in assigning monetary values to environmental characteristics. In a similar way, traditional techniques of graphical comparison have been used in an effort to visually represent the location and degree of impact in a proposed activity. In some cases, color-coded maps have been used to indicate the location of especially sensitive areas. In other instances, similar maps have been used to indicate the locations and quantities of particular effluents from proposed projects. More sophisticated techniques also have been developed to graphically represent environmental impacts. These techniques include the use of map overlays (Ref. 7-12), computer math analysis, value profiles (Ref. 7-13), shaded matrices (Ref. 7-4), relative merit analysis (Ref. 7-1), physical models, and contour mapping of impact severity (Ref. 7-14).

In general, graphical techniques are useful primarily for communication; they are not inherently quantitative, nor are they sufficiently precise to allow accurate representation of the degree of impact. The severity of impact may be indicated at a particular location on a map by shading in a particular color (usually red since red has a traditional connotation of danger or hazard). This representation is a clear indication to the viewer that something detrimental is predicted to occur at that particular location. However, it is difficult to represent the severity of the impact which is predicted. Obviously, a red shading on a map can be used to indicate more severe impact than would be represented by a green or blue shading, but such color representations of impact are not sufficiently precise to allow planners to make informed decisions or to allow members of the public to form accurate impressions of the predicted impact of particular actions.

Because of the difficulties in representing degree of impact in a quantitative fashion, some investigators have developed numerical systems for environmental analysis. These quantitative methods include ranking systems (Ref. 7-4); priority ratings (Ref. 7-7); expected value methods (Ref. 7-16); value matrices (Ref. 7-15); utility theory applications (Ref. 7-4); decision models (Ref. 7-8); etc. Unfortunately, none of the quantitative numerical methods of assessment which have been developed are sufficiently discriminatory. In order to be convenient for use on a wide variety of projects, the systems necessarily have been somewhat oversimplified (Refs. 7-4, 7-6, 7-9, 7-2).

None of the methods mentioned above is sufficiently comprehensive to be considered applicable to all environmental assessments, for all types of projects. However, some of the methods are useful in accomplishing some of the goals of the total assessment task. For example, matrices can be utilized to identify proposed activities which are predicted to create significant impacts. One of the most important benefits to be gained from the use of matrices is a thorough examination of proposed actions, with identification of all types of secondary activities which may be included in the main project. In using a matrix, it is easy for an investigator to remember a great number of possible interactions among segments of the environment and possible construction activities.

One of the matrices which is quite useful in preliminary investigations, and in conveying information to the public, is that developed by the United States Geological Survey (Ref. 7-11). However, the USGS matrix should not be considered a quantitative analysis tool. The critical operation in the procedure recommended by the developers of the matrix is the determination of a quantitative estimate of impact. The procedure recommended by the USGS personnel is not inherently flexible. It is not possible to assign proper weighting to the various aspects of the location, construction, and operation of particular facilities. In the USGS procedure, 100 possible actions are listed along the horizontal axis of the matrix and 88 types of impact are listed along the vertical axis. Thus, 8,800 possible interactions can be represented in this matrix. This matrix was intended to be merely a guide to the identification of possible impacts. Only first-order, direct impacts are considered, and probably no more than a dozen activities would be significant for any one project (according to the developers of the matrix) (Ref. 7-11).

Each element (box) in the matrix consists of an upper half, in which the degree of impact of a particular action is expressed as a number from one (least) to ten (greatest), and a lower half in which the overall importance of that particular impact is expressed in a similar fashion. Each number inset into a box in the matrix is obtained as a result of a subjective value judgement. Because the matrix representation thus includes many subjective evaluations of impact, it is open to criticism. The developers of the matrix do not give adequate criteria for making these highly subjective judgements. Furthermore, secondary effects are ignored and no systematic analysis of interactions among possible impacts is considered.

Another method which has been utilized in preparing environmental impact assessments can be classified as a type of "checklist" (Refs. 7-10, 7-18, 7-19, 7-22, 7-23). In this type of method, a long list of possible impacts is compiled, and the investigator reviews this list of potential impacts while examining the possible effects of the proposed action. The analyst simply proceeds through the checklist in order to uncover any possible areas of impact. This type of method is useful primarily as a memory aid. It certainly is not a quantitative investigation tool. Nevertheless, some checklists do have worthwhile features. For example, the Atomic Energy Commission Guidelines for power plant assessment (Ref. 7-18) are very comprehensive. They require a number of descriptive evaluations of a proposed plant: purpose; site character; plant character; environmental effects of construction; impacts of normal operation; effluent control and monitoring; effects of accidents; socioeconomic effects; alternative energy sources and sites; plant design alternatives; benefit/cost analysis; a listing of environmental approvals from and consultations with other agencies; and a listing of documentary references. It is not possible to quantify all of these factors, or even to quantify environmental factors in a summary benefit/cost ratio. Consequently, an investigator using such a comprehensive checklist may produce a great mass of confusing, detailed information. Such an unwieldy compilation of baseline data is of little use unless some selective criteria are used to identify significant topics and areas of impact.

In comparison to the AEC guidelines, the guidelines developed by the Department of Housing and Urban Development (Ref. 7-22) incorporate a valuable concept of "threshold" values of environmental parameters for impact assessment. However, these guidelines are not sufficiently comprehensive to include all types of activities undertaken by federal agencies.

Perhaps the best type of "checklist" method for impact assessment is that developed by personnel of the Army Construction Engineering Research Laboratory, at Champaign, Illinois (Ref. 7-10). It is not possible to obtain quantitative evaluations of impact through the use of this method, but the method does incorporate index matrices for identification of impacts, and provides descriptive material on the evaluation of such impacts in many phases of the environment. Also, it features good reference lists for each area of impact assessment.

In addition to the methods described in the paragraphs above, several highly quantitative methods have been developed. For example, a system has been developed by the personnel of the Battelle Memorial Laboratories, for use in the evaluation of water resources projects (Ref. 7-3). In this method, a somewhat complicated weighting procedure is used to obtain numerical values for the degree of impact. A long list of possible impact areas has been prepared by the developers of this method. A numerical value of impact is produced for each area of possible impact. The developers of this method claim that a quantitative evaluation can be obtained, and comparision of alternatives can be accomplished in a quantitative fashion. Four basic areas of impact are included in this analysis: ecology, environmental pollution, aesthetics, and human interests. A total of 100 parameter importance units (PIU) are distributed among 78 different environmental parameters. "Environmental quality" of each parameter is predicted for a given project, on the basis of a value from 0 (worse) to 1 (best). This value of environmental quality is then multiplied by the PIU value for that parameter, to yield a number of environmental importance units (EIU). The summation of all environmental importance units for all parameters is assumed to be a measure of total relative impact for a proposed action. Values of environmental importance units can be obtained in the same way for all alternatives to a proposed action, and the alternatives can then be compared on the basis of the total EIU scores. Also, in this system, especially controversial or delicate environmental elements are singled out for particular attention.

This method was designed to furnish a quantitative evaluation of impact; however, the method is somewhat inflexible, since it contains inadequate provision for giving appropriate weight to those parameters which are more important in particular situations. The shortcoming of this system consists essentially of the value judgements concerning parameters for analysis, and value of particular parameters which have been made by the developers of the system. However, almost all of the other quantitative systems which have been developed are subject to much the same criticism.

A review of the systems for impact assessment which have been developed to date indicates that no one given system is generally applicable to all types of projects. In order to improve this situation, especially with respect to the evaluation of water resources plans and water resources management, it is suggested that various types of water resources projects be identified and grouped into major categories. These categories of activity could be analyzed in an effort to develop generalized impact assessments. These generalized impact assessments would be directed toward the analysis of types of activity, rather than particular projects in specific locations. The value of this general assessment would be the development of basic evaluations of particular types of water resources projects, and the elimination of much duplicated effort associated with individual assessments of particular projects.

A matrix can be developed as a preliminary analysis tool, and vehicle for communication, in the analysis of various types of projects. Then, a comprehensive checklist could be assembled by water resources experts for use in assessing impacts from various types of water resources activities. This development of a matrix and a checklist for analysis of classes of projects could greatly simplify environmental assessments of water resources projects. A good beginning for such a general assessment is the Battelle system, described in a previous paragraph. However, if a truly general method of analysis for various types of water resources projects could be developed, much effort which has been expended in duplicate investigations, from one project to another, could be saved (Ref. 7-9). Investigators assigned to a particular project could concentrate on identification of particular attributes of the environment in the area under investigation, and could concentrate on particular features of the proposed action, without wasting time in describing the general effects and characteristics of water resources projects.

7-3.3 Organizing the Assessment Team

One of the operations which is not described in NEPA or in the federal agency guidelines for impact assessments is the organization of the investigation team assigned the task of impact evaluation. This is a very important step in any environmental impact assessment. One of the most practicable methods for organizing such a team is to define the assessment tasks to be performed, and then to secure appropriate individuals to perform those tasks. The assessment team should be structured so that it is possible to utilize a systematic, interdisciplinary approach during completion of the assessment. In order to achieve this interdisciplinary approach, every effort should be made to integrate the natural and social sciences with the environmental design arts in the choice of personnel engaged in project planning and in the later implementation of detailed designs.

Interdisciplinary investigations are not identical with multidisciplinary studies; members of assessment teams must be willing to recognize the values and constraints pertinent to fields outside their own areas of expertise. Members of an investigative team must develop an appreciation for experts in other disciplines. In this regard, it may be helpful to include, on the assessment team, some individuals who have expertise in more than one discipline. The authors' experience has shown that an interdisciplinary approach is preferable over all other modes of operation (Ref. 7-24). The assessment can be accomplished most successfully if individuals possessing specialized knowledge in a variety of appropriate disciplines are brought together to determine societal needs, to identify environmental impacts, and also to search for the best possible alternative solutions (Ref. 7-26).

An interdisciplinary team will not function well unless it is led by a conscientious and capable overall coordinator. This coordinator will be called upon to oversee the efforts of the many specialists involved in the assessment. A wide range of specialist disciplines, including everything from demography to aquatic biology to noise control, will be required for a comprehensive assessment of water resources management. The coordinator of an assessment of such activities must be capable of developing at least a working knowledge of and a familiarity with each of the disciplines represented in the study team. Of course, this familiarization must be carried out within the constraints of time and money associated with a particular project, but in a major assessment effort sufficient time should be available for the coordinator to establish at least a minimal working knowledge of many of the disciplines represented by the members of the study team. The coordinator must be able to communicate clearly and concisely with the members of the investigative team. Additionally, the coordinator must see that effective communication is established among all members of the assessment study group.

One of the most neglected activities, and perhaps one of the most worthwhile undertakings in any assessment, is a comprehensive description of the proposed action by the personnel of the initiating agency. Too often, the members of the investigative team are left to discover their own facts about the proposed action, at times in a misguided effort to maintain objectivity. The assessment group should be adequately informed about any physical constraints on either the project location or the project operation. Most of the alternatives to a particular action would probably have been considered during the initial planning of any proposed facility. Therefore, these considerations should be described by the personnel of the initiating agency, so that the assessment team can gain an appreciation of the practical difficulties faced by the project designers. Conversely, the personnel of the initiating agency must make every effort to develop an appreciation for difficulties associated with comprehensive impact assessments. The coordinator of the assessment effort must establish frequent and candid communications between the project initiators and the project evaluators. Such communication can assist greatly in producing a more realistic evaluation of the impact for a particular facility.

Finally, in organizing an assessment team and in supervising an assessment study, the coordinator must establish communication, at a very early stage in the assessment study, with the public. If possible, professional writers and communications experts should be employed in this effort. These professionals should present the detailed and complex information on project characteristics to the lay public in a fashion comprehensible to the average citizen. This may be exceedingly difficult in some cases, but an attempt must be made. One of the chief failings of early impact assessments and impact statements was a virtually complete lack of communication with the general public during the assessment process. If members of the public do not understand the characteristics of a proposed water resources management plan, nor the facilities included in such a management plan, this ignorance may lead to hostility, or at best, apathy toward the proposed activities. These hostile feelings can be eliminated, to a great extent, through educational programs designed to inform the public about the features of the proposed management plans, and to elicit suggestions and comments for possible alternative activities from the public.

7-4. ENVIRONMENTAL ASSESSMENTS OF WATER MANAGEMENT ACTIVITIES

Many of the impact assessments completed to date in the United States have been directed toward identifying the impacts on the water and air environment which could be produced by particular proposed actions. This early emphasis on water and air pollution has resulted from the relatively high state of development in water and air pollution control technology, because of the capability of investigators to monitor parameters which indicate the physical and chemical status of air and water, and because of the attention which has been focused upon natural ecosystems. In the future, it is likely that emphasis will shift from detailed quantitative descriptions of particular systems, in terms of traditional parameters, to the evaluation of large-scale management activities. This is particularly important with respect to the development of water resources plans and water resources management in general.

7-4.1 NEPA—Planning Implications

The lawmakers who developed the National Environmental Policy Act included a requirement in the law for the utilization of systematic, interdisciplinary studies by federal agencies in developing comprehensive environmental plans.

According to the stated ideals of these lawmakers, this systematic type of study was to be implemented on all levels of federal agency activities. Several distinct categories of environmental planning have been affected by this requirement for comprehensive interdisciplinary environmental analysis:

a. Policy Planning—The formulation of agency policies and operational procedures which could have overall effects on environmental quality;
b. Program Planning—The application of available technology to maintenance and improvement of environmental quality in specific types of locations or under specific types of activities;
c. Functional Planning—The development of construction activities undertaken by a specific agency within a limited territory to serve a specific need (e.g., yearly dredging activities carried out on a particular stream in a given locality by a particular local office or district office of the Corps of Engineers);
d. Project Planning—The implementation of plans for specific activities or projects, such as the construction and operation of a particular facility (e.g., design and construction of a wastewater treatment facility at a specific location).

The Senators and Congressmen who developed NEPA called for the correlation of all of these levels of planning, in order to identify possible environmental impacts at each level of planning, and in order to minimize any identified detrimental consequences of particular plans. With respect to water resources management, federal policy, especially policy established through legislative activity, should be subjected to environmental assessment, in order to determine the possible impacts on the water environment of contemplated changes in governmental policies, and to identify possible alternatives to existing governmental policies (Ref. 7-17). This type of assessment, directed toward governmental policies, is much more difficult to undertake than is the assessment of the impacts of a particular facility at a particular location. The best approach to assessments of federal policies with respect to water resources management may be the establishment of national commissions composed of representatives of federal agencies, environmentalist groups, consulting engineers, and members of the public to undertake such controversial and comprehensive evaluation of far-reaching national policies. In the meantime, the bulk of the environmental assessment activity which will be carried out on a day-to-day basis in the United States will be directed toward the investigation of the particular effects and ramifications of specific water resources management activities.

7-4.2 Assessments for Specific Projects

In the evaluation of environmental impacts associated with specific water resources management projects, assessment activities generally will fall into one of several categories: the monitoring of water quality in the existing environment in order to quantitatively characterize that environment without the proposed project or activity; the prediction of impacts from a proposed activity on the basis of modeling that action and its generation of pollutants; and the evaluation of the impacts identified in the previous step, on the basis of direct and indirect effects, short-term and long-term impacts, secondary effects, and effects on the individual and community. The best way to illustrate the assessment of impacts for a specific project may be to describe an example of a water resources management activity and the assessment study which was completed on that activity.

In a midwestern community located along a major river, a study was undertaken, according to the guidelines of the United States Environmental Protection Agency, to develop a 201 Facilities Plan for each of the three counties in the particular metropolitan area. An environmental assessment of the proposed facilities plan was prepared in compliance with the requirements of the Environmental Protection Agency. In essence, the assessment served to summarize much of the work which had been done in developing the facilities plan. In the early stages of plan development, a comprehensive inventory had been conducted to characterize the physical system in the area subject to the facilities plan. The geology, soils, topography, geography, climate, and hydrology of the 3-county area were analyzed as the initial step in developing the facilities plan.

In this description of the physical environment, particular water resources problem areas throughout the study region were identified. In some areas, the low permeability of surface soils and underlying rock layers, for example, precluded the use of septic tanks for wastewater management. In other areas, the presence of cavernous limestone beneath the surficial soils caused particular problems in surface water and wastewater management. The major source of groundwater recharge in the region was the river flowing through the three counties under study. This river was flowing in a channel filled with granular materials, so that much of the groundwater resource in the area was contained in the geologic materials adjacent to the river. Most of the other streams in the study region other than the major river had a 7-day, 10-year low flow of nearly zero, which greatly limited their capacity to receive wastewaters. For this reason, much of the facilities planning was directed toward the treatment of the wastewaters, with subsequent discharge of treated waters into the major river running through the area.

In addition to analyzing the physical system in the study area, the facilities planners studied the social, economic, and land-use characteristics of the region. Population projections for a 20-year study period were obtained, based upon data furnished by the Environmental Protection Agency and the population study agency of the metropolitan area under investigation. Furthermore, an industrial inventory was prepared for the study region so that estimates of quantities and types of industrial wastewaters could be obtained. Finally, since the

Environmental Protection Agency Guidelines for 201 Facilities Plans stipulate that the plans should be based upon acceptable or approved land-use plans, local planning agencies were consulted in order to develop a comprehensive land-use plan for the three counties included in this study. Land-use data important to this study were collected, including the future distribution of residential population and industrial facilities. These projections were based upon capability of the terrain to support certain uses and the rational development and growth of existing community structures.

In order to characterize the state of the environment in the study area, an effort was made to identify and characterize all existing or permitted sources of wastewater in the three counties under study. These wastewater sources were categorized in three major divisions: sources producing more than 100,000 gallons per day; sources producing from 10,000 to 100,000 gallons per day; and sources producing less than 10,000 gallons per day. As might be expected, about two-thirds of the identified sources fell into the last category, producing less than 10,000 gallons per day. More than 55 percent of the total wastewater generated in the study area was generated in residences, apartments, and mobile home parks, and was produced through municipal and package wastewater treatment facilities. About 40 percent of the total flow of wastewater was industrial wastewater, which was processed either by the industries themselves or through municipal wastewater treatment facilities. Commercial, recreational, institutional, and miscellaneous sources of wastewater accounted for about only 2 percent of the total flow.

The water quality in surface streams and in groundwater reserves within the three counties in the study region also was analyzed during the development of the 201 Facilities Plan. More than 100 sampling stations were established in the three counties under study in order to determine not only the existing water quality in the region, but also to identify locations which would allow the most representative sampling of water quality in the future. Many of the stream segments in the 3-county area exhibited high bacterial levels showing probable fecal contamination. Also, water quality analysis during the development of the facilities plan showed that the levels of dissolved oxygen, pH, and total dissolved solids in the sampled streams did not satisfy the drinking water requirements of the federal government. The major river flowing through the study area also had poor water quality characteristics, reflecting the poor state of wastewater treatment in the study area. Two of the largest tributary streams in the area were consistently found to have very poor water quality (insufficient dissolved oxygen content, presence of excessive fecal and total coliform bacteria, and high dissolved solids contents). At this stage in the development of the facilities plan, a model of the surficial streams in the study area was prepared to allow a determination of the assimilative capacity of the major receiving streams in the area. The model was used in a preliminary investigation which showed that the major river in the study area was the only stream that could accept significant wastewater loadings, if the wastewater was subjected only to secondary treatment. Many of the tributary streams in the 3-county area, because of their dry-weather, low-flow characteristics, could accept only wastewaters which had been subjected to a high level of treatment.

Next, in the development of the facilities plan for the study area, the existing wastewater treatment facilities were analyzed. Three of the largest facilities in the study area included only primary treatment of wastewaters. However, one of the largest facilities was to be expanded in the near future to include secondary treatment units. Some of the smaller treatment facilities in the study area were equipped to provide secondary treatment. However, they were found to be discharging into streams with very low assimilative capacities. Thus, many of the small treatment facilities were providing secondary treatment, but the streams into which these facilities were discharging could accept only wastewaters which had been processed in advanced treatment systems. A large number of package treatment plants had been constructed to serve residential subdivisions in the study area; less than 10 percent of these package treatment plants were equipped to provide tertiary treatment, although the majority of these plants were discharging into very sensitive streams.

In order to determine the potential for development of a comprehensive wastewater management plan, the project investigators next analyzed the political, jurisdictional, and administrative situation in the three counties under study. Since the Water Pollution Control Act of 1972 (as amended) called for consolidation of existing facilities with the development of regional wastewater management agencies, the developers of the 201 Facilities Plan for the region under study recommended the development of county-wide sewer districts for each of the three counties under study. It was felt that this development would help to eliminate point sources of water pollution, would help to improve water quality, and would help to obtain economy of scale in the design and construction of treatment facilities. In order to determine the capabilities for facilities design and construction, the developers of the plan investigated the fiscal resource capabilities of the three counties of the study area. This investigation of resources included a determination of federal funds available for water quality management activities. Additionally, revenue-sharing funds and community development revenue-sharing funds were investigated. State programs which provide matching funds to assist small communities in water quality management were studied. All of these sources of money were analyzed in order to determine the total financial capability of the three counties in the study region.

After the planners had gathered the information described in the preceding paragraphs, they employed systematic optimization techniques to determine the alternatives for wastewater management that would require the lowest capital and operating expenses. Alternative sewer system configurations, alternative treatment facility locations, alternative degrees of treatment, alternative degrees of infiltration/inflow treatment and/or correction requirements were included in this phase of the study. More than 20 alternative schemes for wastewater management were developed for the 3 counties. On the basis of an evaluation of these alternatives, it was determined that total flow and waste flow reductions could be achieved through correction of infiltration and inflow into existing and planned sewer systems. The sewer system configurations were evaluated on the basis of the identified trends in land use and the locations of sewers required to consolidate the treatment facilities in the area. Evaluation of the various treatment alternatives indicated that all new plants located on the major river in the study area would require secondary treatment capabilities. The smaller plants located on tributary streams in the 3-county area would require advanced treatment capabilities, tertiary treatment capabilities, or land application techniques. In general, in this metropolitan region, it was found that land application and wastewater reuse were not feasible. However, because of the existing poor water quality of many of the small streams in the area, and because of the limited capacity of these small streams to receive wastewaters, it was decided that a regional treatment scheme would be very advantageous. Of course, such a regional scheme would have cost advantages, as well as advantages of centralized facilities and the possible capability of extending services into remote areas of the 3-county region.

In order to include public participation in the planning process, an evaluation committee composed of representatives of various public and private organizations was formed to evaluate the various alternate wastewater management plans. This committee spent more than 50 hours in conference in the identification of problem areas and in the evaluation of the alternatives which had been proposed by the facilities planners. Members of the committee also proposed modifications to several of the alternatives.

A total of 12 separate plans had been identified as the most cost-effective alternatives for the region to the north of the major river, which bisected the study area. Eight separate plans had been identified as alternatives for the area to the south of the river. These plans included the possible location of plants at 8 sites north of the river and 6 sites south of the river. Additionally, a variety of different degrees of treatment were included in each of the proposed locations. A further alternative was the possible correction of infiltration/inflow outside combined sewers, treatment of infiltration/inflow outside combined sewers, correction of infiltration/inflow for combined and sanitary sewers, and correction of infiltration/inflow for combined sewers with an addition of detention basins for combined sewers. Obviously, a large number of alternatives were considered. In evaluating these alternatives, the community committee developed a matrix of parameters for use in judging each particular scheme. The parameters which were used in this evaluation step included overall effect on water quality, impact on land use, environmental effects, financial capabilities within the study area related to each of the plans, the reliability/flexibility/operability of each of the plans, capabilities of governmental units to manage the facilities and to enforce regulations, project timing for each of the alternatives, success in eliminating existing wastewater sources, coverage of the study area with services, capabilities of upgrading existing facilities, total project cost, energy availability and conservation, availability of total system resources, capabilities to recover and reuse natural resources, and compatibility with water supply requirements for the metropolitan area. These parameters were given quantitative values in a scheme similar to the Battelle system mentioned previously in this chapter. A total effectiveness value was developed for each alternative plan. The cost effectiveness of each of the plans was determined. On the basis of these determinations, the community committee selected optimal plans for the 3 counties in the study region. Subsequently, a public hearing was held to describe the selected plans and no objections to the plans were presented.

At this point, it was possible to develop an environmental assessment of the proposed management plan for the three counties in the study area, and to include the alternatives which had been identified as feasible by the project initiators and by the community evaluation committee. Reference was made in preparing the assessment to the numerous inventories of data which had been gathered in preparing the description of the physical system and the man-made environment in the study area. Next, the proposed water quality management plan was described. The location of the wastewater treatment facilities and the configuration of the proposed sewer systems were described in this section of the assessment. The schedules for construction and completion of the proposed water quality management efforts in the region were described. Reference was made to the material obtained during the development of the 201 Facilities Plan, to describe the environmental setting and the population trends in the planning area. Maps, photographs, and renderings were used to graphically characterize the environmental setting in the study area.

Each of the alternate water quality management plans was described in considerable detail. This was possible because the development of the facilities plan had included a detailed evaluation of alternatives by a community evaluation committee. Much of this evaluation and description was facilitated through dividing the study region into selected water quality management areas. These areas were identified on the basis of drainage basins and groundwater recharge areas within the 3 counties under study.

As required under the Guidelines for environmental assessments prepared by the Council on Environmental Quality, the relationship of the proposed water quality management plan to land-use planning in the 3-county area was analyzed and described. The expected future development of the 3-county study area was described on the basis of the existing land-use plans for the area. These future developments were translated into probable water demands and resultant wastewater discharges. Thus, the land-use planning and population growth characteristics of the area were used to develop an understanding of the demands on the water resources in the study area. It is significant to note that, in the evaluation process, the community evaluation committee considered the compatibility of the 201 Facilities Plan with existing and projected land use to be a very significant consideration. The committee employed 15 criteria in evaluating the proposed management plan; compatibility with existing and projected land uses was considered to be the second most important of the 15 evaluation criteria, exceeded only by overall water quality. This emphasis on compatibility with land use merely emphasized the possible positive and negative impacts which could result from the design, construction, and operation of various sewer configurations and treatment facilities as proposed in the alternative management plans.

In the evaluation of environmental impact of the proposed water quality management plan, the plan initiators utilized a system very similar to that developed by the personnel of Battelle Memorial Laboratories. Four general categories of impact were identified: ecological impacts; impacts on the physical environment; socio-economic impacts; and aesthetic impacts. With regard to the ecological environment, 4 environmental parameters were considered: effects on aquatic flora; effects on terrestrial flora; effects on aquatic fauna; and effects on terrestrial fauna. The physical systems impacts were quantified with regard to impacts on water quality, land quality, air quality, hydrology, geology, and noise levels. The socioeconomic factors which were utilized in identifying and quantifying socioeconomic impacts included existing and future land use, economic development in the study region, archeological resources, historical sites, recreation facilities, existing and planned transportation facilities, and cultural features in the study area. One of the most difficult categories of impact to assess is impact on the aesthetics of a particular area. In this assessment, the investigators studied the physical factors in the study area, the spatial composition of various physical factors, the man-made objects in the 3-county region, and the rather unique aesthetic factors located throughout the area. Environmental quality values were applied to each of the parameters described above. A mechanism similar to the parameter importance units proposed by the Battelle investigators was used to quantify impacts in each of the 4 categories of environmental quality. With respect to the ecological environment, a total of 250 environmental quality units were assigned to the four parameters describing the quality of the ecological environment. Physical environmental quality, represented by the 6 parameters previously mentioned, was assigned 400 environmental quality units. The 7 socioeconomic factors which were investigated to determine impact were assigned a total of 250 environmental quality units, and the aesthetic factors were assigned 100 environmental quality units. Thus, a total of 1,000 environmental quality units were assigned to the 21 various parameters which were used to describe the degree of impact. In determining the degree of impact, the project investigators used a rating from −5 to +5 to evaluate the effect of the operations on any given environmental parameter included in each alternative management plan. The rating of each of the factors was then multiplied by the number of environmental quality units assigned to that factor to determine a summary overall rating for each alternative.

In accordance with the requirements of the National Environmental Policy Act, the project investigators also determined the environmental impact of the null alternative. This assessment of impact was based upon the data accumulated in the previous portions of the investigation which served to describe the existing treatment facilities, existing water quality in the project region, predicted trends in population and socioeconomic growth, and the characteristics of the physical environment in the study region. After obtaining ratings of impact for each of the environmental parameters described above, it was found that the null alternative for the study region would have a significant detrimental impact on water quality. This impact was found to be most severe in the heavily industrialized area to the south of the major river which bisected the study region. In contrast, the proposed management plan was found to have a significant beneficial effect, as determined by the evaluators. In terms of the evaluation system used by the investigators, the selected management plan had a total environmental quality rating in excess of 1,200 while the null alternative had an overall rating of approximately −100.

The potential impacts of the proposed plan included improvement in the overall water quality of the study region, as well as decreases in the possible levels of air pollution in the study region through the elimination of sludge incineration at several facilities. The construction of centralized treatment facilities was considered to have a beneficial effect through the elimination of many small, disruptive treatment facilities throughout the 3 counties in the study area. The detrimental impacts associated with the proposed plan included disruption of the physical system during construction and operation of the treatment facilities and the sewer systems proposed in the plan. The system evaluators enumerated these impacts in a number of different categories. For example, the known archeological sites in the study area were investigated, and it was discovered that several sites would be disturbed to a certain extent by the proposed construction activities. Also, recreational areas including parks, golf courses, and other facilities in the 3 counties of the study area would be affected by the construction of treatment facilities and sewer lines. These facilities were depicted on maps and the locations of treatment plants and sewer lines were shown to illustrate the locations and severity of the impacts anticipated. A similar review and evaluation was made for the wooded areas in the study region to identify the type and degree of impact anticipated through construction and operation of the proposed facilities. The extensive construction associated with this plan was predicted to produce only slight, short-term impacts on the terrestrial flora and fauna in the project area. It was decided that all terrestrial flora in the path of construction activities would be replaced wherever possible. No significant habitats were found to be altered significantly by the proposed construction activities. In general, effects on the ecological environment were predicted to be rather temporary.

A number of historical sites were found in the study area and the possibility of disruption of these sites was investigated by the project evaluators. Several of these sites had already been listed on the National Register of Historical Sites. Other sites were found to be eligible for registration but were not found on the register. Equal importance was attached to these eligible sites, in comparison to the registered sites, in assessing the impact of the proposed facilities. No major impacts on the historical sites in the area were identified, with the exception of some disruption to cemeteries. This was considered to be a rather minor detrimental impact of the proposed plan.

With respect to the physical environment in the study area, a major benefit to water quality was identified as an impact of the proposed plan. However, some detrimental impacts also were identified, including a minor long-term detrimental impact on air quality in the study region as a result of the operation of a sludge incinerator as part of the proposed facilities. Some impact on land quality in the area was identified through the evaluation of landfills for sludge disposal. With respect to short-term impacts of the construction activities, the most significant detrimental impact was found in the portions of the study area where the bedrock topography was very steep and irregular. In these areas, the surficial soil layers were shallow and highly subject to erosion. For the proposed activities, an unavoidable impact would be created through erosion of these areas during construction.

With respect to groundwater resources of the study area, a possible detrimental impact was identified in regions where major sewer lines were proposed to pass through significant groundwater recharge areas. These areas were identified and shown on maps in the environmental impact assessment report. The possible degradation of groundwater was evaluated for the proposed plan and the alternates which were considered feasible for the region. In the same way, the effects of the proposed facility on surface water resources, wetlands, floodplain areas, prime agricultural lands, areas of high population density, and heavily used corridors in the study area were evaluated. In all of these investigations, major attention was directed toward the quantitative evaluation of impact from the proposed plan, while qualitative estimates were developed for the alternatives. However, a quantitative estimate of overall impact was obtained on the basis of the evaluation of alternatives which had been prepared by the community evaluation committee.

The impact assessment was completed with a discussion of the relationship between local, short-term uses of the environment and the maintenance and enhancement of long-term productivity which would be caused by the implementation of the proposed water quality management plan. Since the plan was designed primarily to enhance long-term water quality, the short-term disruption of the environment necessary for construction of the proposed facilities was considered to be necesssary and justifiable in view of the long-term benefits to be obtained. Finally, the irreversible and irretrievable commitments of resources which would be involved in the proposed management plan were investigated. These commitments included allocations of land, buildings, equipment, chemicals, manpower, and energy involved in the construction and operation of the proposed wastewater treatment facilities and collection systems. However, it was considered that most of the land committed to this system would be affected only temporarily in terms of long-term impact, since many of the facilities could be altered or removed in future operations.

7-5. SUMMARY

The National Environmental Policy Act of 1969 has proven to be one of the most significant legislative tools for environmental protection and resource conservation yet created in the United States. This law requires comprehensive evaluation of the predicted impacts of proposed actions, and of predicted effects of all feasible alternatives to the proposed action. Included in the feasible alternatives to the proposed action must be the null alternative, so that attention must be given to the alternative plan of doing nothing to change the existing conditions in a project area. To date, the effectiveness of the National Environmental Policy Act has been developed principally through judicial interpretation of the text and meaning of the Act. Federal judges have insisted on strict adherence to the letter of the law, and have demanded strict compliance on the part of all federal agencies with all of the provisions of NEPA.

Some of the most important provisions of the National Environmental Policy Act include the requirement for an impact statement directed toward all federally funded activities considered to have a significant effect on the environment. Additionally, sufficient time for agency and citizen reaction to statements of predicted impacts is guaranteed by the provisions of NEPA. However, the text of the law does not indicate how investigators are to assess the various impacts. Furthermore, the language of the law gives no direction concerning the ways and means of organizing an assessment team, other than specifying that the impact assessment should be done by an interdisciplinary group of specialists.

A number of different methodologies have been developed for use in impact assessment. Some of these methods are simply graphical representation techniques which are useful primarily as communication tools. Other methods consist of checklists, which are merely memory aids. On the other hand, some extremely complex quantitative methods have been developed. Many of the extremely complicated methods are not sufficiently flexible to be applicable for assessments in all types of water resources projects. A combination of the use of matrices to identify areas of impact, checklists to insure a comprehensive review, and assessment of types of projects and classes of impact may offer the most effective means of accomplishing meaningful environmental assessments. With regard to specific methodologies for use in investigating water resources planning activities, the system developed by the Battelle Memorial Laboratories may serve as a good beginning to the development of a system directed toward the evaluation of a particular project in a specific location.

With respect to water resources planning and management activities, environmental impact statements will have great significance in overall environmental planning and with regard to public participation in the development of water management plans. Impact statements will be required for legislative actions, for development of new regulations, and for the design, construction, and operation of water resources management facilities. With regard to specific water resources projects, it will be necessary to characterize the water quality in the existing environment, describe existing water treatment facilities, evaluate existing sources of wastewater and of water supplies, and predict future demands for water and future levels of wastewater generation. All feasible alternative plans, including the null alternative, must be included in any comprehensive impact assessment. Furthermore, the impact assessment must include a consideration of both direct and indirect impacts on water resources, as well as short-term and long-term effects. Secondary effects of proposed water management activities also must not be neglected. Emphasis must be given to the effects on the total community, rather than to effects on individuals, or specific population centers. If these considerations are included in the preparation of impact assessments and impact statements on water resources management plans, these statements will prove to be extremely valuable adjuncts to the total plannning process.

For further information on environmental impact assessments and statements, the reader is referred to the following texts directed specifically at this subject: *Handbook of Environmental Impact Analysis*, by R. K. Jain, L. Y. Urban, and G. S. Stacey, Van Nostrand Reinhold, New York, 1977. *Environmental Assessments and Statements*, by J. E. Heer, Jr., and D. J. Hagerty, Van Nostrand Reinhold, New York, 1977.

REFERENCES

7-1. Adkins, W. G., *et al.*, "Social, Economic, and Environmental Factors in Highway Decision Making," *148 Progress Report*, Study 2-1-71-148, Texas Highway Department and Fed. Hwy. Adm., Washington, 1971.

7-2. Amlung, D. K. and Eith, A. W., "An Assessment of Methodologies of Environmental Planning," M. Eng. Thesis, University of Louisville, Louisville, May, 1975.

7-3. Battelle Memorial Institute, Columbus Laboratories, "Final Report on an Environmental Evaluation System for Water Resource Planning to the Bureau of Reclamation, Department of the Interior," January, 1972.

7-4. Carter, Everett C., Haefner, Lonnie E., and Hall, Jerome W., "An Informational Report on Techniques for Evaluating Factors Relevant to Decision Making on Highway Locations," DOT, Fed. Hwy. Admin., Off. Env. Policy, Washington, Nov., 1972.

7-5. Council on Environmental Quality, "Preparation of Environmental Impact Statements: Guidelines," *38 Federal Register*, 20550, Aug. 1, 1973.

7-6. Dries, David A., "Environmental Impact Assessment Methodology," M. Eng. Thesis, University of Louisville, May, 1974.

7-7. Engelen, R. E. and Stuart, D. G., "Transport Technologies and Urban Structure: A Framework for Evaluation," paper presented at the ASCEASME National Meeting on Transportation Engineering, Seattle, July, 1971.

7-8. Haefner, L. E. and Redding, M. J., "An Analytical Structure of Community Public Works Decision Process," *Papers and Proceedings*, Joint Conf. Env. Res. Assoc. and Amer. Inst. Arch., Irvine, California, January 1972.

7-9. Hagerty, D. J. and Heer, J. E., Jr., "Impact Assessment of Power Plants," *Journal, Power Division*, ASCE **102**, No. 1, January, 1976.

7-10. Jain, R. K., Urban, L. V., and Stacey, G. S., *Handbook for Environmental Impact Analysis*, Dept. of Army, CERL, Champaign, Illinois, (Draft), April, 1974.

7-11. Leopold, Luna B., Clarke, Frank E., Hanshaw, Bruce B., and Balsley, James R., "A Procedure for Evaluating Environmental Impact," Geological Survey Circular 645, US Govt. Printing Office, Washington, 1971.

7-12. McHarg, Ian L., *Design With Nature*, Natural History Press, Garden Press, Garden City, New York, 1969.

7-13. Oglesby, C. H., Bishop, B., and Willeke, C. E., "A Method for Decisions Among Freeway Location Alternatives Based on User and Community Consequences," *Hwy. Res. Record*, 305, Highway Res. Board, Washington, 1970.

7-14. Redding, M. J., "The Quality of Residential Environments, Preference for Accessibility to Residential Opportunities," Ph.D. Thesis, Northwestern University Technological Institute, Evanston, June, 1970.

7-15. Schimpeler, C. C. and Grecco, W. L., "Systems Evaluation: An Approach Based on Community Structure and Values," *Hwy. Res. Record,* 238, Highway Res. Board, Washington, 1968.

7-16. Schlager, K., "The Rank-Based Expected Value Method of Plan Evaluation," *Hwy. Res. Record*, 238, Highway Res. Board, Washington, 1968.

7-17. Twiss, Robert H., "Linking the EIS to the Planning Process," *Environmental Impact Assessment: Guidelines and Commentary*, ed. Thomas G. Dickert, Universtiy Extension, University of California, (Berkeley, 1974).

7-18. United States Atomic Energy Commission, "Preparation of Environmental Reports for Nuclear Power Plants," Regulatory Guide 4.2, US Govt. Printing Office, Washington, March, 1973.

7-19. United States Department of Agriculture, "Environmental Impact Statements: Proposed Guidelines," 38 *Federal Register*, 31904, Nov. 19, 1973.

7-20. U. S. Environmental Protection Agency, "Preparation of Environmental Impact Statements–Interim Regulation," Washington, D. C. Federal Register, **38**, No. 11, Part II, Jan. 17, 1973, p. 1709.

7-21. U. S. Environmental Protection Agency, Office of Federal Activities, "Manual for Preparation of Environmental Impact Statements for Wastewater Treatment Works, Facilities Plans, and 208 Areawide Waste Treatment Management Plans," Washington, July, 1974.

7-22. United States Department of Housing and Urban Development, "Handbook of Departmental Policies, Responsibilities and Procedures for Protection and Enhancement of Environmental Quality," 38 *Federal Register*, 19182, July 18, 1973.

7-23. Werner, Robert R., "Environmental Impact and the US Army Corps of Engineers," paper presented at Ecological Soc. of America Symposium, Amer. Assoc. Adv. Sci., Washington, Dec., 1972.

7-24. "Environmental Quality as an Objective of Water Resources Planning," Special Printing, Comm. Int. and Ins. Affairs, US Senate, US Govt. Printing Office, Washington, 1972.

7-25. Hearings on S. 1075, S. 237 and S. 1752 Before the Senate Committee on Interior and Insular Affairs, 91st Congress, 1st Session, at 117.

7-26. *Proceedings*, Panel Discussion on the Systematic Interdisciplinary Approach in Highway Planning and Design, ed. MIT Trans. and Comm. Values Proj., Urban Sys. Lab. with Fed. Hwy. Adm., Cambridge, March, 1973.

7-27. *Calvert Cliffs' Coordinating Committee v. Atomic Energy Commission*, 449 F. 2d 1109, 1 ELR 20346 (D.C. Cir. 1971). *cert denied*, 404 U.S. 942 (1972).

7-28. *Environmental Defense Fund v. Ruckelshaus*, 439 F. 2d at 597, 1 ELR at 20064 (D.C. Cir. 1971).

7-29. *Environmental Defense Fund v. Hardin*, 325 F. Supp. 1401, 1 ELR 2027 (D.D.C. 1971).

7-30. *Environmental Defense Fund v. Corps of Engineers*, 325 F. Supp. 728, 1 ELR 20130 (E.D. Ark. 1970-71), ---F. Supp. ---2 ELR 20260 (E.D. Ark.) 342 F. Supp. 1211, 2 ELR 20353 (E.D. Ark.), *Aff'd*, 470 F. 2d 289, 2 ELR 20740 (8th Cir. 1972).

7-31. *Hanly v. Kleindienst*, F. 2d, 2 ELR 20717 (2d Cir. 1972).

7-32. *Hanly v. Mitchell*, ---F Supp. ---2 ELR 20181 (S.D. N.Y.), 460 F. 2d. 640, 2 ELR 20216 (2d. Cir.), *cert. denied*, 41 U.S.L.W. 3247 (Nov. 7, 1972), sub. nom. *Hanly v. Kleindienst*, 471 F. 2d. 822, 4 ERC 1785 (2nd Cir. Dec. 5, 1972), *cert. denied*, 484 F. 2d 448, 5 ERC 1416 (May 21, 1973).

7-33. *Natural Resources Defense Council v. Morton*, 337 F. Supp. 165, 2 ELR 20028 (D.C.C. 1971), 337 F. Supp. 167, 2ELR 20089 (D.D.C. 1971), *motion for summary reversal denied*, 458 F. 2d 827, 2 ELR 20029 (D.C. Cir.), *dismissed as moot*, 337 F. Supp. 170, 2 ELR 20071 (D.D.C. 1972).

7-34. *SCRAP v. United States*, 346 F. Supp. 189, 195 note 8, 2 ELR 20486 20488 note 8 (D.C.C.), *application for stay pending appeal denied sub. nom. Aberdeen & Rockish Ry. Co. v. SCRAP*, ---U.S.---, 2 ELR 20491 (1972) (Burger, Circuit Judge), *cert. granted* 41 U.S.L.W. 3339 (Dec. 19, 1972).

7-35. *SCRAP v. United States*, 346 F. Supp. 189, 2 ELR 20486 (D.D.C. 1972), *rev'd on other grounds*, 5 ERC 1449 (S. Ct. June 18, 1973).

Index

Index